Heat Shock Proteins in the Nervous System

NEUROSCIENCE PERSPECTIVES

Titles in this series

Roger Horton and Cornelius Katona (eds), Biological Aspects of Affective Disorders
Trevor Stone (ed), Adenosine in the Nervous System
Judith Pratt (ed), The Biological Bases of Drug Tolerance and Dependence
Michel Hamon (ed), Central and Peripheral 5-HT_3 Receptors
John Waddington (ed), D_1 : D_2 Dopamine Receptor Interactions
Thomas Barnes (ed), Antipsychotic Drugs and Their Side-Effects
Eva Giesen-Crouse (ed), Peripheral Benzodiazepine Receptors
David Nicholson (ed), Anti-Dementia Agents
Alan Harvey (ed), Natural and Synthetic Neurotoxins
John Wood (ed), Capsaicin in the Study of Pain
Yossef Itzhak (ed), Sigma Receptors

Heat Shock Proteins in the Nervous System

edited by

R. J. Mayer
Department of Biochemistry,
University of Nottingham Medical School,
Queen's Medical Centre, Nottingham, UK

and

I. R. Brown
Department of Zoology,
University of Toronto, Scarborough Campus,
West Hill, Ontario, Canada

ACADEMIC PRESS
Harcourt Brace & Company, Publishers
London San Diego New York
Boston Sydney Tokyo Toronto

ACADEMIC PRESS LIMITED
24/28 Oval Road,
London NW1 7DX

United States Edition published by
ACADEMIC PRESS INC.
San Diego, CA 92101

This book is printed on acid free paper

A catalogue record for this book
is available from the British Library

ISBN 0–12–480960–X

Typeset by Servis Filmsetting Ltd, Manchester, UK
Printed and bound in Great Britain
by T J Press, Padstow, Cornwall.

Contents

Contributors

Andre-Patrick Arrigo Laboratoire du Stress Cellulaire, Centre de Génétique Moleculaire et Cellulaire, CNRS UMR-106, Claude Bernard University, Lyon-I, 69622 Villeurbanne, France

Jean-Pierre Brion Laboratoire d'Anatomie Pathologique, Université Libre de Bruxelles, 808 Route de Lennik, 1070 Brussels, Belgium

Ian R. Brown Department of Zoology, University of Toronto, Scarborough Campus, West Hill, Ontario, Canada M1C 1A4

Michael E. Cheetham Department of Neuroscience, Institute of Psychiatry, De Crespigny Park, Denmark Hill, London SE5 8AF, UK

Dennis W. Dickson Department of Pathology, Albert Einstein College of Medicine, 1300 Morris Park Avenue, Bronx, New York 10461, USA

Barney E. Dwyer Department of Veterans Affairs, Medical Center, 16111 Plummer Street, Sepulveda, CA 91343, USA

Marshall Edwards Mammalian Department B19, Department of Veterinary Clinical Sciences, University of Sydney, New South Wales, NSW 2006, Australia

Lawrence E. Hightower Department of Molecular and Cell Biology, The University of Connecticut, Storrs, CT 06269-3044, USA

M. Landon Department of Biochemistry, University of Nottingham Medical School, Queen's Medical Centre, Nottingham, NG7 2UH, UK

Karen Li Mammalian Department B19, Department of Veterinary Clinical Sciences, University of Sydney, New South Wales, NSW 2006, Australia

Ting Li Department of Molecular and Cell Biology, The University of Connecticut, Storrs, CT 06269-3044, USA

J. Lowe Department of Pathology, University of Nottingham Medical School, Queen's Medical Centre, Nottingham, NG7 2UH, UK

R. J. Mayer Department of Biochemistry, University of Nottingham Medical School, Queen's Medical Centre, Nottingham, NG7 2UH, UK

Patrick Mehlen Laboratoire du Stress Cellulaire, Centre de Génétique Moleculaire et Cellulaire, CNRS UMR-106, Claude Bernard University, Lyon-I, 69622 Villeurbanne, France

Robert N. Nishimura Department of Veterans Affairs, Medical Center, 16111 Plummer Street, Sepulveda, CA 91343, USA

Thaddeus S. Nowak, Jr Department of Neurology, University of Tennessee, 855 Monroe Avenue, Link Bldg, Memphis, TN 38163, USA

Nobuhito Saito Department of Neurosurgery, Tokyo University School of Medicine, Tokyo, Japan

Sadao Suga Department of Neurosurgery, Keio University School of Medicine, Tokyo, Japan

Michael Tytell Department of Neurobiology and Anatomy, and Department of Ophthalmology Programme in Neuroscience, Bowman Gray School of Medicine of Wake Forest University, Winston-Salem, North Carolina 27157, USA

David Walsh Mammalian Department B19, Department of Veterinary Clinical Sciences, University of Sydney, New South Wales, NSW 2006, Australia

Keith D. Wilkinson Department of Biochemistry, 4017 Rollins Research Building, Emory University School of Medicine, Atlanta GA 30322, USA

Shui-Hui C. Yen Department of Pathology, Albert Einstein College of Medicine, 1300 Morris Park Avenue, Bronx, New York 10461, USA

Li Zhe Mammalian Department B19, Department of Veterinary Clinical Sciences, University of Sydney, New South Wales, NSW 2006, Australia

Frank Zeng Mammalian Department B19, Department of Veterinary Clinical Sciences, University of Sydney, New South Wales, NSW 2006, Australia

Series Preface

The neurosciences are one of the most diverse and rapidly changing areas in the biological sphere. The need to understand the workings of the nervous system pervades a vast array of different research areas. By definition research in the neurosciences encompasses anatomy, pathology, biochemistry, physiology, pharmacology, molecular biology, genetics and therapeutics. Ultimately we are striving to determine how the human brain functions under normal circumstances and perhaps more importantly how function changes in organic disease and in altered states of mind. The key to many of these illnesses will unlock one of the major therapeutic challenges remaining in this era.

The difficulty lies in the vastness of the subject matter. However I try, I find it almost impossible to keep abreast of the changes occurring in my immediate sphere of interest, let alone those exciting advances being made in other areas. The array of journals dealing with neurosciences is enormous and the flow of reprints needed to keep me updated is daunting. Inevitably piles of papers accumulate on my desk and in my briefcase. Many lie there unread until sufficient time has passed for their content to be overtaken by yet more of the ever rising tide of publications.

There are various approaches that one can take to deal with this problem. There is the blinkered approach in which you pretend that literature outside your area does not exist. There is the ignore it totally option. Indeed, one colleague of mine has ceased to read the literature in the belief that, if there is a publication of critical importance to his research, someone will tell him about it. I am not that brave and instead I arrived at what I thought was the ideal solution. I started to read critical reviews of areas of current interest. But I soon came unstuck as I realized that, for many subjects of importance to the neurosciences, such authoritative works did not exist.

Instead of simply moaning that the literature was incomplete, I conceived the idea of *Neuroscience Perspectives*. For purely selfish reasons I wanted to have available a series of individual edited monographs dealing in depth with issues of current interest to those working in the neuroscience area. Already a number of volumes have been published which have been well received and the series is thriving with books on a range of topics in preparation or in production. Each volume is designed to bring a multidisciplinary approach to the subject matter by pursuing the topic from the laboratory to the clinic. The editors of the individual volumes are producing balanced critiques of each topic to provide the reader with an up-to-date, clear and comprehensive view of the state of the art.

As with all ventures of this kind, I am simply the individual who initiates a chain of events leading to the production of the series. In reality, it is key individuals at Academic Press who are really responsible for the success of *Neuroscience Perspectives*. In particular, Dr Carey Chapman and Leona Daw have the uneviable task of recruiting editors and authors and keeping the ship on an even keel.

Finally, I hope that *Neuroscience Perspectives* will continue to be enjoyed by my colleagues in the neurosciences. Already the series is being read, understood and enjoyed by a wide audience and it is fast becoming a reference series in the field.

Peter Jenner

Preface

Cells respond to temperature elevation and other traumas by inducing a set of genes encoding cell stress (heat shock) proteins which may play important roles in cellular repair and/or protective mechanisms. This new book surveys the current state of knowledge concerning the induction of heat shock genes and proteins in the nervous system following such traumatic events as hyperthermia, ischaemia and tissue wounding. The possible neuroprotective effects of the heat shock response are discussed. In addition to reviewing progress in animal model systems, this book discusses heat shock proteins such as ubiquitin in relation to human neurodegenerative diseases. Each chapter presents an overview of a specific subject area and includes current results from each author's laboratory and a viewpoint on future research directions.

This book on the expression of stress genes in the nervous system in response to traumatic events and during neurodegenerative diseases will be of value to researchers and graduate students interested in areas of neuroscience, ranging from the molecular to the clinical level.

This book is organized in two parts, the first dealing with cell stress genes and their protein products and the second dealing with ubiquitin and the nervous system. The chapters are grouped so that each of the two sections begins with a description of the basic molecular cell biology of heat shock proteins or ubiquitin. In the first section of the book the chapters deal with *Functional roles of heat shock proteins, Injury* and *Neuroprotection.* The final two chapters of the first section then emphasize the role of heat shock proteins in the nervous system during *Development* and *A novel neuronal heat shock protein.*

The basic cellular roles of ubiquitin are emphasized in the first chapter of the second section of the book, followed by a chapter in which the importance of protein ubiquitination in *Aging* (Chapter 10) is contrasted with the importance of ubiquitin and heat shock proteins in *Neurodegenerative diseases* (Chapter 11).

R. J. Mayer and I. R. Brown

CHAPTER 1

STRUCTURE AND FUNCTION OF THE MAMMALIAN HSP70 FAMILY

Lawrence E. Hightower and Ting Li

Department of Molecular and Cell Biology, The University of Connecticut, Storrs, Connecticut 06269-3044, USA

Table of Contents

1.1 Overview of the hsp70 family of molecular chaperones

1.1.1 Introduction

The 70-kDa family of heat shock proteins (hsp70s) is the most abundant and conserved subset of eukaryotic stress proteins. The mammalian family consists of at least four distinct classes present in different cellular compartments: the strictly stress-inducible hsp70 and constitutively expressed heat shock cognate protein hsc70 (or as it is often called, clathrin-uncoating ATPase), both located in the cytoplasm and nucleus; the 78-kDa glucose-regulated protein grp78 (also called BiP) located

Heat Shock Proteins in the Nervous System
ISBN 0-12-480960-X

on the luminal surface of the rough endoplasmic reticulum (ER) and a recently identified member, mt-hsp70 (also called grp75) in the matrix of mitochondria, both of which are normally abundant but also stress-inducible. All of the hsp70 family proteins can be operationally described as peptide/unfolded protein-stimulated ATPases and they are classified as molecular chaperones. Molecular chaperones have been defined as cellular proteins that mediate the correct folding of polypeptides and their assembly into oligomeric structures, but which are not part of the mature protein or complex (Ellis and van der Vries, 1991). Whether stress-inducible chaperones serve only to limit further protein damage or direct unrepairable proteins to degradation in stressed cells or are directly involved in renaturation, are open questions. In this chapter, the focus is primarily on constitutive hsc70, about which the most is known structurally and functionally, and emphasis is given to work with particular relevance to responses of nervous tissue. For detailed recent reviews of the hsp70 family, see (Morimoto *et al.*, 1990; Nover, 1991; Lee, 1992; Welch, 1992).

The mammalian hsps were discovered along with the avian hsps (Kelley and Schlesinger, 1978) four years after Tissieres and Mitchell reported their discovery of the *Drosophila* hsps (Tissieres *et al.*, 1974). In the same paper, Kelley and Schlesinger showed that hsps were the same proteins induced in avian and mammalian cells by amino acid analogues (Hightower and Smith, 1978; Hightower, 1980). By the first international meeting devoted to the heat shock response, held at Cold Spring Harbor on the twentieth anniversary of Ferruccio Ritossa's discovery (Ritossa, 1962), Ian Brown and Fredric White reported on pioneering studies from their respective laboratories using mammalian brain tissue, studies which increased the appreciation of the heat shock response as a more general stress response (Brown *et al.*, 1982; White and Currie, 1982). The studies from Ian Brown's laboratory are summarized in Chapter 2. As an introduction, Fred White's contributions to our understanding of the hsp70 family of stress proteins in mammalian brain are summarized.

Studies of incubated brain slices were begun by White with the original intent of establishing a more amenable system to study normal patterns of brain protein synthesis. However, the processes of preparing and/or incubating the slice as an explant led to the rapid induction of a small set of proteins dominated by hsp70 (then called SP71) and it soon became clear that the incubated slices were more suitable to the study of tissue trauma. Less obvious was the identity of SP71 and the nature of the induction signal. Enter serendipity in the form of a mutual colleague, Dr Barry Hall, probably the only biologist at the time aware of both trauma-induced rat brain SP71 and the 70-kDa protein known in the Hightower laboratory as a protein induced by the arginine analogue, canavanine, in cultured chicken embryo cells. Barry arranged the introductions, and the ensuing collaboration resulted in the identification of trauma-induced and analogue-induced hsp70 as the same proteins; the discovery of the constitutive member of the mammalian hsp70 family, originally called P73 and now known as hsc70; and the introduction of the terms 'cellular stress response' and 'stress proteins' into the literature (Hightower and White, 1981). Thus, amino acid analogues provided the bridge between trauma and the heat shock response, with Milton Schlesinger's laboratory showing that amino acid analogues and heat induced the same proteins, and with the work described above showing that amino acid analogues and tissue trauma induced the same proteins.

Using incubated slices of rat telencephalon and whole body heating of rats, White found evidence of developmental regulation of hsp70 induction, which occurred in the brains of heat shocked rats and incubated brain slices from control rats only after three weeks of postnatal development (White, 1981a; Currie and White, 1983). A developmental link to the maturation of the cerebral microvasculature and the establishment of the blood–brain barrier was proposed based on the finding of high amounts of hsp70 synthesis in cells associated with cerebral microvessels (White, 1980b, 1981a). Further efforts to identify the stress-responsive cells in incubated brain slices led to a detailed characterization of brain microvessels (White *et al.*, 1981; White, 1981b) and to a study of stress protein induction of developing cell cultures derived from rat cerebellum (Pearce *et al.*, 1983). The bulk of the evidence supported the conclusion that neither neurons nor astrocytes responded strongly to the trauma of explantation, even though stress proteins were heat-inducible in both cell types. More recent studies have confirmed and extended these observations on the heat shock response of cultured neurons and astrocytes, reporting that the response is substantially greater in astrocytes than in neuronal cells (Dwyer *et al.*, 1991; Nishimura *et al.*, 1991). Cerebral endothelial cells appeared to be most sensitive to trauma-induced synthesis of stress proteins in brain slices. The specific stressor(s) was never identified, however the damage to the vascular endothelium that triggered the induction of stress proteins may well have been protein damage since the addition of polyhydroxyl alcohols, like glycerol, to the incubation medium blocked the induction of SP71 in brain slices (White and White, 1987). More extensive studies of the effects of both deuterium oxide and glycerol as cellular protectors were used to provide additional support for the abnormal protein hypothesis of hsp induction, which states that inducers of the heat shock or stress response share the ability to damage cellular proteins and that cells respond to such damage by inducing hsps (Hightower, 1980; Hightower *et al.*, 1985; Edington *et al.*, 1989).

Early in his work with incubated brain slices, White obtained biochemical evidence of a protein translocation system which could be inhibited by vinblastine, colchicine and low calcium ion concentrations (White, 1979). SP71 turned out to be one of the proteins translocated away from capillaries (White, 1980a). Two explanations were suggested: (i) synthesis of SP71 in astrocyte end-feet which abut microvessels and intracellular movement away from the capillaries; and (ii) synthesis of SP71 in endothelial cells followed by extracellular transfer to astrocyte glial cells and translocation away from the capillaries. The second possibility, seemingly heretical at a time when hsps were known only as cytoplasmic proteins, is appealing now in the light of the discoveries of glia-axon transfer proteins, which include SP71

(Tytell *et al.*, 1986; Sprang and Brown, 1987) and the selective release of hsps from cultured mammalian cells by a secretory pathway-independent mechanism (Hightower and Guidon, 1989).

Using our current understanding of the molecular chaperoning functions of SP71 or hsp70, one can now propose a tissue pathway in which hsc70 in normal tissue and both hsp70 and hsc70 in stressed tissue serve as intracellular *and extracellular* carriers of peptides and partially folded proteins, for example, from vascular endothelial cells to astrocytes and from these cells to neurons. The stimulated synthesis of radioactively labelled hsp70 in microvascular cells during incubation of the brain slice provided a tagged population of molecular chaperones that could be followed through the pathway. Left to their own devices, neurons distribute newly synthesized molecular chaperones by slow axonal transport, which may be too slow to meet the needs of stressed or even normal axons, as described in greater detail by Tytell in Chapter 4.

1.1.2 Hsp induction pathway

A flurry of recent papers have added considerable detail to our understanding of the heat shock induction pathway. At least three distinct heat shock factors (HSFs) have been found in higher vertebrates and the following description applies to mammalian HSF1. Lindquist and colleagues originally suggested that hsp70 may autoregulate the heat shock response in *Drosophila* (DiDomenico *et al.*, 1982; Solomon *et al.*, 1991). In order to explain the negative, autoregulatory nature of the response, it has been proposed that one or more members of the hsp70 family associate either transiently, or as a stable complex, with HSF1 in non-stressed cells, keeping the transcription factor in an inactive form (Craig and Gross, 1991; Baler *et al.*, 1992; Morimoto, 1993). Although no evidence of a stable complex between hsp70/hsc70 and the inactive form of HSF1 has been reported, Morimoto and coworkers (Abravaya *et al.*, 1992) showed that exogenous hsp70 inhibits activation of HSF1 *in vitro*. Some models use the very low basal amounts of inducible hsp70 present in unstressed cells as the negative regulator whereas others invoke a shift in the proportion of free and bound hsc70. In the latter scenario, shown in Figure 1, stressors cause the accumulation of misfolded or damaged proteins that bind hsc70, depleting the free pool of hsc70 and consequently favouring the dissociation of HSF1-hsc70 complexes. In support of this hypothesis, the accumulation of aberrant proteins in cells has been correlated positively with the heat shock transcriptional response (Abravaya *et al.*, 1992), and cycloheximide blocks the activation of transcription at moderate temperature elevations in human cells, presumably by removing newly synthesized proteins as targets for thermal denaturation (Zimarino *et al.*, 1990; Baler *et al.*, 1992). In stressed cells, HSF1 forms trimers which accumulate in the nucleus. Trimer formation is dependent upon several leucine zippers located near the DNA binding domain at the amino terminus of HSF1. Recently it has been suggested that another leucine zipper located in the carboxy terminal domain of HSF1 may inhibit intermolecular zipper

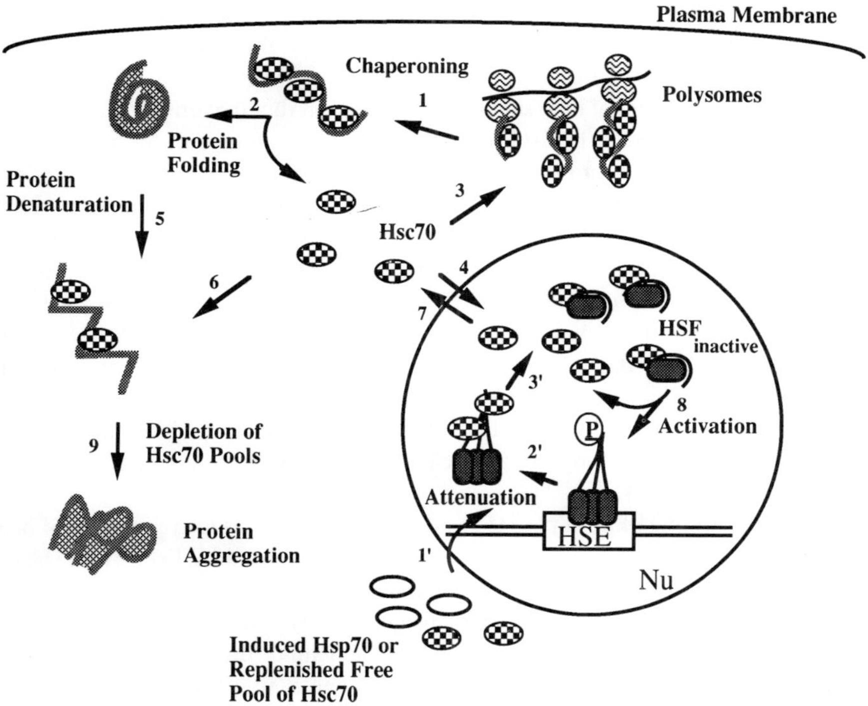

Figure 1 Illustration of the induction pathway. Steps 1–4 represent normal chaperoning activities of hsc70 in unstressed cells. The interaction between hsc70 and $HSF_{inactive}$ may be transient. As a consequence of a heat shock or interactions with other stressors that cause proteotoxicity, proteins unfold and rebind hsc70, thus depleting the free pools of this chaperone, steps 5 and 6. This results in depletion of nuclear hsc70 pools as well, step 7, and the subsequent activation of HSF. Further reduction in hsc70 levels would result in the cellular accumulation of sufficient amounts of unfolded proteins to allow aggregation, step 9. The functioning of newly induced hsp70 would eventually restore protein homeostasis and bound hsc70 would be released again into free pools as denatured proteins were either degraded or refolded. Either free hsc70 or induced hsp70 could then interact with active HSF trimers to release them from their binding sites on HSEs and dissociate them to inactive monomers, steps 1′–3′, causing attenuation or down-regulation of transcription.

interactions by forming an intramolecular zipper in monomers (Rabindran *et al.*, 1993). Possibly hsp70/hsc70 facilitates intramolecular zipper formation, keeping HSF1 inactive. Once formed, HSF1 trimers bind to arrays of inverted repeats of the element NGAAN located in the 5′-flanking region of heat shock-responsive genes (Perisic *et al.*, 1989; Baler *et al.*, 1993; Sarge *et al.*, 1993). HSF1 undergoes phosphorylation in stressed cells as well and although this modification may alter its transcriptional activity, recent studies suggest that phosphorylation is not essential for its activation (Sarge *et al.*, 1993). In an alternative model, HSF monomers act as the sensors of stress and unfold to a form that requires stabilization by hsc70. The chaperoned HSFs can then form active trimers (Schlesinger and Ryan, 1993).

Another major aspect of heat shock-gene regulation that may involve interactions between HSF1 and hsp70/hsc70 is the relatively rapid attenuation of transcription following induction, at least in response to moderate stress levels. Extreme levels of stress result in prolonged periods of elevated rates of transcription of heat shock genes. During attenuation, HSF trimers are released from DNA, dephosphorylated, and returned to the pool of inactive monomers. Since hsp70 has been detected in association with trimers of HSF1 in stressed cell extracts, current models include a role for this protein in the dissociation of trimers (Abravaya *et al.*, 1992; Baler *et al.*, 1992). Thus, to close the autoregulatory loop, one could propose that inducible hsp70 accumulates in stressed cells to a level at which hsp70 causes dissociation of active HSF1 trimers. Alternatively, hsp70 may remove and/or help refold denatured proteins, resulting in the release of bound hsc70, which may then dissociate HSF1 trimers.

The recent finding of multiple HSFs adds new possibilities for regulation of heat shock genes. For example, HSF1 and not HSF2 is activated by heat shock, heavy metal ions, amino acid analogues and oxidative stress, i.e. it follows the rule that stressors which cause protein damage or proteotoxicity activate HSF1 (Hightower, 1991). However, HSF2 and not HSF1 is activated during haemin-induced differentiation of erythroleukemia cells, which suggests that the abnormal protein hypothesis may not apply for HSF2 (Sistonen *et al.*, 1992). Recently, a third factor, avian HSF3, was isolated which is not induced under conditions that induce either HSF1 or HSF2 (Nakai and Morimoto, 1993). There are plenty of new inducers with potential relevance as physiological effectors and biological response modifiers as well as hints of other signalling pathways for the induction of heat shock genes to go around. For example, salicylates, prostaglandins and arachadonic acid are all potent activators of HSFs and have roles in inflammatory responses. Some inducers stimulate phospholipase C and phospholipase A_2 activity in mammalian cells and roles for Ca^{2+} in HSF activation have been proposed (Calderwood and Stevenson, 1993). There may also be an hsc70/hsp70 connection in these signalling pathways since both of these proteins, including hsc70 purified from rat brain, are known to bind non-esterified palmitic and stearic acids (Guidon and Hightower, 1986a,b). Intracellular concentrations of fatty acids increase dramatically in the adult mammalian brain and other tissues to a lesser extent at the beginning of ischaemia, presumably due to the activation of phospholipases, that degrade membrane phosphoglycerides (Rodriguez de Turco and Bazan, 1983). Although purely speculative at this point, it is possible that fatty acids released during this response may affect hsc70/hsp70 interactions with HSFs resulting in factor activation. The fatty acid binding sites on these hsps may bind inducers such as arachadonic acid and prostaglandins as part of an HSF-activation pathway as well. Inducers and induction conditions have been divided into three general classes (Morimoto *et al.*, 1992): (i) environmental stress, which includes the stressors that cause activation of HSF1; (ii) non-stressful conditions which do not activate HSF1, including cell cycle events, growth factors, development, differentiation, and oncogenes; and (iii) pathophysiological states including fever, inflammation, ischaemia, neuronal injury, etc. Obviously, this last class includes complex tissue responses that could potentially result in activation of all three known HSFs.

As noted above for class II inducers, not all cases of up-regulation of heat shock gene expression involve HSF1 activation. Recent findings of hsp70 family proteins associated with nascent polypeptide chains indicate that molecular chaperoning starts at the polysomes (Beckmann *et al.*, 1990, 1992). This implies a close link between rates of protein synthesis, i.e. cellular demand for chaperones, and intracellular pools of hsc70 (see Figure 1). Confirming this expectation, transforming growth factor-β1 which is a regulator of protein synthesis in chicken embryo cells causes a rapid post-transcriptional increase in hsp70-family gene expression that precedes an increase in overall protein synthesis (Takenaka and Hightower, 1992, 1993). The suggestion is that cells expand their pools of molecular chaperones in preparation for increasing production of new proteins. This form of up-regulation of chaperones may be particularly relevant to brain injury since TGF-β is one the cytokines released by platelets at wound sites and it has been implicated in wound healing. The inducers of hsps in such complex processes as tissue injury in brain (Brown *et al.*, 1989), or in the incubated brain explants described earlier, have not been identified, but cytokines such as TGF-β and mediators of inflammation are good candidates.

1.1.3 Cellular locations

In unstressed mammalian cells in culture, hsc70 exhibits diffuse fluorescence over both nucleus and cytoplasm in immunolocalization studies. Hsc70 as well as the inducible hsp70 are found to localize in nuclei and concentrate heavily in nucleoli after heat shock (Pelham, 1984; Velazquez and Lindquist, 1984; Welch and Feramisco, 1984; Lewis and Pelham, 1985; Welch and Mizzen, 1988). Nucleolar accumulation correlates with marked alterations in the integrity of the nuclei and particularly the nucleoli after heat shock. Immunoelectron microscopy showed that most of hsc70 and hsp70 is specifically localized to the granular region of the nucleolus, the site of biogenesis and assembly of ribosomal subunits and small ribonucleoproteins (snRNPs) (Welch and Suhan, 1986). The alterations in nuclear and nucleolar morphology coincide with impaired functioning of these heat-sensitive targets, such as inhibition of mRNA splicing/transport in the nuclear matrix (Mayrand and Pederson, 1983; Yost and Lindquist, 1986; Sadis *et al.*, 1988), and inhibition of rRNA synthesis, pre-ribosome assembly and snRNP assembly in the nucleoli (Ellgard and Clever, 1971; Heine *et al.*, 1971; Robin and Hogness, 1975; Sadis *et al.*, 1988). The binding of hsp70 and hsc70 to the nucleoli is reversible. During recovery from heat shock, nucleoli slowly regain their normal morphology, with hsp70 initially losing its association with nucleoli but mostly remaining in the nucleoplasm in discrete areas, and gradually regaining its normal diffuse nuclear (excluding nucleoli) and cytoplasmic distribution (Pelham, 1984; Velazquez and Lindquist, 1984; Welch and Feramisco, 1984; Welch and Suhan, 1986; Welch and Mizzen, 1988). Identical behaviour is observed for hsc70 using antibodies specific for this protein (Pelham *et al.*, 1985).

In mammalian cells recovering from stress, much of the perinuclear and cell perimeter hsp70 co-localizes with translationally active cytoplasmic ribosomes by immunofluorescence analysis (Welch and Suhan, 1986). This finding set the stage for understanding more recent work showing that hsp70/hsc70 bind to nascent polypeptides in polysomes (Beckmann *et al.*, 1990). As illustrated in Figure 1, the chaperoning pathway actually begins as soon as a nascent polypeptide displaying an appropriate amino acid sequence for binding to the cytosolic hsp70 family members emerges from ribosomes. Thus, in addition to the nuclear localization of hsp70/hsc70 in stressed cells, another potentially very interesting site of accumulation occurs at or near the plasma membrane. Some of the initial autoradiographic and immunohistologic studies showed localization of a portion of hsp70 at this site in *Drosophila* cells (Velazquez *et al.*, 1980; Velazquez and Lindquist, 1984). Later came studies showing the localization of an hsp70 family member in the ruffling membranes of motile fibroblasts (LaThangue, 1984), the immunoprecipitation of hsp70/hsc70 as a complex with a solubilized cell surface glycoprotein, gp90 (Hughes and August, 1982), the release of hsc70 in vesicles containing terminal transferrin receptors during reticulocyte maturation (Davis *et al.*, 1986), and association of hsc70 with synaptosomal membranes in rat brain (Lim *et al.*, 1984; Whatley *et al.*, 1986). In addition, a member of the hsp70 family has been implicated in antigen presentation. Antisera raised against this protein blocks presentation of the processed antigen cytochrome *c* displayed on the surface of antigen presenting cells to an antigen-specific T-cell hybrid (Vanbuskirk *et al.*, 1989).

There is evidence of interactions between hsc70 and cytoskeletal elements. Hsc70 is identical to β-internexin, which is a microtubule-associated protein (MAP) that has a lower binding affinity for microtubules than most MAPs (Green and Liem, 1989) and also the same as a ubiquitous intermediate filament-associated protein (Napolitano *et al.*, 1985, 1987). Two other cytoskeleton-associated proteins are highly related if not identical to hsc70, based on biochemical and immunological evidence. These include the 68-kDa microtubule-associated protein (MAP68) (Weatherbee *et al.*, 1980; Lim *et al.*, 1984; Welch *et al.*, 1985 a,b), and the thermin A protein which associates with cytoskeleton (Wang *et al.*, 1980, 1981; Weatherbee *et al.*, 1980). The recent observation that Chinese hamster ovary (CHO) cell mutants, resistant to microtubule inhibitors contained an additional new isoform of hsc70 and a new isoform of hsp60 suggests a role of hsc70 and hsp60 in the assembly and function of microtubules *in vivo* (Ahmad *et al.*, 1990; Gupta, 1990). However, the functional significance of hsc70 association with cytoskeleton is not clear at this point.

1.1.4 Molecular activities

Serendipity struck twice in linking a 73-kDa brain protein to what we now call hsc70. In addition to Barry Hall's role as messenger of serendipity in the demonstration that bovine brain P73 is a heat shock cognate protein, Milton Schlesinger played the second messenger role in a 1983 visit to the Rothman laboratory (Schlesinger, 1986). This laboratory had been busily purifying clathrin from bovine brain and had discovered an associated 70-kDa protein that dissociated clathrin baskets using ATP hydrolysis. Milton recognized the similarities in the biochemical properties of clathrin-uncoating ATPase and the hsp70 family, which led to a collaboration among several labs in the USA and an independent demonstration by Ungewickell of the relationship (Ungewickell, 1985; Chappell *et al.*, 1986). As part of his contribution, William Welch improved the purification of hsp70 proteins by incorporating an ATP-affinity column step in the protocol (Welch and Feramisco, 1985a,b). So, approximately ten years after the discovery of hsps, there was an *in vitro* assay for the hsp70s, but the *in vivo* significance of this activity in clathrin-mediated processes remains unclear. Very recently, Black and co-workers used non-denaturing immunoprecipitations to identify proteins that interact with clathrin during axonal transport and demonstrated for the first time a stable association of clathrin and hsc70 in intact cells (Black *et al.*, 1991), suggesting a direct involvement of hsc70 in intraneuronal transport of molecules.

Both genetic and biochemical evidence appeared together in support of a role for hsp70 in the translocation of proteins across membranes of the endoplasmic reticulum and mitochondria using yeast as a model system (Chirico *et al.*, 1988; Deshaies *et al.*, 1988). Since these observations in yeast, several studies have now shown that mammalian hsc70 is also involved in protein import into organelles. For example, purified hsc70 can maintain a mitochondrial precursor protein in a non-aggregated form in an ATP-dependent manner and can also retard the folding of the precursor protein during renaturation (Sheffield *et al.*, 1990). Using a permeabilized cell system, Shi and Thomas demonstrated that import of nucleoplasmin into HeLa cell nuclei requires hsc70 and hsp70, as well as other cytosolic factors that are NEM-sensitive. The nuclear transport activity of the HeLa extract is abolished when depleted of hsc70 and hsp70, and is fully restored by the addition of purified hsc70 or hsp70 (Shi and Thomas, 1992). Using antibodies against hsc70 to inhibit nuclear import of karyophilic proteins into mammalian nuclei, Imamoto *et al.* (1992) obtained evidence that hsc70 interacts directly with proteins containing nuclear localization sequences in the cytoplasm prior to import. These sites probably represent the continuation of a chaperoning pathway that begins with hsc70 binding to nascent polypeptides to block premature folding.

The lysosomal degradation of certain cellular proteins containing the N-terminal peptide sequence KFERQ, or similar sequences, is increased during serum withdrawal. The cytosolic protein that specifically recognizes the KFERQ sequence for such lysosomal targeting, the 73-kDa peptide recognition protein (prp73), is hsc70 based on immuno-crossreactivity, sequence identity of two internal peptides, and a comparison of their abilities to stimulate lysosomal uptake of proteins *in vitro* (Chiang *et al.*, 1989; Terlecky *et al.*, 1992). Although a detailed discussion is beyond the scope of this review, it is important to note that the chaperoning pathways do not end at membrane boundaries. For translocation across the endoplasmic reticulum, grp78 is positioned on the luminal side to accept the translocated polypeptide and mt-hsp70 is positioned in the matrix of mitochondria for the same purpose. Furthermore, many polypeptides are passed from hsp70 family chaperones to members of the hsp60 chaperone family located in both mitochondria and the cytoplasm for folding.

The results described above imply that hsp70 family proteins may preferentially bind to unfolded proteins and peptides. In direct tests of this hypothesis, it has been shown that hsc70 and grp78 can bind short peptides (Flynn *et al.*, 1989), and that purified hsc70 can distinguish between folded and unfolded forms of the same protein such as native cytochrome *c* and its unfolded precursor apocytochrome *c* as a substrate for peptide/unfolded protein-stimulated ATPase activity (Palleros *et al.*, 1991; Sadis and Hightower, 1992). A milestone in structural studies of this protein family was achieved by McKay and colleagues with the crystallization and determination of the three-dimensional structure of the N-terminal 44-kDa ATPase fragment of bovine hsc70 (Flaherty *et al.*, 1990). The molecule consists of a two-lobed, four-domain structure with ATP bound in a deep cleft between the lobes. The tertiary folding resembles the nucleotide-binding domain of hexokinase and is also very similar to actin. These similarities in structure were unexpected because there are no regions of significant amino acid sequence identity among these proteins. The similarities in structure perhaps explain how hsc70 oligomerizes (like actin) and undergoes substrate-driven conformational changes (like hexokinase).

Much less is known about the carboxy terminal peptide-binding domain of hsp70 family proteins. Using deletion mutants and assays based on localization of hsp70 to nuclei in stressed cells, Milarski and Morimoto provided evidence that the C-terminal 160 amino acids have an important role in polypeptide ligand binding (Milarski and Morimoto, 1989). Using circular dichroism spectroscopy and secondary structure prediction, Sadis and coworkers predicted that the C-terminal domain has a high content of α-helix and they suggested that two of the predicted helices might form a peptide-binding groove, similar to the structure of the human class I major histocompatibility antigen (Sadis *et al.*, 1990). Subsequently, Rippmann *et al.* (1991) constructed computer models of an hsp70 peptide-binding domain using the known structural coordinates of the HLA protein as a template, and these models will no doubt be among those tested in fine-structure mutational analyses of this domain. Major unanswered questions include the nature of the binding site on unfolded proteins and are there different building sites for different hsp70 family members. In the peptide-binding study carried out by Rothman and coworkers, the K_ms varied over a 100-fold range, suggesting preferences for some sequences, but the sampling of peptides was too small to determine a consensus binding sequence. In a recent comparison of the ability of DnaK and grp78 to substitute for hsc70 in an *in vitro* assay based on the stimulation of translocation of precursor proteins in mammalian microsomes, it was concluded that neither of these proteins can substitute for hsc70 and that polypeptide substrate specificity is likely (Wiech *et al.*, 1993). In further support of substrate preferences, two yeast hsp70 isoforms that differ by only 3% in their amino acid sequences have marked differences in their clathrin-uncoating activities (Gao *et al.*, 1992).

Experiments based on lysate-mediated reconstitution studies of protein complexes are providing convincing evidence of a requirement for hsc70 in producing functional heterocomplexes. Rabbit reticulocyte lysate reconstitutes the functional and untransformed state of the progesterone and glucocorticoid receptors by generating a heterocomplex of the receptor with both hsp90 and hsc70, as well as hsp56 (or hsp59) and some other proteins. The formation of these large heterocomplexes of progesterone receptors (Smith *et al.*, 1990, 1992) and of glucocorticoid receptors (Scherrer *et al.*, 1990; Hutchinson *et al.*, 1992) with hsp90 and hsc70 is necessary to present the receptor to the hormone. The binding of hormone to the receptors triggers the dissociation of hsp90 and hsc70 from the receptor and the hormone-receptor complex now proceeds to its nuclear binding sites. Like hsp90, hsc70 seems to bind directly to the hormone-binding domain of the receptor and the reconstitution requires ATP and certain monovalent cations. It is therefore proposed that hsc70 would bind to and unfold the receptor to a hormone binding-competent state, which is further stabilized by the association of hsp90 (Hutchinson *et al.*, 1992). The binding of hormone to the receptor then induces the dissociation of hsps and the activation of the receptor to a DNA binding conformation.

At the cellular level, mammalian hsp70/hsc70 proteins function in thermoprotection. Micro-injection of rat fibroblasts with monoclonal antibodies against both hsp70 and hsc70 proteins blocks the migration of these proteins to the nucleus and nucleolus following heat shock, and impairs the injected cells' ability to survive a heat challenge (Riabowol *et al.*, 1988). In a different approach, Chinese hamster ovary cells were transfected with a plasmid containing a 5′-control region of the hsp70 gene with multiple binding sites for the heat shock transcription factor (HSF) linked to a dihydrofolate reductase gene. DHFR-coamplification of this heat shock element (HSE)-containing region by over 20 000 fold resulted in greater than 90% reduction in the level of heat-inducible hsp70 after heat shock, presumably due to titration of HSF by the large excess of HSEs present in the cell. These cells exhibited a marked increase in thermosensitivity and showed extensive vacuolation, lysis and detachment after heat stress (Johnston and Kucey, 1988). Recently, Gloria Li and coworkers developed stably transfected rat cell lines that constitutively express the human inducible hsp70 and showed that the constitutively expressed human hsp70 is capable of nucleolar–cytoplasmic shuttling in rat cells after heat shock (Li *et al.*, 1991). The expression of human hsp70 conferred heat resistance to rat cells and the degree of thermoprotection correlated with the constitutive level of the transfected hsp70.

1.2 Antisense RNA strategies for studying hsc70 function

1.2.1 Introduction

Although considerable progress has been made toward understanding the biochemical and biophysical properties of purified hsc70 in the past few years, elucidation of its biological functions *in vivo*, particularly in mammalian cells, has just begun. Our approach to this problem involves application of the recently developed antisense RNA technology to generate hsc70-deficient mammalian cells, and analyses of the resulting pseudomutant phenotypes. Among existing antisense RNA and DNA strategies, nuclear expression of antisense RNA from an engineered antisense gene was chosen for this study because of the long half-life and abundance of the hsc70 protein *in vivo*.

Antisense gene expression has been used in a number of experimental systems to inhibit exogenous and endogenous gene expression (van der Krol *et al.*, 1988; Helene and Toulme, 1990; Takayama and Inouye, 1990; Murray and Crockett, 1992). In mammalian cells, antisense RNA has been most successful in inhibiting expression of exogenous viral or bacterial or endogenous genes encoding low abundance regulatory proteins, such as protooncogenes, while reports on inhibition of endogenous genes of abundant proteins are few and with variable degrees of success. Examples of successful suppression of endogenous genes in mammalian cells by antisense RNA include the inhibition of human c-*myc* (Yokoyama and Imamoto, 1987), human c-*fos* (Holt *et al.*, 1986), mouse c-*fos* (Nishikura and Murray, 1987), rat $pp60^{c\text{-}src}$ (Amini *et al.*, 1986) human c-*raf-1* (Kasid *et al.*, 1989), human EGF receptor (Hunts *et al.*, 1986; Moroni *et al.*, 1992), mouse tissue inhibitor of metalloproteinases (TIMP) (Khokha *et al.*, 1989), mouse HPRT (Stout and Caskey, 1987), endoB cytokeratin (Trevor *et al.*, 1987), human creatine kinase B (Ch'ng *et al.*, 1989), rat S100β-calcium binding protein (Selinfreund *et al.*, 1990), mouse β-actin (Izant and Weintraub, 1985), and in transgenic mice, the myelin basic protein (Katsuki *et al.*, 1988) and the glucocorticoid receptor (Pepin *et al.*, 1992). There are also examples of experiments where antisense RNA failed to inhibit target gene suppression in cells such as human γ-actin and β-tubulin (Gunning *et al.*, 1987), mouse H-*ras* (Salmons *et al.*, 1986), monkey CAT transgene (Kerr *et al.*, 1988) and HPRT in transgenic mice (Munir *et al.*, 1992).

The mechanism of antisense RNA regulation of gene expression is not well understood. It is generally believed that antisense RNA hybridizes to the complementary endogenous RNA to interfere with nuclear processing such as splicing and export, and with mRNA translation in the cytoplasm. It is also assumed that RNA:RNA duplexes formed in the nucleus (and possibly cytoplasm) are rapidly degraded by double-stranded-specific ribonuclease (dsRNase) or inactivated by duplex unwinding–modifying activities present in the nucleus, both of which contribute to the major effects of inhibition by antisense RNA expressed in the nucleus. In many cases, a large excess of antisense RNA over sense RNA is needed to achieve efficient inhibition. The effectiveness of antisense RNA also depends on the accessibility of both the antisense and endogenous RNA for efficient hybridization which is determined by their secondary structure, protein associations and other factors. The fact that antisense RNA can potentially block at different steps during gene expression, makes interpretation of results and particularly rational design of strategies for a given target, rather difficult. None-the-less, successful inhibition of gene expression in mammalian cells by antisense RNA expressed from artificial antisense genes has been demonstrated in many studies (van der Krol *et al.*, 1988; Takayama and Inouye, 1990).

The antisense RNA approach has been used sparingly in the study of stress proteins. McGarry and Lindquist were the first to describe the use of antisense RNA to inhibit heat shock gene expression (McGarry and Lindquist, 1986). Heat-inducible expression from an hsp70 promoter of antisense RNA complementary to mRNA for the small heat shock protein hsp26 in *Drosophila* cells resulted in varying degrees of inhibition of hsp26 induction after heat shock. In clones with high-level antisense RNA expression, the hsp26 mRNA level was significantly reduced (presumably due to degradation of duplex RNA) and the rate of hsp26 protein synthesis was decreased by more than 80%. However, no apparent changes in the patterns of protein synthesis during recovery from heat shock were observed in these cells, suggesting that either hsp26 is not an important determinant in this process or other members of the family can compensate for its reduction. Using the same system, these researchers examined the effect of inhibition of *Drosophila* hsp70 synthesis during heat shock (Lindquist *et al.*, 1988). Heat-inducible expression of antisense hsp70 RNA suppressed the heat shock induction of hsp70 synthesis to below 25% of that observed in the vector control. In compensation, hsp70 synthesis was extended for a longer period of time during recovery. Furthermore, there was an increase in the rate and duration of synthesis of hsp68, hsp83 and small hsps during recovery. The recovery of normal protein synthesis was also delayed in antisense cells after a severe heat shock. These results support the role of hsp70 in the regulation of the heat shock response.

There are only two examples of using antisense RNA to study mammalian stress proteins, both of which target the ER protein grp78 in Chinese hamster ovary (CHO) cells. Dorner and coworkers selected CHO cells that constitutively expressed high levels of antisense RNA against grp78, resulting in significant reductions in levels of grp78 mRNA and protein. No obvious phenotype was observed in these cells with regard to cell growth and viability. However, reduction of grp78 in mutant cells producing tPA led to reduced grp78 binding and increased secretion of tPA (Dorner *et al.*, 1988). Another study by an independent group using a similar strategy provided a detailed analysis on the molecular phenotype associated with the expression of grp78 antisense RNA (Li *et al.*, 1992). CHO cells expressing high levels of antisense RNA against grp78 from the strong constitutive promoter (RSV LTR) were selected through coamplification with the DHFR gene. Despite high levels of antisense RNA present, these cells have a normal basal level of grp78 protein synthesis but produce a 10-fold higher level of endogenous grp78 mRNA,

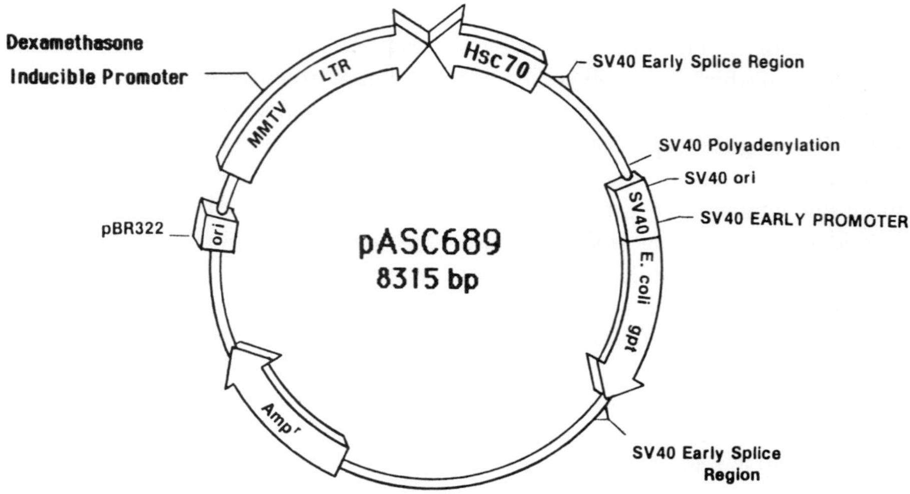

Figure 2 Antisense hsc70 expression vector pASC689. The diagram is drawn to scale and the structural features are described in the text.

suggesting the presence of a compensatory response to maintain a basal level of grp78 necessary for normal cell growth. However, the stress induction of grp78 synthesis by the calcium ionophore A23187 was blocked and the cells became sensitive to A23187 treatment, indicating the importance of grp78 in the ER stress response and perhaps calcium homeostasis. These studies demonstrate the feasibility of the antisense approach in the study of stress proteins, as well as the uncertainty of the outcome for any one target gene, cell and expression system.

1.2.2 Development of cell lines expressing antisense hsc70 RNA

Since inhibition of hsc70 synthesis may be lethal to the cells, an inducible antisense expression vector pASC689 was constructed (Figure 2). The mammalian transcription unit consists of a glucocorticoid hormone-inducible promoter from the mouse mammary tumour virus long terminal repeat (MMTV LTR), a rat hsc70 cDNA fragment with 22bp 5′UTR and one-third of the 5′ coding sequence in antisense orientation to the LTR, and the 3′SV40 early splicing region and SV40 polyadenylation sequences. A dominant selectable marker, the *E. coli* gpt gene, is expressed from the SV40 early promoter. Introduction of the construct into NIH 3T3 fibroblasts and selection using the linked marker resulted in stable transformants, from which clonal cell lines were developed. Control lines were developed using the parental vector pMSG lacking the antisense sequence. Cells were induced for antisense hsc70 RNA expression with dexamethasone and their levels of hsc70 protein synthesis were measured. Typically, each transformed line was cultured in the presence and absence of 1–2 μM dexamethasone for 10 days and the steady-state level of hsc70 synthesis was determined by silver staining of total cell protein extracts separated on one-dimensional polyacrylamide gels under denaturing conditions. This assay was used as a method of primary screening for promising clones, since the ultimate objective was to isolate cell lines with reduced hsc70 protein, after induction of antisense expression by dexamethasone.

A number of clones were identified and further analysed to determine levels of accumulation of antisense RNA and endogenous hsc70 RNA by Northern blot analysis and proteins were studied in greater detail using two-dimensional gels. The results showed that the basal expression of antisense hsc70 RNA reduced the level of endogenous hsc70 RNA but did not significantly affect the accumulation of hsc70 protein, and that the dexamethasone-induced expression of antisense RNA caused a small reduction in hsc70 protein. Different levels of regulation or a possible compensatory mechanism may be responsible for these observations.

An important clue that led to improvements in our antisense strategy came quite unexpectedly from an experiment which showed a more effective cross-inhibition of the highly inducible hsp70 by antisense hsc70 RNA in heat shocked cells, suggesting that antisense RNA was more efficient in targeting newly synthesized transcripts. To test whether antisense hsc70 RNA could block protein synthesis directed by newly synthesized hsc70 RNA, cells were first made quiescent by serum deprivation and then stimulated with serum to re-enter the cell cycle. The serum deprivation served to reduce the synthesis of hsc70 mRNA and protein and to build up a pool of antisense RNA, in part because cell division was prevented. Serum stimulation would then cause a transient, rapid increase in total protein synthesis as well as hsc70 synthesis directed by newly transcribed hsc70 RNA, which would be an effective target for antisense hsc70 RNA according to our prediction. The results using two independently isolated antisense clones support this hypothesis. Upon serum stimulation, the rate of hsc70 synthesis increased rapidly during the first 5 h and was blocked effectively in cells induced for antisense expression by dexamethasone. As shown in the scanning profiles of the 70-kDa region from one-dimensional gels (Figure 3), the increase in hsc 70 synthesis upon serum stimulation (panels A and C) was substantially reduced in the antisense clone (panel D) but not in the control (panel B) when dexamethasone was present in cultures during serum starvation and stimulation. Interestingly, a reduction in the rate of grp78 synthesis was also noted in dexamethasone-induced antisense clones (panel D). These results were confirmed and quantified using two-dimensional polyacrylamide gels. Quantification of these gels showed that hsc70 synthesis was reduced by 52% after dexamethasone induction in the highest antisense RNA-expressing clone but was only reduced by 13% in the control clone. They indicated that the antisense RNA was effective in reducing a transient and rapid increase of hsc70 synthesis by targeting newly synthesized and possibly nuclear hsc70 transcripts. The quiescent state was important for these observations since we found that actively dividing cells contain high levels of hsc70 RNA and protein, rendering antisense RNA less effective.

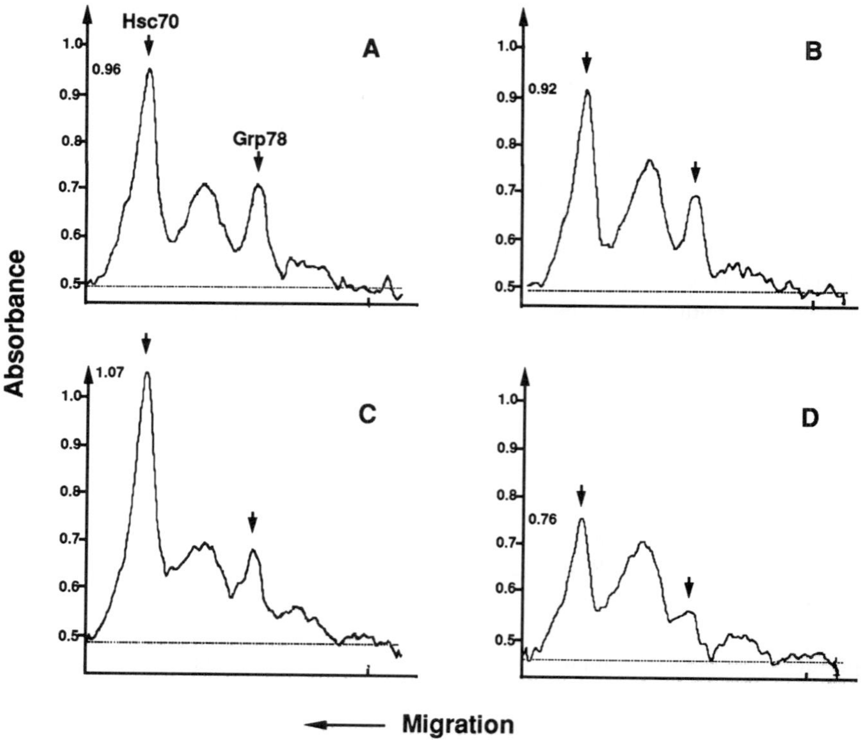

Figure 3 Scan profiles of the 70-kDa region of a one-dimensional gel. The 70-kDa regions of fluorographs of one-dimensional polyacrylamide gels (SDS-PAGE) were scanned using an LKB UltroScan densitometer. The absorbance values, offset from the *y*-axis, for the maximum peak heights of hsc70 are indicated and the dotted lines mark the background absorbance. Arrows indicate the hsc70 and grp78 peaks: (A) vector-transformed control cells minus dexamethasone; (B) vector transformed control cells plus dexamethasone; (C) antisense-transformed cells minus dexamethasone; (D) antisense-transformed cells plus dexamethasone.

1.2.3 Phenotype of hsc70-deficient mammalian cells

It was anticipated that hsc70 would be an essential protein in mammalian cells, so the ultimate phenotype of the hsc70-deficient cell should be inviability. Indeed, we found that prolonged serum deprivation was lethal to cells expressing basal levels of antisense hsc70 RNA, that is, cells that had not been treated with dexamethasone. Ironically, dexamethasone treatment appeared to protect the viability of cells expressing antisense RNA. When cells were put under extended periods of serum deprivation, we noticed that cells from an uninduced (minus dexamethasone) antisense clone were killed, while cells from a control clone remained quiescent. We found that after three days of serum deprivation both control cells and antisense cells were quiescent as shown in part by the accumulation of cytoplasmic stress fibres. Control cells continued to accumulate stress fibres and remained quiescent beyond day six of serum deprivation (Figure 4C). However, localized cell death appeared in antisense cells at day four of serum deprivation and extended to the entire culture by day 6 (Figure 4A). Two major morphological changes were associated with antisense cell death. The main class of dead cells remained attached to the culture substratum and had contracted cytoplasm with darkened or empty nuclei. The other type of dead cell rounded up to form a sac of fine particles of degraded structures in the cytoplasm before it detached or lysed. In either case, we often observed distinct and thickened nuclear boundaries, especially during the early stage of cell death, suggesting that nuclei might be an early target during such cell killing. The lethality observed here was due to the basal expression of antisense hsc70 RNA since no dexamethasone was used. The basal expression of antisense RNA together with serum deprivation apparently reduced the amount of hsc70 below the level that is required for cell survival under such conditions. Other explanations for the lethality have not been ruled out. For example, double-stranded RNA itself might trigger a lethal response under conditions of serum deprivation. And recently, it was discovered that the mouse U14 snRNA is transcribed as part of hsc70 pre-mRNA and is excised directly from the introns to produce mature U14 snRNA (Leverette *et al.*, 1992). Reduction of hsc70 RNA levels by antisense RNA may result in depletion of U14 snRNA, which is probably a lethal event, based on previous mutational analyses of yeast (Jarmolowski *et al.*, 1990; Li *et al.*, 1990).

Since prolonged serum deprivation was lethal during basal antisense hsc70 RNA expression, further induction of antisense RNA by dexamethasone was expected to accelerate cell killing. The results showed just the opposite. No cell death was observed when dexamethasone was added to cultures during prolonged serum deprivation and antisense cells remained quiescent and contained stress fibres (Figure 4B). Control cells treated with dexamethasone appeared less quiescent than untreated cells and accumulated fewer stress fibres during serum deprivation (Figure 4D). These results demonstrated that dexamethasone protected 3T3 cells from the lethal effects of antisense hsc70 expression during serum deprivation, presumably by altering the physiological state of the cells.

1.2.4 Antisense cell death

A straightforward but not unique interpretation for the cell killing observed here is necrosis or cytotoxicity, i.e. reduction of hsc70 caused deficiencies in cellular functions leading to lethal metabolic or structural abnormalities. Deficiency in hsc70 may cause accumulation and aggregation of unfolded polypeptides, particularly unstable nascent polypeptide chains (Beckmann *et al.*, 1990) and thermally sensitive proteins (Lepock *et al.*, 1988), leading to general proteotoxicity (Hightower, 1991). Early morphological changes in the nucleus may suggest structural or functional damage due to reduced chaperoning functions. It has been shown that hsc70 binds

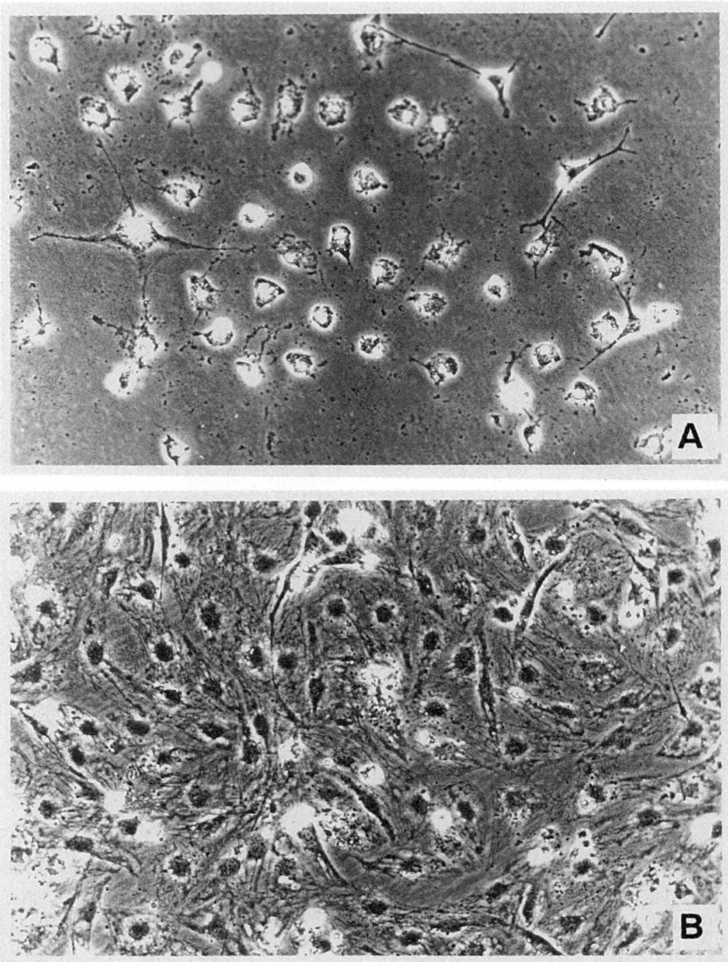

Figure 4 Light photomicrographs of NIH 3T3 cells showing the lethal effects of basal antisense hsc70 RNA expression at day six of serum deprivation and its protection by dexamethasone: (A) antisense-transformed cells, minus dexamethasone, showing death of the entire culture; (B) antisense-transformed cells, plus dexamethasone, showing a quiescent cul-

to and facilitates lysosomal degradation of selected cytosolic proteins containing the KFERQ-motif, which increases during serum withdrawal (Chiang *et al.*, 1989). Antisense inhibition of hsc70 synthesis in cells under prolonged serum deprivation may affect this response. Another possibility is that perturbed hsc70-cytoskeleton interactions may eventually affect the structural integrity of cellular organelles. Protein import into organelles and clathrin-mediated membrane trafficking are additional processes in which an hsc70 deficiency might cause cytotoxicity. As

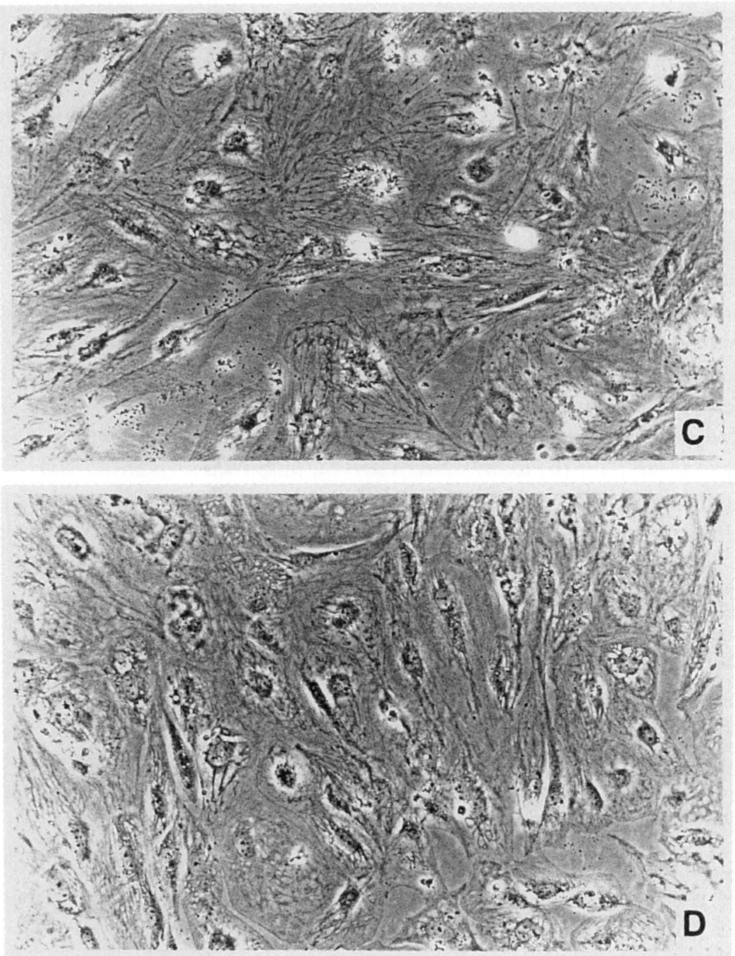

ture with abundant stress fibres; (C) vector-transformed control cells, minus dexamethasone, shown in deep quiescence with numerous stress fibres and filaments; (D) vector-transformed control cells, plus dexamethasone, showing quiescence with fewer stress fibres.

discussed above, other possible causes of cell killing such as depletion of U14 snRNA have not been ruled out.

The kinetics of cell death and the associated morphologies, particularly those of the cells that remained attached and contracted after death, suggest an alternative interpretation, programmed cell death or apoptosis (Bursch *et al.*, 1990; Ellis *et al.*, 1991). Antisense cells appeared to die rapidly at a critical time of growth factor depletion and showed certain morphological characteristics of apoptosis such as cytoplasm contraction, darkened nuclei and thickened nuclear boundaries (possible signs of chromatin condensation around the periphery of the nucleus), eventual rounding and detachment. As seen in apoptotic cell death, many dead cells became contracted and did not have the uncontrolled swelling and total membrane breakdown seen in necrosis. We do not have evidence as yet as to whether the diagnostic DNA fragmentation (internucleosomal DNA cleavage) found in classical apoptosis is present in these cells. Further experiments are required to clarify whether apoptosis or necrosis (cytotoxicity) is involved in the cell killing process.

Dexamethasone protected 3T3 cells from the lethal effects of antisense hsc70 expression under prolonged serum deprivation, despite the increased accumulation of antisense hsc70 RNA. We observed that cells treated with dexamethasone were less quiescent and contained fewer stress fibres. We also found dexamethasone did not directly affect the expression of hsc70 or other stress proteins with the exception of hsp90, but did alter the expression of several unknown proteins in NIH 3T3 cells. This is consistent with earlier observations that dexamethasone induced hsp90 but has no effect on the synthesis of grps or other hsps (Kasambalides and Lanks, 1983; Kasambalides and Lanks, 1985; Fisher *et al.*, 1986), and that dexamethasone induced the synthesis of some non-heat shock proteins (Maytin and Young, 1983). Thus the effects of dexamethasone were demonstrated at both the morphological and molecular levels. Finally, it has been reported very recently that dexamethasone induces the accumulation of αB-crystallin, a low molecular weight stress-protein, which led to the development of a thermoresistant state in NIH 3T3 cells (Aoyama *et al.*, 1993). It is therefore possible that αB-crystallin may also be responsible for the protective effect of dexamethasone against the antisense induced cell death that we have observed.

1.3 Future directions

1.3.1 Molecular chaperoning

The mammalian hsp70 family will continue to feature prominently in studies of the new concept of molecular chaperoning and *in vivo* protein folding and assembly. Ironically, we know much more about the structure and activities of the constitutive members of this family, hsc70 and grp78, than we do about stress-inducible hsp70. The question remains as to whether hsp70 is simply more hsc70, i.e. the two are functionally identical but stressed cells need increased chaperoning capacity, or whether they have different functions. In particular, can one or both of these proteins facilitate the renaturation of denatured proteins and/or solubilize protein aggregates? Of the two major studies done with the *E. coli* homologue of these proteins, DnaK, to answer this question, one was retracted (Gaitanaris *et al.*, 1990,

1992) and the other may involve reversible association of core RNA polymerase subunits, not actual denaturation and aggregation (Skowyra *et al.*, 1990). Hsp70 and hsc70 co-localize to nucleoli in stressed cells and there is now evidence that these two proteins may form ATP-sensitive complexes in cells, favouring the idea that they may perform very similar if not identical functions (Brown *et al.*, 1993). However, as discussed earlier, Hsc70 and grp78 are not functionally interchangeable in microsomal transport assays and even two highly related yeast hsp70 isoforms differ in protein substrate recognition. Studies underway in several laboratories using random peptide phage display libraries to determine consensus recognition sequences for various hsp70 family members should aid in resolving this issue.

Another major effort underway is the search for accessory proteins in eukaryotic cells analogous to the bacterial DnaJ and GrpE proteins which regulate the activity of DnaK [see Chapter 8 by Cheetham *et al.* on *Neuronal homologues of the bacterial heat shock protein DNAJ*]. DnaJ homologues have been found in yeast and the cytoplasmic homologue YDJ1 interacts with a yeast hsp70 family member to stimulate its ATPase activity and affect its interactions with unfolded proteins (Cyr *et al.*, 1992). So far, no eukaryotic equivalent to bacterial GrpE, which stimulates adenine nucleotide exchange, has been reported. However, ATP/ADP exchange is rate-limiting for the ATPase activity of purified bovine hsc70 *in vitro*, so this can be a likely point of regulation for this protein as well (Sadis and Hightower, 1992). It is becoming clear that molecular chaperones like the hsp70 family do not work alone and are part of protein shuttling pathways in cells that include steps at which folding is blocked and others at which folding is facilitated.

1.3.2 Hsp70 proteins as carriers of peptide-mimetic drugs

Based upon our knowledge of the interaction of molecular chaperones with short peptides, it seems likely that peptides are usually bound to these proteins within cells and possibly in extracellular fluids as well. This may be of particular relevance to nervous tissue in which neurotropic and psychoactive peptides play important physiological roles and to which access is limited by structures such as the blood–brain barrier. Once the consensus sequences for peptide-bonding to chaperones have been determined, it should be possible to design modified peptides or peptide-mimetic drugs that could follow chaperoning pathways, both within cells and perhaps also across boundary cell layers and into tissues. In addition, as information develops on possible roles for hsp70 family proteins in antigen processing and presentation, it may be possible to use peptide-mimetic drugs to modulate immune responses. For example, hsc70 binds to deoxyspergualin (DSG), a synthetic analogue of a bacterial compound with strong immunosuppressive activities and which bears structural similarity to several basic amino acid side chains. Nadler and colleagues have suggested several possible modes of action for DSG, including direct inhibition of hsc70, transport of DSG along intracellular chaperoning pathways to sites of immunosuppressive activity, and modulation of glucocorticoid receptor activity through interactions with hsc70 in the receptor complex (Nadler *et al.*, 1992). Also, one of the yeast hsp70 homologues binds to a cyclosporin A photo-affinity reagent (Moss *et al.*, 1992). Cyclosporin A is a cyclic undecapeptide and the most frequently used immunosuppressive drug for the prevention of tissue graft rejection. Understanding the significance of the interaction of these drugs with molecular chaperones may further our understanding of the mechanism of action of these drugs, suggest ways to increase their effectiveness, and provide clues to reducing undesirable side-effects.

Acknowledgements

This chapter is dedicated to Fredric P. White, an excellent explorer and travelling companion on the frontiers of science. I would like to acknowledge the many contributions of our former research technician Carol N. White and former graduate students Peter Guidon, Jr, Sandra Whelan, Brent Edington, Seth Sadis, and Ivone Takenaka. The antisense studies described are part of Ting Li's PhD dissertation and benefited from the input of current graduate students Judy Boice, Mary Brown, May Leung, John Ryan and Carol White. We thank Mary Brown for critically reading the manuscript.

Work in the authors' laboratory was supported by the National Science Foundation, National Institutes of Health, and benefited from the use of the University of Connecticut Biotechnology Center.

References

Abravaya, K., Myers, M.P., Murphy, S.P., & Morimoto, R. (1992) The human heat shock protein hsp70 interacts with HSF, the transcription factor that regulates heat shock-gene expression. *Genes & Devel.* **6**, 1153–1164.

Ahmad, S., Ahuja, R., Venner, T.J., & Gupta, R.S. (1990) Identification of a protein altered in mutants resistant to microtubule inhibitors as a member of the major heat shock-protein (hsp70) family. *Mol. Cell. Biol.* **10**, 5160–5165.

Amini, S., DeSeau, V., Reddy, S., Shalloway, D., & Bolen, J.B. (1986) Regulation *pp60*$^{c\text{-}src}$ synthesis by inducible RNA complementary to c-*src* mRNA in polyomavirus-transformed rat cells. *Mol. Cell. Biol.* **6**, 2305–2316.

Aoyama, A., Frohli, E., Schafer, R., & Klemenz, R. (1993) αB-Crystallin expression in mouse NIH 3T3 fibroblasts: Glucocorticoid responsiveness and involvement in thermal protection. *Mol. Cell. Biol.* **13**, 1824–1835.

Baler, R., Dahl, G., & Voellmy, R. (1993) Activation of human heat shock genes is accompanied by oligomerization, modification, and rapid translocation of heat shock transcription factor HSF1. *Mol. Cell. Biol.* **13**, 2486–2496.

Baler, R., Welch, W.J., & Voellmy, R. (1992) Heat shock-gene regulation by nascent polypeptides and denatured proteins: hsp70 as a potential autoregulatory factor. *J. Cell. Biol.* **117**, 1151–1159.

Beckmann, R.P., Lovett, M., & Welsh, W.J. (1992) Examining the function and regulation of hsp70 in cells subjected to metabolic stress. *J. Cell Biol.* **117**, 1137–1150.

Beckmann, R.P., Mizzen, L.A., & Welch, W.J. (1990) Interaction of hsp70 with newly synthesized proteins: Implications for protein folding and assembly. *Science* **248**, 850–854.

Black, M.M., Chestnut, M.H., Pleasure, I.T., & Keen, J.H. (1991) Stable clathrin-uncoating protein (hsc70) complexes in intact neurons and their axonal transport. *J. Neuroscience* **11**, 1163–1172.

Brown, C.R., Martin, R.L., Hansen, W.J., Beckmann, R.P., & Welch, W.J. (1993) The constitutive and stress inducible forms of hsp70 exhibit functional similarities and interact with one another in an ATP-dependent fashion. *J. Cell Biol.* **120**, 1101–1112.

Brown, I.R., Cosgrove, J.W., & Clark, B.D. (1982) Physiologically relevant increases in body temperature induce the synthesis of a heat shock protein in mammalian brain and other organs. In *Heat Shock from Bacteria to Man* (eds Schlesinger, M.J., Ashburner, M. and Tissieres, A.), pp. 361–368. New York, Cold Spring Harbor Laboratory.

Brown, I.R., Rush, S.J., & Ivy, G.O. (1989) Induction of a heat shock gene at the site of tissue injury in the rat brain. *Neuron* **2**, 1559–1564.

Bursch, W., Kleine, L., & Tenniswood, M. (1990) The biochemistry of cell death by apoptosis. *Biochem. Cell Biol.* **68**, 1071–1074.

Calderwood, S.K., & Stevenson, M.A. (1993) Inducers of the heat shock response stimulate phospholipase C and phospholipase A_2 activity in mammalian cells. *J. Cell. Physiol.* **155**, 248–256.

Ch'ng, J.L.C., Mulligan, R.C., Schimmel, P., & Holmes, E.W. (1989) Antisense RNA complementary to 3′ coding and non-coding of creatine kinase is a potent inhibitor of translation *in vivo*. *Proc. Natl Acad. Sci. USA* **86**, 10 006–10 010.

Chappell, T.G., Welch, W.J., Schlossman, D.M., Palter, K.B., Schlesinger, M.J., & Rothman, J.E. (1986) Uncoating ATPase is a member of the 70-kilodalton family of stress proteins. *Cell* **45**, 3–13.

Chiang, H.-L., Terlecky, S.R., Plant, C.S., & Dice, J.F. (1989) A role for a 70-kilodalton heat shock protein in lysosomal degradation of intracellular proteins. *Science* **246**, 382–385.

Chirico, W.J., Walters, M.G., & Blobel, G. (1988) 70K heat shock related proteins stimulate translocation into microsomes. *Nature* **332**, 805–810.

Craig, E.A., & Gross, C.A. (1991). Is hsp70 the cellular thermometer? *TIBS (Trends Biochem. Sci.)* **16**, 135–140.

Currie, R.W., & White, F.P. (1983) Characterization of the synthesis and accumulation of a 71-kilodalton protein induced in rat tissues after hyperthermia. *Can. J. Biochem. and Cell Biol.* **61**, 438–446.

Cyr, D.M., Lu, X.Y., & Douglas, M.G. (1992) Regulation of hsp70 function by a eukaryotic DnaJ homologue. *J. Biol. Chem.* **267**, 20 927–20 931.

Davis, J.Q., Dansereau, D., Johnstone, R.M., & Bennett, V. (1986) Selective externalization of an ATP-binding protein structurally related to the clathrin-uncoating ATPase/heat shock protein in vesicles containing terminal transferrin receptors during reticulocyte maturation. *J. Biol. Chem.* **261**, 15 368–15 371.

Deshaies, R., Koch, B., Werner-Washburne, M., Craig, E.A., & Scheckman, R. (1988) 70kD stress-protein homologues facilitate translocation of secretory and mitochondrial precursor polypeptides. *Nature* **332**, 800–805.

DiDomenico, B.J., Bugaisky, G.E., & Lindquist, S. (1982) The heat shock response is self-regulated at both the transcriptional and posttranslational levels. *Cell* **31**, 593–603.

Dorner, A.J., Krane, M.G., & Kaufman, R.J. (1988) Reduction of endogenous GRP78 levels improves secretion of a heterologous protein in CHO cells. *Mol. Cell. Biol.* **8**, 4063–4070.

Dwyer, B.E., Nishimura, R.N., de Vellis, J., & Clegg, K.B. (1991) Regulation of heat shock protein synthesis in rat astrocytes. *J. Neurosci. Res.* **28**, 352–358.

Edington, B.V., Whelan, S.A., & Hightower, L.E. (1989) Inhibition of heat shock (stress) protein induction by deuterium oxide and glycerol: Additional support for the abnormal protein hypothesis of induction. *J. Cell. Physiol.* **139**, 219–228.

Ellgard, E.G., & Clever, U. (1971) RNA metabolism during puff induction in *Drosophila melanogaster. Chromosoma* **36**, 60–78.

Ellis, R.E., Yuan, J., & Horvitz, R. (1991) Mechanisms and functions of cell death. *Annu. Rev. Cell Biol.* **7**, 663–698.

Ellis, R.J., & van der Vries, S.M. (1991) Molecular chaperones. *Annu. Rev. Biochem.* **60**, 321–347.

Fisher, G.A., Anderson, R.L., & Hahn, G.M. (1986) Glucocorticoid-induced heat resistance in mammalian cells. *J. Cell. Physiol.* **128**, 127–132.

Flaherty, K.M., DeLuca-Flaherty, C., & McKay, D.B. (1990) Three-dimensional structure of the ATPase fragment of a 70K heat shock cognate protein. *Nature* **346**, 623–628.

Flynn, G.C., Chappell, T.G., & Rothman, J.E. (1989) Peptide binding and release by proteins implicated as catalysts of protein assembly. *Science* **245**, 385–390.

Gaitanaris, G.A., Papavassiliou, A.G., Rubock, P., Siverstein, S.J., & Gottesman, M.E. (1990) Renaturation of denatured λ-repressor requires heat shock proteins. *Cell* **61**, 1013–1020.

Gaitanaris, G.A., Papavassiliou, A.G., Rubock, P., Siverstein, S.J., & Gottesman, M.E. (1992) Retraction: Renaturation of denatured λ-repressor requires heat shock proteins. *Cell* **70**, 714.

Gao, B., Biosca, J., Craig, E.A., Green, L.E., & Eisenberg, E. (1992) Uncoating of coated vesicles by yeast hsp70 proteins. *J. Biol. Chem.* **226**, 19 565–19 571.

Green, L.A.D., & Liem, R.K.H. (1989) β-internexin is a microtubule-associated protein identical to the 70-kDa heat shock cognate protein and the clathrin-uncoating ATPase. *J. Biol. Chem.* **264**, 15 210–15 215.

Guidon, P.T., Jr., & Hightower, L.E. (1986a) Purification and initial characterization of the 71-kDa rat heat shock protein and its cognate as fatty acid binding proteins. *Biochemistry* **25**, 3231–3239.

Guidon, P.T., Jr., & Hightower, L.E. (1986b) The 73-kilodalton heat shock cognate protein purified from rat brain contains nonesterified palmitic and stearic acids. *J. Cell. Physiol.* **128**, 239–245.

Gunning, P., Leavitt, J., Muscat, G., Ng, S.-U., & Kedes, L. (1987) A human β-actin expression vector system directs high-level accumulation of antisense transcripts. *Proc. Natl Acad. Sci. USA* **84**, 4831–4835.

Gupta, R.S. (1990) Microtubules, mitochondria, and molecular chaperones: a new hypothesis for *in vivo* assembly of microtubules. *Biochem. Cell Biol.* **68**, 1352–1363.

Heine, U., Sverak, L., Kondratick, J., & Bonar, R.A. (1971) The behavior of HeLa-S3 cells under the influence of supranormal temperatures. *J. Ultrastruct. Res.* **34**, 375–396.

Helene, C., & Toulme, J.-J. (1990) Specific regulation of gene expression by antisense, sense and antigene nucleic acids. *Biochem. Biophys. Acta* **1049**, 99–125.

Hightower, L.E. (1980) Cultured animal cells exposed to amino acid analogues or puromycin rapidly synthesize several polypeptides. *J. Cell. Physiol.* **102**, 407–427.

Hightower, L.E. (1991) Heat shock stress proteins, chaperones, and proteotoxicity. (Meeting Review) *Cell* **66**, 191–197.

Hightower, L.E., & Guidon, P.T., Jr. (1989) Selective release from cultured mammalian cells of heat shock (stress) proteins that resemble glia-axon transfer proteins. *J. Cell. Physiol.* **138**, 257–266.

Hightower, L.E., Guidon, P.T., Jr., Whelan, S.A., & White, C.N. (1985) Stress responses in avian and mammalian cells. In *Changes in Eukaryotic Gene Expression in Response to Environmental Stress* (eds Atkinson, B.G. and Walden, D.B.), pp 197–210. New York, Academic Press.

Hightower, L.E., & Smith, M.D. (1978) Effects of canavanine on protein metabolism in Newcastle disease virus-infected and uninfected chicken embryo cells. In *Negative strand viruses and the host cell* (eds Mahy, B.W.J. and Barry, R.D.), pp 395–405. London, Academic Press.

Hightower, L.E., & White, F.P. (1981) Cellular responses to stress: comparison of a family of 71–73 kilodalton proteins rapidly synthesized in rat tissue slices and canavanine-treated cells in culture. *J. Cell. Physiol.* **108**, 261–275.

Holt, J.T., Gopal, T.V., Moulton, A.D., & Nienhuis, A.W. (1986) Inducible production of c-fos antisense RNA inhibits 3T3 cell proliferation. *Proc. Natl Acad. Sci. USA* **83**, 4794–4798.

Hughes, E.N., & August, J.T. (1982) Coprecipitation of heat shock proteins with a cell surface glycoprotein. *Proc. Natl Acad. Sci. USA* **79**, 2305–2309.

Hunts, J., Merlino, G., Pastan, I., & Shimizu, N. (1986) Reduction of EGF receptor synthesis by antisense RNA vectors. *FEBS Lett.* **206**, 319–322.

Hutchinson, K.A., Czar, M.J., Scherrer, L.C., & Pratt, W.B. (1992) Monovalent cation selectivity for ATP-dependent association of the glucocorticoid receptor with hsp70 and hsp90. *J. Biol. Chem.* **267**, 14 047–14 053.

Imamoto, N., Matsuoka, Y., Kurihara, T., Kohno, K., Miyagi, M., Sakiyama, F., Okada, Y., Tsunasawa, S., & Yoneda, Y. (1992) Antibodies against 70-kD heat shock cognate protein inhibit mediated nuclear import of karyophilic proteins. *J. Cell Biol.* **119**, 1047–1061.

Izant, J.G., & Weintraub, H. (1985) Constitutive and conditional supression of exogenous and endogenous genes by antisense RNA. *Science* **229**, 345–352.

Jarmolowski, A., Zagorski, J., Li, H., & Fournier, M.J. (1990) Identification of essential elements in U14 RNA of *Saccharomyces cereviciae. EMBO J.* **9**, 4503–4509.

Johnston, R.N., & Kucey, B.L. (1988) Competitive inhibition of hsp70 gene expression causes thermosensitivity. *Science* **242**, 1551–1554.

Kasambalides, E.J., & Lanks, K.W. (1983) Dexamethasone can modulate glucose-regulated and heat shock protein synthesis. *J. Cell. Physiol.* **114**, 93–98.

Kasambalides, E.J., & Lanks, K.W. (1985) Antagonistic effects of insulin and dexamethasone on glucose-regulated and heat shock protein synthesis. *J. Cell. Physiol.* **123**, 283–287.

Kasid, U., Pfeifer, A., Brenna, T., Beckett, M., Weichselbaum, R.R., Dritschilo, A., & Mark, G.E. (1989) Effect of antisense c-raf-1 on tumorigenicity and radiation sensitivity of a human squamous carcinoma. *Science* **243**, 1354–1356.

Katsuki, M., Sato, M., Kimura, M., Yokoyama, M., Kobayashi, K., & Nomura, T. (1988) Conversion of normal behavior to shiverer by myelin basic protein antisense cDNA in transgenic mice. *Science* **241**, 593–595.

Kelley, P.M., & Schlesinger, M.J. (1978) The effect of amino acid analogues and heat shock on gene expression in chick embryo fibroblasts. *Cell* **15**, 1277–1286.

Kerr, S.M., Stark, G.R., & Kerr, I.M. (1988) Excess antisense RNA from infectious recombinant SV40 fails to inhibit expression of a transfected, inteferon-inducible gene. *Eur. J. Biochem.* **175**, 65–73.

Khoka, R., Waterhouse, P., Yagel, S., Lala, P.K., Overall, C.M., Norton, G., & Denhardt, D.T. (1989) Antisense RNA-induced reduction in murine TIMP levels confers oncogenicity on Swiss 3T3 cells. *Science* **243**, 947–950.

LaThangue, N.B. (1984) A major heat shock protein defined by a monoclonal antibody. *EMBO J.* **3**, 1871–1879.

Lee, A.S. (1992) Mammalian stress response: induction of the glucose-regulated protein family. *Current Opinion in Cell Biology* **4**, 267–273.

Lepock, J.R., Frey, H.E., Rodahl, A.M., & Kruuv, J. (1988) Thermal analysis of CHL V79 cells using differential scanning calorimetry: Implications for hyperthermic cell killing and the heat shock response. *J. Cell. Physiol.* **137**, 14–24.

Leverette, R.D., Andrews, M.T., & Maxwell, E.S. (1992) Mouse U14 snRNA is a processed intron of the cognate hsc70 heat shock pre-messenger RNA. *Cell* **71**, 1215–1221.

Lewis, M.J., & Pelham, H.R.B. (1985) Involvement of ATP in the nuclear and nucleolar functions of the 70-kDa heat shock protein. *EMBO J.* **4**, 3137–3143.

Li, G.C., Li, L., Liu, Y., Mak, J.Y., Chen, L., & Lee, W.M. (1991) Thermal response of rat fibroblasts stably transfected with human 70-kDa heat shock protein-encoding gene. *Proc. Natl Acad. Sci. USA* **88**, 1681–1685.

Li, H., Zagorski, J., & Fournier, M.J. (1990) Depletion of U14 small nuclear RNA (snR128) disrupts production of 18S rRNA in *Saccharomyces cereviciea. Mol. Cell. Biol.* **10**, 1145–1152.

Li, L.-J., Li, X., Ferrario, A., Rucker, N., Liu, E.S., Wong, S., Gomer, C.J., & Lee, A.S. (1992) Establishment of a Chinese hamster ovary cell line that expresses grp78 antisense transcripts and suppresses A23187 induction of both GRP78 and GRP94. *J. Cell. Physiol.* **153**, 575–582.

Lim, L., Hall, C., Leung, T., & Whatley, S. (1984) The relationship of the rat brain 68-kDa microtubule-associated protein with synaptosomal plasma membranes and with the *Drosophila* 70-kDa heat shock protein. *Biochem. J.* **224**, 677–680.

Lindquist, S., McGarry, T.J., & Golic, K. (1988) Use of antisense RNA in studies of the heat shock response. In *Antisense RNA and DNA* (ed Melton, D.A.), pp 71–77. Cold Spring Harbor, New York, Cold Spring Harbor Laboratory Press.

Mayrand, S., & Pederson, T. (1983) Heat shock alters nuclear ribonucleoprotein assembly in *Drosophila* cells. *Mol. Cell. Biol.* **3**, 161–171.

Maytin, E.Y., & Young, D.A. (1983) Separate glucocorticoid, heavy metal, and heat shock domains in thymic lymphocytes. *J. Biol. Chem.* **258**, 12 718–12 722.

McGarry, T.J., & Lindquist, S. (1986) Inhibition of heat shock protein synthesis by heat-inducible antisense RNA. *Proc. Natl Acad. Sci. USA* **83**, 399–403.

Milarski, K.L., & Morimoto, R.I. (1989) Mutational analysis of the human hsp70 protein: Distinct domains for nucleolar localization and adenosine triphosphate binding. *J. Cell Biol.* **109**, 1947–1962.

Morimoto, R.I. (1993) Cells in Stress: Transcriptional Activation of Heat Shock Genes. *Science* **259**, 1409–1410.

Morimoto, R.I., Sarge, K.D., & Abravaya, K. (1992) Transcriptional regulation of heat shock genes. *J. Biol. Chem.* **267**, 21 987–21 990.

Morimoto, R.I., Tissieres, A., & Georgopoulos, C. (1990) Stress Proteins in Biology and Medicine. Cold Spring Harbor, New York, Cold Spring Harbor Laboratory Press.

Moroni, M.C., Willingham, M.C., & Beguinot, L. (1992) EGF-R antisense RNA blocks expression of the epidermal growth factor receptor and suppresses the transforming phenotype of a human carcinoma cell line. *J. Biol. Chem.* **267**, 2714–2722.

Moss, M.L., Palmer, R.E., Kuzmic, P., Dunlap, B.E., Henzel, W., Kofron, J.L., Mellon, W.S., Royer, C.A., & Rich, D.H. (1992) Identification of actin and HSP-70 as cyclosporin-A binding proteins by photoaffinity labelling and fluorescence displacement assays. *J. Biol. Chem.* **267**, 22 054–22 059.

Munir, M.I., Rossiter, B.J.F., & Caskey, C.T. (1992) Antisense RNA production in mammalian fibroblasts and transgenic mice. In *Antisense RNA and DNA* (ed Murray, J.A.H.), pp 97–108. New York, Wiley-Liss, Inc.

Murray, J.A.H., & Crockett, N. (1992) Antisense techniques: an overview. In *Antisense RNA and DNA* (ed Murray, J.A.H.), pp 1–49. New York, Wiley-Liss Inc.

Nadler, S.G., Tepper, M.A., Schacter, B., & Mazzucco, C.E. (1992) Interaction of the immunosuppressant deoxyspergualin with a member of the hsp70 family of heat shock proteins. *Science* **258**, 484–486.

Nakai, A., & Morimoto, R.I. (1993) Characterization of a novel chicken heat shock transcription factor, heat shock factor-3, suggests a new regulatory pathway. *Mol. Cell Biol.* **13**, 1983–1997.

Napolitano, E.W., Pachter, J.S., Chin, S.S.M., & Leim, R.K.H. (1985) β-internexin, a ubiquitous intermediate filament-associated protein. *J. Cell. Biol.* **101**, 1323–1331.

Napolitano, E.W., Pachter, J.S., & Liem, R.K.H. (1987) Intracellular distribution of mammalian stress proteins. *J. Biol. Chem.* **262**, 1493–1504.

Nishikura, K., & Murray, J.M. (1987) Antisense RNA of proto-oncogene c-fos blocks renewed growth of quiescent 3T3 cells. *Mol. Cell. Biol.* **7**, 639–649.

Nishimura, R.N., Dwyer, B.E., Clegg, K., Cole, R., & de Vellis, J. (1991) Comparison of the heat shock response in cultured cortical neurons and astrocytes. *Mol. Brain Res.* **9**, 39–45.

Nover, L. (1991) Heat Shock Response. Boca Raton, CRC Press, Inc.

Palleros, D.R., Welch, W.J., and Fink, A.L. (1991) Interaction of hsp70 with unfolded proteins: Effects of temperature and nucleotides on the kinetics of binding. *Proc. Natl Acad. Sci. USA* **88**, 5719–5723.

Pearce, B.R., Dutton, G.R., & White, F.P. (1983) Induction of a stress protein in developing cell cultures of the rat cerebellum. *J. Neurochem.* **41**, 291–294.

Pelham, H., Lewis, M., & Munor, S. (1985) A nucleolar function for mammalian and *Drosophila* 70-kDa heat shock proteins. In *Sequence Specificity in Transcription and Translation* (eds Calendar, R. and Gold, L.), pp 681–688. New York, Alan R. Liss.

Pelham, H.R.B. (1984) Hsp70 accelerates the recovery of nucleolar morphology after heat shock. *EMBO J.* **3**, 3095–3100.

Pepin, M., Pothier, M.F., & Barden, N. (1992) Impaired type II glucocorticoid-receptor function in mice bearing antisense RNA transgene. *Nature* **355**, 725–728.

Perisic, O., Xiao, H., & Lis, J.T. (1989) Stable binding of *Drosophila* heat shock factor to head-to-head and tail-to-tail repeats of a conserved 5 bp recognition unit. *Cell* **59**, 797–806.

Rabindran, S.K., Haroun, R.I., Clos, J., Wisniewski, J., & Wu, C. (1993) Regulation of heat shock factor trimer formation – role of a conserved leucine zipper. *Science* **259**, 230–234.

Riabowol, K.T., Mizzen, L.A., & Welch, W.J. (1988) Heat shock is lethal to fibroblasts micro-injected with antibodies against hsp70. *Science* **242**, 433–436.

Rippmann, F., Taylor, W., Rothbard, J., & Green, N. (1991) A hypothetical model for the peptide binding domain of hsp70 based on the peptide domain of HLA. *EMBO J.* **10**, 1053–1059.

Ritossa, F. (1962) A new puffing pattern induced by temperature shock and DNP in *Drosophila. Experientia* **18**, 571–573.

Robin, G.M., & Hogness, D.S. (1975) Effects of heat shock on low molecular weight RNA's in *Drosophila melanogaster*: Accumulation of a novel form of 5S RNA. *Cell* **6**, 207–213.

Rodriguez de Turco, E.B., & Bazan, N.G. (1983) Changes in free fatty acids and diglycerides in mouse brain at birth and during anoxia. *J. Neurochem.* **41**, 794–800.

Sadis, S., Hickey, E., & Weber, L. (1988) Effects of hyperthermia on RNA metabolism in HeLa cells. *J. Cell. Physiol.* **135**, 377–386.

Sadis, S., & Hightower, L.E. (1992) Unfolded proteins stimulate molecular chaperone hsc70 ATPase by accelerating ADP/ATP exchange. *Biochemistry* **31**, 9406–9412.

Sadis, S., Raghavendra, K., & Hightower, L.E. (1990) Secondary structure of the mammalian 70-kDa heat shock cognate protein analysed by circular dichroism spectroscopy and secondary structure prediction. *Biochemistry* **29**, 1899–8206.

Salmons, B., Groner, B., Friis, R., Meullener, D., & Jaggi, R. (1986) Expression of anti-sense mRNA in H-*ras* transfected NIH/3T3 cells does not suppress the transformed phenotype. *Gene* **45**, 215–220.

Sarge, K.D., Murphy, S.P., & Morimoto, R.I. (1993) Activation of heat shock gene transcription by heat shock factor-1 involves oligomerization, acquisition of DNA-binding activity and nuclear localization and can occur in the absence of stress. *Mol. Cell Biol.* **13**, 1392–1407.

Scherrer, L.C., Dalman, F.C., Massa, E., Meshinchi, S., & Pratt, W.B. (1990) Structural and functional reconstitution of the glucocorticoid receptor-hsp90 complex. *J. Biol. Chem.* **265**, 21 397–21 400.

Schlesinger, M. (1986) Heat shock proteins: the search for functions. *J. Cell Biol.* **103**, 321–325.

Schlesinger, M.S., & Ryan, C. (1993) An ATP and hsc70 dependent oligomerization of nascent heat shock factor (HSF) polypeptide suggests that HSF, itself, could be a 'sensor' for the cellular stress response. *Protein Science*, in press.
Selinfreund, R.H., Barger, S.W., Welsh, M.J., & Van, E.L.J. (1990) Antisense inhibition of glial S100β production results in alterations in cell morphology, cytoskeletal organization, and cell proliferation. *J. Cell Biol.* **111**, 2021–2128.
Sheffield, W.P., Shore, G.C., & Randall, S.K. (1990) Mitochondrial precursor protein. Effects of 70-kilodalton heat shock protein on polypeptide folding, aggregation, and import competence. *J. Biol. Chem.* **265**, 11 069–11 076.
Shi, Y., & Thomas, J.O. (1992) The transport of proteins into the nucleus requires the 70-Kilodalton heat shock protein or its cytosolic cognate. *Mol. Cell. Biol.* **12**, 2186–2192.
Sistonen, L., Sarge, K.D., Phillips, B., Abravaya, K., & Morimoto, R.I. (1992) Activation of heat shock factor-2 during haemin-induced differentiation of human erythroleukemia cells. *Mol. Cell. Biol.* **12**, 4104–4111.
Skowyra, D., Georgopolous, C., & Zylicz, M. (1990) The *E. coli* dnaK gene product, the hsp70 homologue, can reactivate heat-inactivated RNA polymerase in an ATP hydrolysis-dependent manner. *Cell* **62**, 939–944.
Smith, D.F., Schowater, D.B., Kost, S.L., & Toft, D.O. (1990) Reconstitution of progesterone receptor with heat shock proteins. *Mol. Endocrinol.* **4**, 1704–1711.
Smith, D.F., Stensgard, B.A., Welch, W.J., & Toft, D.O. (1992) Assembly of progesterone receptor with heat shock proteins and receptor activation are ATP-mediated events. *J. Biol. Chem.* **267**, 1350–1356.
Solomon, J.M., Rossi, J.M., Golic, K., McGarry, T., & Lindquist, S. (1991) Changes in hsp70 alter the acquisition of thermotolerance and regulation of the heat shock response in *Drosophila. New Biologist* **3**, 1106–1220.
Sprang, G.K., & Brown, I.R. (1987) Selective induction of a heat shock gene in fibre tracts and cerebellar neurons of the rabbit brain detected by *in situ* hybridization. *Mol. Brain Res.* **3**, 89–93.
Stout, J.T., & Caskey, C.T. (1987) Antisense RNA inhibition of endogenous genes. *Meth. Enzymol.* **151**, 519–530.
Takayama, K.M., & Inouye, M. (1990) Antisense RNA. *Crit. Rev. Biochem. Mol. Biol.* **25**, 155–184.
Takenaka, I.M., & Hightower, L.E. (1992) Transforming growth factor-β1 rapidly induces hsp70 and hsp90 molecular chaperones in cultured chicken embryo cells. *J. Cell. Physiol.* **152**, 568–577.
Takenaka, I.M., & Hightower, L.E. (1993) Regulation of chicken hsp70 and hsp90 family gene expression by transforming growth factor-β1. *J. Cell. Physiol.* **155**, 54–62.
Terlecky, S.R., Chiang, H.-L., Olson, T.S., & Dice, J.F. (1992) Protein and peptide binding and stimulation of *in vitro* lysosomal proteolysis by the 73-kDa heat shock cognate protein. *J. Biol. Chem.* **267**, 9202–9209.
Tissieres, A., Mitchell, H.K., & Tracy, U. (1974) Protein synthesis in salivary glands of *Drosophila melanogaster*. Relation to chromosome puffs. *J. Mol. Biol.* **84**, 389–398.
Trevor, K., Linney, E., & Oshima, R.g. (1987) Suppression of endo B cytokeratin by its antisense RNA inhibits the normal coexpression of endo A cytokeratin. *Proc. Natl Acad. Sci. USA* **84**, 1040–1044.
Tytell, M., Greenberg, S.G., & Lasek, R.J. (1986) Heat shock-like protein is transferred from glia to axon. *Brain Res.* **363**, 161–164.
Ungewickell, E. (1985) The 70-kD mammalian heat shock proteins are structurally and functionally related to the uncoating protein that releases clathrin triskelion from coated vesicles. *EMBO J* **4**, 3385–3391.
van der Krol, A.R., Mol, J.N.M., & Stuitje, A.R. (1988) Modulation of eukaryotic gene expression by complimentary RNA or DNA sequences. *BioTechniques* **6**, 958–976.

Vanbuskirk, A., Crump, B.L., Margoliash, E., & Pierce, S.K. (1989) A peptide binding protein having a role in antigen presentation is a member of the hsp70 heat shock family. *J. Exp. Med.* **170**, 1799–1809.
Velazquez, J.M., Di Domenico, B.J., & Lindquist, S. (1980) Intracellular location of heat shock proteins in *Drosophila*. *Cell* **20**, 679–689.
Velazquez, J.M., & Lindquist, S. (1984) Hsp 70: nuclear concentration during environmental stress and cytoplasmic storage during recovery. *Cell* **36**, 655–662.
Wang, C., Asai, D.J., & Lazarides, E. (1980) The 68 000-dalton neurofilament-associated polypeptide is a component of non-neuronal cells and of skeletal myofibrils. *Proc. Natl Acad. Sci. USA* **77**, 1541–1545.
Wang, C., Gomer, R., & Lazarides, E. (1981) Heat shock proteins are methylated in avian and mammalian cells. *Proc. Natl Acad. Sci. USA* **78**, 3531–3535.
Weatherbee, J.A., Lufting, R.B., & Weihung, R.R. (1980) Purification and reconstruction of HeLa cell microtubules. *Biochemistry* **19**, 4116–4123.
Welch, W.J. (1992) Mammalian stress response – cell physiology, structure/function of stress proteins, and implications for medicine and disease. *Physiol. Rev.* **72**, 1063–1081.
Welch, W.J., & Feramisco, J.R. (1984) Nuclear and nucleolar localization of the 72,000-dalton heat shock protein in heat shocked mammalian cells. *J. Biol. Chem.* **259**, 4501–4513.
Welch, W.J., & Feramisco, J.R. (1985a) Rapid purification of mammalian 70-kilodalton stress proteins: affinity of the proteins for nucleotides. *Mol. Cell. Biol.* **5**, 1229–1237.
Welch, W.J., Feramisco, J.R., & Blose, S.H. (1985b) The mammalian stress response and the cytoskeleton: Alterations in intermediate filaments. *Ann. NY Acad. Sci.* **455**, 57–67.
Welch, W.J., & Mizzen, L.A. (1988) Characterization of the thermotolerant cell. II. Effects on the intracellular distribution of heat shock protein 70, intermediate filaments, and small nuclear ribonucleoprotein complexes. *J. Cell Biol.* **106**, 1117–1130.
Welch, W.J., & Suhan, J.P. (1986) Cellular and biochemical events in mammalian cells during and after recovery from physiological stress. *J. Cell Biol.* **103**, 2035–2052.
Whatley, S.A., Leung, T., Hall, C., & Lim, L. (1986) The brain 68-kilodalton microtubule-associated protein is a cognate form of the 70-kilodalton mammalian heat shock protein and is present as a specific isoform in synaptosomal membranes. *J. Neurochem.* **47**, 1576–1583.
White, F.P. (1979) Subcellular particles involved in the translocation of proteins in rat brain. *J. Neurobiol.* **10**, 591–607.
White, F.P. (1980a) The synthesis and possible transport of specific proteins by cells associated with brain capillaries. *J. Neurochem.* **35**, 88–94.
White, F.P. (1980b) Differences in protein synthesized *in vivo* and *in vitro* by cells associated with the cerebral microvasculature. A protein synthesized in response to trauma? *Neuroscience* **5**, 1793–1799.
White, F.P. (1981a) The induction of 'stress' proteins in organ slices from brain, heart and lung as a function of postnatal development. *J. Neurosci.* **1**, 1312–1319.
White, F.P. (1981b) Protein and RNA synthesis in cerebral microvessels: A radioautographic study. *Brain Res.* **229**, 43–52.
White, F.P., & Currie, R.W. (1982) A mammalian response to trauma: the synthesis of a 71-kDa protein. In *Heat Shock from Bacteria to Man* (eds Schlesinger, M.J., Ashburner, M. and Tissieres, A.), pp 379–388. New York, Cold Spring Harbor Laboratory Press.
White, F.P., Dutton, G.R., & Norenberg, M.D. (1981) Microvessels isolated from rat brain: localization of astrocyte processes by immunohistochemical techniques. *J. Neurochem.* **36**, 328–332.
White, F.P., and White, S.R. (1987) Effects of polyhydroxyl alcohols on protein synthesis in brain slices. *J. Neurochem.* **48**, 1560–1565.
Wiech, H., Buchner, J., Zimmermann, M., Zimmermann, R., & Jakob, U. (1993) Hsc70, immunoglobulin heavy chain binding protein, and hsp90 differ in their ability to stimulate transport of precursor proteins into mammalian microsomes. *J. Biol. Chem.* **268**, 7414–7421.

Yokoyama, K., & Imamoto, F. (1987) Transcriptional control of the endogenous MYC protooncogene by antisense RNA. *Proc. Natl Acad. Sci. USA* **84**, 7363–7367.

Yost, H.J., & Lindquist, S. (1986) RNA splicing is interrupting by heat shock and is rescued by heat shock-protein synthesis. *Cell* **45**, 185–193.

Zimarino, V., Wilson, S., & Wu, C. (1990) Antibody-mediated activation of *Drosophila* heat shock factor *in vitro*. *Science* **249**, 546–549.

CHAPTER 2

INDUCTION OF HEAT SHOCK GENES IN THE MAMMALIAN BRAIN BY HYPERTHERMIA AND TISSUE INJURY

Ian R. Brown

Department of Zoology, University of Toronto, Scarborough Campus, West Hill, Ontario, M1C 1A4, Canada

Table of Contents

Heat Shock Proteins in the Nervous System
ISBN 0-12-480960-X

2.1 Introduction

2.1.1 The heat shock response and its relevance to the nervous system

Cells respond to sudden increases in temperature by exhibiting an inhibition of ongoing RNA and protein synthesis and an induction of a set of genes encoding heat shock proteins (hsps), which may play important roles in cellular repair and protective mechanisms (Lindquist and Craig, 1988; Pardue *et al.*, 1989; Morimoto *et al.*, 1990; Hightower, 1991; Nover, 1991; Georgopoulos and Welch, 1993). Numerous studies on heat shock proteins have utilized organisms including *E. coli*, yeast, *Drosophila* and vertebrate cells grown in culture. However, less work has been carried out on intact thermoregulating animals despite the potential significance of the response in that context. As will be shown in this article, the heat shock response is physiologically relevant because heat shock genes are turned on in the mammalian nervous system following fever-like temperatures and tissue injury. Three of the main classes of heat shock proteins are the hsp90s, hsp70s, and the low-molecular-weight hsp20s, the categories being governed by the approximate molecular mass (in kilodaltons, kDa) of the principal proteins which are expressed. This review will focus on expression of the mammalian hsp70 class which has been the centre of most of the studies in the nervous system, however, existing studies on the induction of the hsp90s and hsp20s will also be covered. The term hsp70 will be used to denote stress-inducible members of the 70-kDa family while constitutively-expressed members are referred to as hsc70. An overview of the structure and function of the mammalian hsp70 family of molecular chaperones, including mechanisms of induction, is provided in the Chapter 1 in this volume (recent reviews: Hightower, 1991; Gething and Sambrook, 1992; Hartl and Martin, 1992; Morimoto, 1993; Georgopoulos and Welch, 1993).

The brain is a complex organ composed of many cell types. Do they all respond similarly to heat shock or tissue injury concerning expression of heat shock genes? What is the response in glial cells compared to neuronal cells, for example? The mammalian hsp70 class is a multigene family composed of constitutively expressed and stress-inducible members (Welch, 1990; Morimoto *et al.*, 1990; Hightower, 1991; Nover, 1991). How do the levels of constitutive hsc70 proteins and stress-inducible hsp70 members compare in various neural cells types of the mammalian brain? Since heat shock proteins have been proposed to exhibit protective properties in tissue culture systems (Johnston and Kucey, 1988; Riabowol *et al.*, 1988; Li *et al.*, 1991; Solomon *et al.*, 1991), do they offer strategies for tissue protection in the nervous system? These and other questions will be addressed.

2.2 Neural induction of hsp70 by physiologically relevant increases in body temperature

2.2.1 Initial studies on the effect of hyperthermia on brain protection synthesis

Early studies in our laboratory demonstrated that increases in body temperature of 2–3°C in the rabbit induced a transient disaggregation of brain polysomes to monosomes with maximal shift, after 30–60 min, and a return to normal levels by 4 h (for review: see Brown *et al.*, 1982b). This transient inhibition of brain protein synthesis results from a lesion at an early step in initiation of translation (Cosgrove *et al.*, 1981, Cosgrove and Brown, 1984). Concomitant with this transient, hyperthermia-induced, disaggregation of brain polysomes, induction of hsp70 and hsp90 has been detected by analysing labelled rabbit brain proteins which were resolved by one- or two-dimensional gel electrophoresis, following either *in vivo* injection of labelled amino acids or cell-free translation of isolated brain polysomes (Freedman *et al.*, 1981; Brown, 1983, 1985a,b; Cosgrove and Brown, 1983). Whole-body hyperthermia (Currie and White, 1981) and amphetamine-induced hyperthermia (Nowak 1988) have also been shown to induce hsp70 in the rodent brain and a prominent induction is observed in the cerebral microvascular system (White, 1980, 1981; Inasi and Brown, 1982). A lower threshold temperature for induction of translatable hsp70 mRNA is apparent in the neonatal brain compared to the adult (Brown, 1983).

Following heat shock, the induction of hsp70 and hsp90 protein has been reported in the rabbit retina both *in vivo* and *in vitro* (Clark and Brown, 1982, 1986a,b, 1987). Hsp70 which is induced in retinal ganglion cells is transported down the optic nerve by slow axonal transport (Clark and Brown, 1985), specifically with slow component b, which includes such transported molecules as actin, spectrin, clathrin, and calmodulin (Tytell and Barbe, 1987). A constitutively expressed hsc70 protein, known to be a clathrin-uncoating ATPase, has also been reported to be axonally transported as part of slow component b (de Waegh and Brady, 1989). Induction of an hsp70 in isolated dorsal root ganglion from bullfrog (Stone and Dougher, 1988, 1989) and from rat (Morandi *et al.*, 1989) has been demonstrated following temperature shock, and in the invertebrate nervous system, heat shock induces hsp70 in adaxonal glia of the squid giant axon (Tytell *et al.*, 1986) and Aplysia (Greenberg and Lasek, 1985). More recently we have continued our studies on the mammalian system in collaboration with the laboratory of Michael Tytell, mapping out the pattern of hsp70 protein and mRNA induction in the rat retina and corneal epithelium, following whole body hyperthermia (Yamaguchi *et al.*, 1990; Tytell *et al.*, 1993, 1994). The photoreceptor layer is the principal site of hsp70 induction in the rat retina and prior whole body temperature protects against the selective degradation of photoreceptors, which is induced by bright light treatment (Barbe *et al.*, 1988; Tytell *et al.*, 1993, 1994; see Chapter 4 for a full discussion).

In the above-mentioned studies, the appearance of a 70-kDa labelled protein on one- or two-dimensional autoradiograms indicated that fever-like temperatures were sufficient to induce expression of hsp70 genes in the mammalian nervous system. These early studies carried out in our laboratory and that of Fredric White, were presented at the First International Meeting on the heat shock response held at the Cold Spring Harbor Laboratory (Brown *et al.*, 1982a; White and Currie, 1982). A detailed account of the early contributions from Fredric White's group is given by Hightower and Li in Chapter 1 of this volume. Subsequent research on neural hsps has focused on mapping out, at the mRNA level, the pattern of expression of constitutive hsc70 and stress-inducible hsp70 genes in the brain, utilizing Northern blotting and *in situ* hybridization. In addition, the distribution of constitutively expressed and stress-inducible proteins has been established by Western blotting and immunocytochemical studies.

2.2.2 Mapping out patterns of expression of hyperthermia-inducible hsp70 mRNA

Early studies at the protein synthesis level indicated that hyperthermia induced the rapid synthesis of hsp70 in the mammalian brain (Brown *et al.*, 1982a; Cosgrove and Brown, 1983). The next task was to determine whether there were regional and cell type differences in the expression of hsp70 genes in the nervous system. Analysis of the pattern of expression of heat shock genes in the mammalian nervous system could further our understanding of the selective vulnerability of certain brain cells to various forms of trauma and neurogenetic diseases.

One hour after elevation of body temperature by 2–3°C, a massive induction of a 2.7 kilobase (kb) mRNA in the rabbit brain is detected by Northern blotting, while a 2.5 kb constitutively expressed species is present in control animals (Brown *et al.*, 1985; Sprang and Brown, 1987). The rabbit genome contains a family of hsp70 genes, including both constitutive and stress-inducible members (Brown *et al.*, 1985). Hsp70 nucleic acid probes have been reported which permit the selective detection of neural transcripts derived from either inducible or constitutively expressed members of the hsp70 gene family (Brown and Rush, 1990; Nowak *et al.*, 1990; and Miller *et al.*, 1991). In the rabbit brain, induction of a 2.7 kb hsp70 transcript is transient and parallels the rise and fall in body temperature, while levels of a 2.5 kb hsc70 mRNA show little change (Brown and Rush, 1990). The inducible transcript was not detected in the control brain. In heat shocked rats or rats made hyperthermic with amphetamine, an inducible hsp70 mRNA species was detected in brain by Northern blotting (Miller *et al.*, 1991). At equivalent total thermal response values, amphetamine-treated animals showed higher levels of hsp70 mRNA compared to heat shocked rats, suggesting that amphetamine enhanced the induction. A two- to three-fold induction of a constitutive hsc70 mRNA was also noted following either hyperthermic treatment. Induction of both the hsp70 and the hsc70 mRNAs was higher in the cerebellum than in the cerebral cortex.

Experimental autoimmune encephalomyelitis (EAE), a model of inflammatory demyelination in the central nervous system has been reported to increase hsc70 mRNA levels but not to induce hsp70 mRNA in the spinal cord at 12 days, the height of inflammation and clinical signs (Aquino *et al.*, 1993), Despite the elevation of hsc70 mRNA, increases in hsc70 protein were not detected. High levels of hsc70 protein were noted in neurons. In the spinal cord, hsc70 comprised 2–3% of the total cellular protein. We have also noted a differential induction of hsc70 mRNA compared to hsp70 mRNA in chondrocytes after treatment with interleukin 1 or phorbol myrisate acetate (Cruz *et al.*, 1991).

Very little work has been carried out on stress induction of the low molecular weight hsps in the mammalian nervous system. In the rat, heat stress was found to induce hsp27 mRNA in lung and skin but not in brain (Blake *et al.*, 1990a). In the same animals, prominent induction of hsp70 mRNA, but not hsp27 mRNA, was noted in brain with high levels present after 35 min of exposure to 40°C. As reviewed by Arrigo and Mehlen in Chapter 7 in this volume, the expression, cellular localization and function of the constitutive low molecular weight heat shock proteins have been analysed during development of the *Drosophila* and mammalian nervous systems. Recent studies from the laboratories of Robert Tanguay and Peter Koetzel have also reported constitutive expression of the low molecular weight hsps in the developing and adult nervous systems of *Drosophila* and mouse (Marin *et al.*, 1993; Gernold *et al.*, 1993; Tanguay *et al.*, 1993). Further work on stress-induction of the low molecular weight hsps is desirable, given that these proteins have been reported to play a role in protective mechanisms (Landry *et al.*, 1989; Weber, 1992).

An age-related impairment in the induction of hsps might contribute to a reduction in thermal tolerance with age. It is thus of interest that in aged rats a lower level of inducible hsp70 mRNA has been reported following heat shock, compared to younger animals in brain, lung and skin (Blake *et al.*, 1991). However, this relative decline of heat-induced hsp70 mRNA levels with age is correlated with an attenuated increase in colonic temperature in the older rats. It was concluded that age-related differences in heat-induced hsp70 mRNA were the result of a deficit in heat generation in the older animals rather than an impairment in the heat shock response with age. In a related study, however, body temperature changes in response to heat stress were found to be no different in adult and aged rats, yet the relative level of inducible hsp70 mRNA was several-fold lower in the aged rats in dentate gyrus granule cells and pyramidal cells of the hippocampus (Pardue *et al.*, 1992). These findings suggest that some cell populations in the hippocampus might be at increased risk for stress-related injury in aged animals.

In situ hybridization has been utilized to map out the pattern of expression of hsp70 genes in the control and hyperthermic mammalian brain. This procedure enables one to identify the cells in a complex tissue which are expressing the gene of interest. One hour after hyperthermia, a strong induction of hsp70 mRNA was noted in fibre tracts throughout the rabbit forebrain and cerebellum, a pattern consistent with a strong glial response to fever-like heat shock (Sprang and Brown, 1987; Brown, 1990). Induction was also noted in the choroid plexus, microvasculature, pia mater and granule cell layer of the cerebellum. Constitutive expression was observed in several neuronal enriched areas, such as hippocampal regions CA1 to CA4, and the Purkinje and granule cellular layers of the cerebellum (Sprang and Brown, 1987). In 1 h hyperthermic rabbits, induction of hsp70 mRNA was not detected in neuronal cells other than granule neurons in the cerebellum. Intravenous injection of LSD has been a convenient means of producing a rapid and predictable increase in body temperature in rabbits (Brown *et al.*, 1982b; Brown, 1985a,b). An *in situ* hybridization study indicated that induction of hsp70 mRNA in the rabbit cerebellum is due to hyperthermic effects of this drug (Manzerra and Brown, 1990). Induction was not present when drug-induced hyperthermia is blocked. The pattern of hsp70 mRNA induction in the cerebellum was similar following heat production generated by either drug (LSD) or non-drug means (placement of animals in a warm incubator).

Our initial *in situ* hybridization studies were carried out using a 35S-labelled riboprobe which hybridized to both constitutively expressed and stress-inducible hsp70 transcripts. Subsequently, these studies were extended using riboprobes which are able to discriminate hyperthermia-inducible 2.7 kb transcripts from constitutively expressed 2.5 kb transcripts in order to better identify reactive cell types in the rabbit cerebellum (Brown and Rush, 1990; Manzerra and Brown, 1992a). In control animals, the constitutive specific riboprobe detected the 2.5 kb hsc70 transcript in several neuronal cell types in the cerebellum, such as Purkinje and granule cells, with little increase in signal in 1 h hyperthermic animals. The inducible-specific riboprobe revealed a prominent induction of the 2.7 kb hsp70 mRNA species at 1 h after a 2–3°C increase in body temperature in Bergmann glial cells in the Purkinje layer, in glial cells in the deep white matter fibre tracts and in granule cells. The inducible transcript was not detected in the cerebellum of control animals. A striking neuronal-glial cell type difference in the localization of the two transcripts is highlighted by observations on the large Purkinje neurons which are surrounded by smaller Bergmann glial cells, a type of astrocyte. The neurons showed high levels of the 2.5 kb constitutive hsc70 transcript in both control and heat-treated animals and no induction of the 2.7 kb species at 1 h, while the adjacent Bergmann glial cells demonstrated a very strong induction of the 2.7 kb species at the same time point and no detectable signal with the constitutive specific probe. As will be discussed in Section 2.2.3 of this chapter, these observations on neuronal/glial differences in constitutive and stress-inducible hsp70 mRNA levels are supported by similar observations at the protein level in the same animal, namely immunocytochemical observations showing that the inducible protein is localized to glial types (Bergmann glial and oligodendrocytes), while high levels of the constitutive protein are found in Purkinje neurons (Manzerra *et al.*, 1993).

To further explore possible neuronal and glial cell type differences in the expression of constitutive and hyperthermia-inducible hsp70 mRNAs, the rabbit spinal cord was selected since it exhibits: clearly defined grey matter, neuronal enriched areas; and white matter, fibre tract areas, rich in glial cells (Manzerra and Brown, 1992b). Abundant levels of the 2.5 kb constitutive mRNA were detected in large motor neurons of both control and hyperthermic animals. Within 1 h after hyperthermia, a prominent induction of the inducible 2.7 kb hsp70 mRNA was noted in fibre tracts of the spinal cord, consistent with a strong glial response to heat shock. Induction was not observed in the large motor neurons in 1 h hyperthermic animals. Experiments on neural cells grown in primary tissue culture have also reported that induction of hsp70 in neuronal cells after heat shock is attenuated or undetectable compared to that seen in glial cells (Nishimura *et al.*, 1988, 1991; Marcuccilli *et al.*, 1992).

In glial cell cultures derived from the adult human brain, a differential expression of hsp70 has been reported using immunocytochemical methods, in that oligodendrocytes, but not astrocytes, showed an up-regulation of hsp70 protein 18 h following heat shock at 43°C for 30 min (Freedman *et al.*, 1992). However, a time course difference in the accumulation of induced hsp70 protein has been reported between the two cell types. In cultured mouse cells, the peak hsp70 protein level in astrocytes was observed at 3 h after a heat shock at 43°C of 90 min, well before peak levels were attained in oligodendrocytes (Satoh *et al.*, 1992). Rat astrocytes grown in primary culture have also been reported to induce hsp70 after treatment with the heavy metal zinc (Swanson and Sharp, 1992). Induction of hsp70 has been reported in cerebellar granule neurons grown in primary culture following heat shock, or treatment with sodium arsenite or a calcium ionophore, but not by excitotoxic injury with glutamate (Lowenstein *et al.*, 1991). As previously mentioned, cerebellar granule cells are the only neuronal cell type in which we have observed hsp70 mRNA induction in the 1 h hyperthermic rabbit (Sprang and Brown, 1987; Brown, 1990; Manzerra and Brown, 1992a).

Why do Purkinje neurons show no detectable induction of hsp70 mRNA in 1 h hyperthermic animals while surrounding Bergmann glial cells show a strong induction? A feed-back control mechanism may exist in which the level of endogenous hsc70 or hsp70 protein regulates the level of subsequent stress protein induction (Morimoto, 1993; see also Figure 1, Chapter 1). The high levels of constitutive hsc70 in Purkinje neurons may enable these postmitotic cells to counteract or buffer the initial effects of the heat stress without the requirement for induction of hsp70. Recently, hsc70 and hsp70 in mammalian cells have been reported to exhibit functional similarities (Brown *et al.*, 1993). Glial cells with comparatively low levels of the constitutive hsc70 protein are observed to rapidly induce hsp70 after hyperthermia. With more severe insults to neurons, the level of stress may exceed the buffering capacity of the constitutive hsc70 protein, triggering the neuronal induction of hsp70.

Recently, we have noted a delayed accumulation of hsp70 mRNA in Purkinje neurons (Manzerra *et al.*, 1993). While the inducible hsp70 mRNA species was not detected in Purkinje neurons of 1 h hyperthermic animals, it was observed at 3 h and particularly at 10 hours post-heat shock, at a time when body temperature had returned to normal. Certain neuronal cell types may accumulate hsp70 mRNA in response to physiologically relevant increases in body temperature, however the response is delayed when compared to the rapid response seen in glial cell types.

The delayed response in Purkinje neurons may be due to after effects of the hyperthermic treatment rather than to direct heat effects. It should be noted that the pattern of expression of inducible hsp70 in different regions and/or cell types in the mammalian brain varies according to the type and severity of stress applied (for a review see Brown, 1990). In ischaemia and kainic acid studies, hsp70 is induced primarily in neurons with a pattern of inducibility matching the known histopathology of the insult (Vass *et al.*, 1988, 1989; Gonzalez *et al.*, 1989; see also Chapter 3 by Nowak *et al.*).

High levels of hsc70 mRNA in neuronal cell populations and much lower levels in glial cells have also been reported in other laboratories (Pardue *et al.*, 1992). In heat shocked adult rats, highest levels of inducible hsp70 mRNA were noted in glia cells, intermediate levels in granule cells of the dendate gyrus and lowest levels in pyramidal cells of Amon's horn. Higher temperatures or longer time intervals of the hyperthermic stress may induce hsp70 mRNA in certain neuronal cell populations (Blake *et al.*, 1990b; Pardue *et al.*, 1992). Hyperthermia in the non-anaesthetized rat induces hsp70 mRNA in glia cells and the microvasculature (Blake *et al.*, 1990b), in agreement with our previous studies (Sprang and Brown, 1987; Brown 1990). In addition, expression was observed in brain regions regulating the endocrine stress response, such as the dentate gyrus and hypothalamus, and the magnitude of this response increased with higher body temperatures.

2.2.3 Analysis of stress-inducible hsp70 protein in neural cells

The pattern of induction of hsp70 protein has been mapped out in the mammalian brain following whole body hyperthermia. These studies were facilitated by the availability of antibodies directed against hsp70 (Welch *et al.*, 1989; Welch, 1990). In agreement with the above mentioned *in situ* hybridization studies (Sprang and Brown, 1987; Manzerra *et al.*, 1993), immunocytochemical studies revealed a strong induction of hsp70 in glial cells either *in vivo* or *in vitro*, following temperature elevation (Nishimura *et al.*, 1988, 1991; Marini *et al.*, 1990). Western blotting revealed peak levels of inducible hsp70 protein at 5–10 h after whole body hyperthermia in the rabbit brain (Manzerra *et al.*, 1993). The inducible hsp70 and constitutively expressed hsc70 protein species can be resolved into separate components by two-dimensional Western blotting. Immunocytochemistry indicated that the induced protein is localized to oligodendrocytes in fibre tracts of the cerebellar deep white matter and to Bergmann glial cells, a type of astrocyte, in the Purkinje layer (Manzerra and Brown, 1992c). At the same time point, Purkinje neurons show no detectable inducible-hsp70 protein, however, they demonstrate high levels of a constitutively expressed hsc70 protein in both the neuronal cell body and in dendritic processes extending into the molecular layer. In summary, our *in situ* hybridization and immunocytochemical studies, which detect mRNA and protein distribution respectively, support each other and indicate hsp70 inducibility in glial cell types and high constitutive hsc70 expression in neuronal cell types.

In other laboratories, immunocytochemical studies in the adult rat have also indicated that following hyperthermia, induction of hsp70 protein is intense in glial cells and vascular elements (Marini *et al.*, 1990). A more recent study reported that hyperthermia induces hsp70 in glial and endothelial cells throughout the rat brain (Li *et al.*, 1992). In addition, neuronal expression of hsp70 was noted particularly in cerebral neuronal populations which control the neuroendocrine stress response. It should be noted however, that these observations were made 24 h after hyperthermia and the animals experienced a transient drop in body temperature immediately after whole body hyperthermia. Body temperature dropped 8.6°C in 1 h from a peak heat shock temperature of 41.5°C to 32.9°C, followed by a gradual rise to normal body temperature (Li *et al.*, 1992). As we and others have noted, induction of hsp70 mRNA in glia precedes that in neurons (Pardue *et al.*, 1992; Manzerra *et al.*, 1993).

Focal electrical stimulation of the rat hippocampus has been utilized to determine whether neuronal excitation, without cell damage, can induce hsp70 (Sloviter and Lowenstein, 1992). Depending on the duration of the afferent stimulation, specific hippocampal cells can be either injured or merely excited with no cellular damage. These studies demonstrate that non-stressful neuronal excitation does not induce hsp70, whereas potentially injurious neuronal stress, caused by repetitive excitation, results in induction of hsp70.

In the guinea-pig auditory system, hsp70 has been detected in the cochlea of both the unstressed (Neely *et al.*, 1991) and heat-stressed animals with no differences in the distribution of the protein between control and experimental animals (Dechesne *et al.*, 1992). In the rat, however, hsp70 was not detected in the unstressed cochlea, while heat shock induced hsp70 in most inner ear tissues (Dechesne *et al.*, 1992). Using both immunocytochemistry and *in situ* hybridization procedures, the greatest induction in the cochlea was observed in the stria vascularis and in the microvascular system. The broad induction response in the auditory periphery contrasts to the more specific induction pattern which is observed in the brain following hyperthermia (Brown, 1990; Manzerra and Brown, 1992a,b).

High intensity acoustic stimulation has been shown to induce hsp70 in the rat cochlea and auditory brainstem (Lim *et al.*, 1992). Immunocytochemistry revealed that noise stimulation induced maximal levels of hsp70 in outer hair cells of the cochlea 6 h after exposure, while little or no signal was observed in inner hair cells or spiral ganglion cells. In the brainstem, noise stimulation induced hsp70 in the cochlear nucleus, medial nucleus of the trapezoid body and lateral superior olive, but not in the inferior colliculus.

2.2.4 Tissue protective effects of heat shock

In mammalian cells grown in tissue culture, functional hsp70 or expression of hsp70 appears to be necessary if cells are to survive temperature shock (Johnston and Kucey, 1988; Riabowol *et al.*, 1988). The protective effect of heat shock has been noted in the nervous system. A brief heat shock protects rat embryos against subsequent teratogenic heat shock effects (see Walsh *et al.*, Chapter 6). Protection against light-induced degeneration of rat retinal photoreceptors is conferred by prior whole-body hyperthermia (Barbe *et al.*, 1988). In purified cultures of cerebellar granule cells (Lowenstein *et al.*, 1991) and in cortical cultures (Rordorf *et al.*, 1991), prior heat shock protects neurons from toxic concentrations of glutamate (see Chapter 5 by Dwyer and Nishimura). Identification of the critical factor(s) which are responsible for these heat-induced tissue protective effects in the nervous system is required. Is the induction of hsp70 in a particular brain cell or import of the protein into that cell, the critical factor which improves the chances of survival of that cell? The retinal system may offer some answers.

In collaboration with the laboratory of Michael Tytell we have carried out a time-course study to map out the cell types in the rat retina which induce hsp70 protein and mRNA, in response to whole-body hyperthermia (Tytell *et al.*, 1993, 1994). These studies show that photoreceptors are the principal site of hsp70 induction and that the time of maximal hsp70 protein accumulation in the photoreceptors corresponds to the time when photoreceptors are maximally protected against degeneration by bright light. Evidence that hsp70 may be the protective agent is suggested by the observation that intra-ocular injection of purified protein confers protection (Tytell *et al.*, 1993; see Chapter 4 for a full discussion).

Up-regulation of hsp70 in selected cells in the nervous system may open up new avenues for enhancing cellular repair and survival, and reducing neuron loss after various traumatic conditions. To test this possibility, an attempt is being made to create transgenic mice in which expression of hsp70 is up-regulated in specific populations of neurons by use of promoter 1 of the rhombotin gene (Uney *et al.*, 1992). This promoter has been shown to direct LacZ expression in specific populations of neurons in the CA1 and CA2 region of the adult mouse hippocampus. Caution is warranted, however, since engineering cells to express hsp70 at normal temperatures has been reported to be deleterious to cell growth. In *Drosophila*, constitutive over-expression of hsp70 results in growth rates that are sharply decreased in both tissue culture cells and *in vivo* in salivary glands (Feder *et al.*, 1992). Attempts to transfect cells with extra copies of the wild-type hsp70 gene or with constructs driven by heterologous promoters, which would increase the expression of wild-type hsp70 at normal temperatures, have usually failed (Solomon *et al.*, 1991; Feder *et al.*, 1992). Nature appears to have evolved regulatory mechanisms to minimize expression of hsp70 at normal temperatures. While induction of hsp70 is beneficial to *Drosophila* cells following heat shock or other traumas, the protein has negative effects on cell growth rates and hence its synthesis is tightly regulated. *Drosophila* cells attempt to repress the activity of hsp70 at normal temperature by sequestering the protein in granules from which it cannot be redeployed (Feder *et al.*, 1992).

These observations point out that while it might be beneficial to increase the level of hsp70 in targeted neural cells to promote cellular repair and/or protect against subsequent trauma, engineering cells, particularly those in the developing nervous system, to carry out constitutive expression of hsp70 may well have negative effects on cell growth. Our studies have shown that hsc70 protein is present endogenously at high levels in large neurons in the mammalian brain, however, these cells are in a postmitotic state. The strategy of developing non-harmful procedures for activating the heat shock response in the brain could have considerable therapeutic value, particularly if the pattern of induction was targeted to encompass the profile of cells which are affected, or at risk, by a particular trauma.

2.3 Effect of tissue injury on the induction of heat shock genes in the nervous system

We have demonstrated that localized tissue injury, namely a small surgical cut in the rat cerebral cortex, induces hsp70 gene expression (Brown *et al.*, 1989). Two hours post-surgery, a highly localized induction of hsp70 mRNA was detected in cells proximal to the lesion site, with a reduction in signal by 12 h. Both neurons and glial cells at the injury site appeared to respond by induction of hsp70 mRNA. Thus, induction of a member of the hsp70 gene family is a physiologically relevant response which is activated at an early stage following tissue injury. The induction pattern serves as a marker to identify a population of reactive cells that respond rapidly to tissue trauma. Peripheral axotomy has also been shown to be a cell stressor which induces hsp70 mRNA (New *et al.*, 1989). In that study, cutting the facial nerve in the adult hamster caused the induction of hsp70 mRNA in facial nerve nuclear groups in the brain-stem. The observed time course of hsp70 mRNA induction paralleled the induction observed in the rat cerebral cortex wounding study (Brown *et al.*, 1989). After concussive spinal cord injury in the rat, induction of hsp70 protein in neuronal and glial cells at the injury site has been reported within 30 min using Western blotting and immunocytochemistry (Gower *et al.*, 1989). Crush injury of the crayfish central nervous system (CNS) has also been reported to induce the synthesis and accumulation of an hsp70 isoform in injured tissue (Xue and Grossfield, 1993).

Recently, mild fluid percussive head injury in the rat has been shown to induce hsp70 (Tanno *et al.*, 1993). In that immunocytochemical study, earliest expression was noted 3 h post-injury in neurons and glia in the cortical area surrounding the necrotic zone at the impact site. Induction was most frequent in neurons. By 6 h, blood vessels in the necrotic area were positive. By 24 h, maximal signal was apparent in neurons, glia and blood vessels in the cortical area immediately adjacent to the necrotic area at the impact site and in other brain regions such as the parasagital cortex and CA3 of the posterior hippocampus. It was noted that the pattern of hsp70-induction paralleled the temporal and regional pattern of breakdown of the blood–brain barrier after similar head injury. Breakdown of the blood–brain barrier results in entry of humoral agents that may injure brain tissue. For example, neurons are known to accumulate plasma proteins at the sites of barrier disruption. Denatured proteins are thought to activate the heat shock response (Ananthan *et al.*, 1986; Pelham, 1986). Areas of the brain that express hsp70 after head injury may therefore reflect the pattern of cellular uptake of plasma proteins following localized breakdown of the blood–brain barrier. The expression of hsp70, which was most frequently associated with neurons, was not associated with neuronal cell loss (Tanno *et al.*, 1993). Thus the protein may define a population of neurons that recover from an injured state. Lateral fluid percussion head injury in the rat has also been shown to induce not only hsp70 mRNA but also grp78 and grp94 mRNA which were increased in the cortex ipsilateral to the impact site (Lowenstein *et al.*, 1994). This report is the first to show changes in mRNAs encoding glucose regulated proteins (grps), following acute injury to the central nervous system.

2.4 Expression of hsp90 in the nervous system

2.4.1 Constitutive expression of hsp90 in the mammalian brain

Hsp90 is an abundant protein in the cytoplasm of unstressed eukaryotic cells comprising 1–2% of the total cellular protein. In steroid hormone-responsive tissues, hsp90 has been found associated with non-activated steroid-hormone receptor complexes, preventing receptor binding to nuclear DNA elements until hormone is bound to the receptor (Catelli *et al.*, 1985; Schuh *et al.*, 1985; Baulieu, 1987; Pratt, 1987; Carson-Jurica *et al.*, 1989; Brunt *et al.*, 1990). Hsp90 has also been found associated with actin (Koyasu *et al.*, 1986), several kinases including the transforming protein $pp60^{src}$ (Opperman *et al.*, 1981) and initiation factor (eIF-2α) kinase (Rose *et al.*, 1989). Hsp90 is thought to play a role in the control of receptor binding and enzyme activity. Higher levels of hsp90 have been reported in the bovine brain compared to other organs (Itoh *et al.*, 1990). Immunocytochemical studies reveal the presence of hsp90 in most neurons in the rat brain (Itoh *et al.*, 1993). The distribution of constitutively expressed hsp90 mRNA has been investigated in the rat brain (Izumoto and Herbert, 1993). High mRNA levels were noted in neurons and expression was low or absent in glial cells. The greatest abundance of hsp90 mRNA was noted in the Purkinje layer of the cerebellum. A correlation was apparent between the distribution of rat brain hsp90 mRNA and that of the glucocorticoid receptor mRNA.

In humans, the hsp90 gene family is comprised of hsp90 α and hsp90 β members (Hickey *et al.*, 1986, 1989; Rebbe *et al.*, 1987) which correspond to hsp86 and hsp84 in mouse (Ullrich *et al.*, 1986; Moore *et al.*, 1989). In rat brain, no differences were found in the distribution of the hybridization signal with human α and β hsp90 probes, suggesting similar constitutive patterns of expression of the equivalent rat hsp90 mRNA species (Izumoto and Herbert, 1993). However, Northern blots were not shown to confirm that the human probes are able to discriminate two separate hsp90 mRNA species in rat brain.

2.4.2 Stress and hormonal induction of hsp90

Both isoforms of hsp90 are present at high levels in unstressed cells and both are heat inducible, however only hsp90 α is inducible by the adenovirus E1A gene product (Hickey *et al.*, 1986, 1989). Hyperthermia in the rabbit has been shown to induce an increase in the synthesis of hsp90 in brain microsomal and synaptic fractions concomitant with an induction of hsp70 (Freedman *et al.*, 1981). *In vivo* labelling studies demonstrated a hyperthermia-induced increase in the synthesis of an hsp90 protein in the rabbit retina, in addition to hsp70 induction (Clark and Brown, 1982). Translation of purified retinal polysomes in a cell-free protein synthesis system and resolution of labelled proteins by two-dimensional gel electrophoresis revealed that hyperthermia increases the relative abundance of polysomal mRNA encoding hsp90 protein (Clark and Brown, 1982). High constitutive levels of the hsp90 mRNA and protein were detected in the control retina. Heat shock of the isolated retinal preparation also induced hsp90 (Clark and Brown, 1986a). Recently, we have investigated the effect of hyperthermia on the induction of hsp90 mRNAs in the rabbit brain. Using riboprobes which are able to discriminate hsp90 α and β mRNA species, we have noted an increase in both forms 1 h after hyperthermic treatment, with levels remaining above control value until 10 h. *In situ* hybridization and immunocytochemical experiments are in progress to map out the distribution of hsp90 mRNAs and protein in regions of the control and hyperthermic rabbit brain.

Hsp90 was thought to be localized to the cytoplasm, however, it has been found to translocate to the nucleus in heat shocked cells (Silver *et al.*, 1983; Schlesinger, 1990). Immunocytochemical studies in mouse N2A neuroblastoma cells suggest that the mouse hsp90 member, hsp84, is present in the cytoplasm and nucleus of control cells, accumulates in the nucleus in heat shocked cells, then returns to cytoplasm during recovery (van Bergan en Henegouwen *et al.*, 1987).

Hsp90 levels in the ventromedial hypothalamus of the female rat have been shown to be modulated by oestradiol administration. At 12 h after injection of steroid hormone, maximal induction of hsp90 was detected in the ventromedial hypothalamus by one-dimensional Western blotting, while in the uterus peak levels were seen at 18 h (Olazabal *et al.*, 1992a). An immunocytochemical analysis in this same study indicated that the hsp90 localization in brain is primarily neuronal. Sex differences were reported in the regulation of hsp90 in the ventromedial hypothalamus. Oestradiol administration increased hsp90 protein levels in ovariectomized female rats but not in castrated males (Olazabal *et al.*, 1992b). In the absence of hormone, hsp90 levels in control female and male animals were similar but after hormone administration, hsp90 levels were higher in females. As the authors point out, since constitutive levels of hsp90 are much more abundant than steroid receptors, it is unclear why oestrogen administration should induce even higher levels of hsp90. These hsps may mediate other effects of oestrogen, and/or newly synthesized steroid receptors may bind only to newly synthesized hsp90. Similar effects of steroid hormone on hsp90 synthesis have been reported in other systems (Ramachandran *et al.*, 1988; Brunt *et al.*, 1990).

2.5 Conclusion: perspectives on hsp70 and hsc70 in neuronal and glial cells

Analysis of the distribution of constitutively expressed and hyperthermia-inducible mRNAs and proteins of the hsp70 multigene family have revealed interesting cell type differences in the mammalian nervous system. High levels of hsc70 mRNA have been found in neuronal cell populations in control animals, particularly in large neurons such as Purkinje neurons in the cerebellum and pyramidal neurons in the hippocampus, whereas expression was not detected in glial cell populations (Sprang and Brown, 1987; Brown, 1990; Manzerra and Brown, 1990, 1992a,b; Pardue *et al.*, 1992). Our recent immunocytochemical studies have demonstrated that high levels of hsc70 protein are apparent in the cell body and dendritic processes of Purkinje neurons, but a signal was not detectable in neighbouring glial cells. Brain tissue has been widely utilized as an abundant source of hsc70 protein where it comprises 1% of soluble brain protein (Schlossman *et al.*, 1984). Brain contains the highest levels of hsc70 protein of the 13 body tissues of young mice which were examined (Tanguay *et al.*, 1993). Some years ago, clathrin-uncoating ATPase isolated from bovine brain was reported to be a constitutively expressed member of the hsp70 multigene family (Ungewickell, 1985; Chappell *et al.*, 1986). Axonal transport of hsc70 in neurons as part of slow component b has been reported (de Waegh and Brady, 1989). Recently, the hsc70 protein has been reported to interact with newly synthesized proteins, retarding folding until synthesis and translocation events have been completed (Beckmann *et al.*, 1990). Observations of high levels of hsc70 in neurons of the hippocampus and cerebellum suggest that this protein plays an important role in the normal metabolic functioning of neurons, perhaps in translocation of proteins between intracellular compartments and in vesicle-mediated cycling events (review: Brown, 1990).

When body temperature is raised to physiologically relevant levels in the rabbit brain, a rapid induction of hsp70 mRNA within 1 h is detected in specific cell types, such as the microvasculature, the choroid plexus, oligodendrocytes in white matter fibre tracts, and Bergman glial cells, a type of astrocyte in the Purkinje layer of the cerebellum (Sprang and Brown, 1987; Manzerra and Brown, 1992a). We have confirmed the induction in glial cells at the hsp70 protein level by immunocytochemistry (Manzerra *et al.*, 1993). At the 1 h time-point, hsp70 mRNA induction is not detectable in neuronal cell populations in the rabbit brain other than in granule cells in the cerebellum. Why is hsp70 induction not detected in large post-mitotic neurons by our conditions of hyperthermia, whereas neighbouring glial cells show a robust induction at 1 h? Are these neurons at risk due to a failure to induce hsp70 or does their high endogenous level of hsc70 compensate for their requirements. Engineering cells to exhibit constitutive over-expression of hsp70 has been reported to dampen subsequent induction of hsp70 in tissue culture systems by reducing the extent of activation of heat shock transcription factor (HSF) after heat shock treatment (Mosser *et al.*, 1993). Recent experiments suggest an interaction of hsp70 with HSF (Abravaya *et al.*, 1992), and a regulatory role for hsp70 in HSF activation has been proposed (Morimoto *et al.*, 1992; Morimoto, 1993). In unstressed cells, HSF exists in a non-DNA binding form through transient interactions with hsp70 (or hsc70, see Figure 1, Chapter 1). During heat shock, the formation of misfolded or aggregated protein complexes competes with HSF for association with hsp70. Released HSF is then free to assemble into trimers and bind to DNA, triggering hsp70 transcription. If high endogenous levels of hsc70 in certain neuronal cells could also dampen HSF activation, a reduced heat shock response would be observed in these cells.

Different types of neural trauma, or variations in the level of a particular stressor, induce characteristic cellular responses in the mammalian brain, with regard to the type of brain cell that reacts by inducing hsp70 and the timing of the induction response (Brown, 1990). As reviewed in this chapter, in the rabbit brain system we observe a rapid glial response to physiologically relevant increases in body temperature. Higher increments of temperature elevation or heat shock followed by a transition to hypothermia, which may cause increased cellular perturbation, appears to cause additional hsp70 induction in select neuronal cell populations in the mammalian brain (Blake *et al.*, 1990a,b; Li *et al.*, 1992; Pardue *et al.*, 1992). Following ischaemia, hsp70 is induced in neuronal cell populations (see Chapter 3). The pattern of induction of hsp70 following various traumatic events may be a useful marker of cellular injury and may identify previously unrecognized areas of vulnerability in the nervous system (Gonzalez *et al.*, 1989; Brown, 1990). The high level of hsc70 in certain neuronal cells and the resultant potential for a dampening of HSF activation in those cells following hyperthermic treatment, highlights the importance of being aware of cell-type differences in hsc70 levels in the brain. Endogenous levels of hsc70 may influence whether induction of hsp70 is triggered in a particular cell type by a given stimulus. As previously discussed, certain neuronal cells appear to have high levels of hsc70 compared to glial cells and this could affect the pattern of the induction response.

Could transport of hsp70 from inducing glial cells to adjacent neuronal cell bodies or neuronal processes, produce a mechanism for rapid hsp70 delivery? In the invertebrate nervous system, hsp70 is synthesized in adaxonal glial cells after heat shock and rapidly exported into axons, thus providing a fast response mechanism for delivery of hsp70 to regions of the neuron which are distant from the cell body (Tytell *et al.*, 1986). An experiment carried out in our laboratory suggests that a similar mechanism may occur in the mammalian nervous system. Following pulse labelling *in vivo* for 1 h by injection of 35S-methionine into the lateral ventricles of the brain, synaptic fractions were purified from control and hyperthermic rabbits (Freedman *et al.*, 1981). One-dimensional gel autoradiography revealed labelled hsp70 in highly purified synaptic fractions isolated from the cerebral hemispheres of hyperthermic animals. We now know from our *in situ* hybridization studies that, at the 1 h time-point under investigation, induction of hsp70 mRNA is observed in glial cells in fibre tracts, and is not detected in neuronal cells in the cerebral hemispheres (Sprang and Brown, 1987; Manzerra and Brown, 1992a,c). The labelled hsp70 protein, which was found associated with the synaptic fractions after the 1 h pulse, would therefore seem to have originated in neighbouring glial cells. With the caveat that the newly synthesized hsp70 was not released from glial cells during tissue homogenization and subsequently absorbed onto synaptic membranes during the isolation procedure, a rapid transport of newly synthesized hsp70 protein from glial cells to neuronal processes could account for the observation. Release of hsp70 from cultured rat embryo cells has been reported (Hightower and Guidon, 1989).

As discussed in Section 2.2.4, hsp70 has been shown to play an important role in the development of thermotolerance in *Drosophila* and in mammalian cells grown in tissue culture (Li *et al.*, 1991; Solomon *et al.*, 1991). Protective effects of a prior heat shock treatment have also been noted in neural cells grown in tissue culture (Lowenstein *et al.*, 1991; Rordof *et al.*, 1991), in the developing nervous system (see Chapter 6) and in the retina (Barbe *et al.*, 1988; Tytell *et al.*, 1993,1994). Evidence for a role of hsp70 as a protective agent in the retinal example is suggested by the observation that prior intra-ocular injection of purified protein confers protection against damage of photoreceptors by bright light damage (Tytell *et al.*, 1993; and see Chapter 4). Might other heat shock proteins also be involved in tissue protective effects in the nervous system? In tissue culture systems, the low-molecular weight hsps have been reported to play a role in the thermotolerance (Landry *et al.*, 1989, Weber, 1992). Recently, these proteins have been shown to be constitutively expressed in the developing and adult nervous system (see Chapter 7), however, virtually no work has been carried out on stress induction of this class of heat shock protein in the nervous system. It should be kept in mind that a set of proteins is induced in the heat shock response and that other members, besides hsp70, may well play a role in protective mechanisms in the nervous system.

Increased synthesis of hsp70 is beneficial to cells reacting to trauma, however, constitutive over-expression has been reported to be detrimental to cell growth in both tissue culture systems and *in vivo* in *Drosophila* salivary gland tissue (Feder *et al.*, 1992). Strategies to engineer neural cells to constitutively over-express hsp70, with the objective of generating cells which are more resistant to various types of trauma, may thus encounter problems with cell growth rates unless the timing of the over-expression is carefully controlled. The development of non-harmful procedures for activating the heat shock response in the nervous system could have considerable therapeutic value in promoting repair and/or tissue protection, particularly if triggering procedures could be selected which result in a pattern of induction which is specific to the profile of cells which are affected, or are at risk by a given trauma. As shown in the various chapters of this book, divergent types of neural stress, ranging from hyperthermia to ischaemia, turn on the heat shock response in the nervous system and result in the induction of hsp70 in specific cell types. Are there common mechanisms of induction in the various types of neural trauma? The elucidation of the mechanism by which divergent types of stress activate the heat shock response in a complex tissue such as the nervous system, remains to be accomplished.

Parallel to our *in vivo* observations on the hyperthermic rabbit brain, it has been reported that hsp70 induction in neuronal cells grown in primary culture is attenuated or undetectable compared to a robust response seen in cultured glial cells (see Chapter 5 and also Marcuccilli *et al.*, 1992). Tissue culture systems offer avenues for dissecting out the molecular basis of variations in hsp induction patterns. Recently, I spent my sabbatical leave in the laboratory of Richard Morimoto at Northwestern University where many advances in the molecular biology of the heat shock response have been achieved using mouse and human tissue culture cells (for a review: see Morimoto *et al.*, 1992; Morimoto, 1993). A human glioblastoma cell line (T98G) and a cervical epithelial cell line (HeLa) showed a robust induction of hsp70 and hsp90 α following heat shock (Mathur *et al.*, 1993). However, a retinoblastoma cell line (Y79), which demonstrates some neuronal-like features, showed an attenuated hsp70 induction, but normal hsp90 α induction, in a nuclear run-on transcription assay. Analysis in the Y79 cells of the stress-inducible heat shock transcription factor, HSFl, by mobility shift assay, Western blot analysis and genomic footprinting revealed that the factor is activated and functional, based on DNA-binding, phosphorylation and occupancy of the hsp90 α promoter. These studies indicate that activation of the heat shock transcriptional response is a complex, multistep process and that activation of HSFl is not sufficient for transcriptional induction of hsp70 in Y79 cells. It should be noted that these observations were made on populations of cells which are undergoing cell division, while previous investigations in the intact animal have been on neurons which are in a post-mitotic state. Studies on neural cells grown in tissue culture are providing new insights into the sequence of molecular events involved in transcriptional induction of hsp70. As studies in this area progress and other types of neural cells are employed, the tissue culture approach should contribute significantly to our understanding of the molecular biology of the heat shock response.

Acknowledgements

I thank Professors Lawrence Hightower, Julie Silver and Michael Tytell for helpful comments on the manuscript. This work was supported by grants from the Medical Research Council of Canada.

References

Abravaya, K., Myers, M.P., Murphy, S.P. & Morimoto, R.I. (1992) The human heat shock protein hsp70 interacts with HSF, the transcription factor that regulates heat shock gene expression. *Genes and Dev.* **6**, 1153–1164.

Ananthan, J., Goldberg, A.L. & Voellmy, R. (1986) Abnormal proteins serve as eukaryotic stress signals and trigger the activation of heat shock genes. *Science* **232**, 522–524.

Aquino, D.A., Kliptel, A.A., Brosnan, C.F. & Norton, W.T. (1993) The 70-kDa heat shock cognate protein (hsc70) is a major constituent of the central nervous system and is

upregulated only at the mRNA level in acute experimental autoimmune encephalomyelitis. *J. Neurochem.* **61**, 1340–1348.
Barbe, M.F., Tytell, M., Gower, D.J. & Welch, W.J. (1988) Hyperthermia protects against light damage in the rat retina. *Science* **241**, 1817–1820.
Baulieu, E.E. (1987) Steroid hormone antagonists at the receptor level: a role for the heat shock protein MW 90,000 (hsp90). *J. Cell Biochem.* **35**, 161–174.
Beckmann, R.P., Mizzen, L.A. & Welch, W.J. (1990) Interaction of hsp70 with newly synthesized proteins: Implications for protein folding and assembly. *Science* **248**, 850–854.
Blake, M.J., Gershon, D., Fargnoli, J. & Holbrook, N.J. (1990a) Disordant expression of heat shock protein mRNAs in tissues of heat stressed rats. *J. Biol. Chem.* **265**, 15 275–15 279.
Blake, M.J., Nowak, T.S. & Holbrook, N.J. (1990b) *In vivo* hyperthermia induces expression of hsp70 mRNA in brain regions controlling the neuroendocrine response to stress. *Mol. Brain Res.* **8**, 89–92.
Blake, M.J., Fargnoli, J., Gershon, D. & Holbrook, N. (1991) Concomitant decline in heat-induced hyperthermia and hsp70 mRNA expression in aged rats. *Am. J. Physiol.* **260** R663–667.
Brown, C.J., Martin, R.L., Hansen, W.J., Beckman, R.P. & Welch, W.J. (1993) The constitutive and stress inducible forms of hsp70 exhibit functional similarities and interact with one another in an ATP-dependent fashion. *J. Cell Biol.* **120**, 1101–1112.
Brown, I.R. (1983) Hyperthermia induces the synthesis of a heat shock protein by polysomes isolated from the faetal and neonatal mammalian brain. *J. Neurochem.* **40**, 1490–1493.
Brown, I.R. (1985a) Effect of hyperthermia and LSD on gene expression in the mammalian brain and other organs. In *Changes in Eukaryotic Gene Expression in Response to Environmental Stress* (eds Atkinson, B.G. and Waldon, C.B.) pp 211–225. New York, Academic Press Inc.
Brown, I.R. (1985b) Modification of gene expression in the mammalian brain after hyperthermia. In *Gene Expression in Brain* (eds Zomzely-Neurath, C. & Walker, W.A.), pp 157–171. New York, John Wiley and Sons Inc.
Brown, I.R. (1990) Induction of heat shock (stress) genes in the mammalian brain by hyperthermia and other traumatic events: a current perspective. *J. Neurosci. Res.* **27**, 247–255.
Brown, I.R. & Rush, S.J. (1990) Expression of heat shock genes (hsp70) in the mammalian brain: Distinguishing constitutively expressed and hyperthermia-inducible species. *J. Neurosci. Res.* **25**, 14–19.
Brown, I.R., Cosgrove, J.W. & Clark, B.D. (1982a) Physiologically relevant increases in body temperature induce the synthesis of a heat shock protein in mammalian brain and other organs. In *Heat Shock from Bacteria to Man* (eds Schlesinger, M.J., Ashburner, M. & Tissieres, A), pp 361–367. New York, Cold Spring Harbor Laboratory Press.
Brown, I.R., Heikkila, J.J. & Cosgrove, J.W. (1982b) Analysis of protein synthesis in the mammalian brain using LSD and hyperthermia as experimental probes. In *Molecular Approaches to Neurobiology* (ed. Brown, I.R.), pp 221–253. New York, Academic Press Inc.
Brown, I.R., Lowe, D.G. & Moran, L.A. (1985) Expression of a heat shock gene in faetal and maternal rabbit brain. *Neurochem. Res.* **10**, 1277–1284.
Brown, I.R., Rush, S.J. & Ivy, G.O. (1989) Induction of a heat shock gene at the site of tissue injury in the rat brain. *Neuron* **2**, 1559–1564.
Brunt, S.A., Riehl, R. & Silver, J.C. (1990) Steroid hormone regulation of the *Achlya ambisexulis* 85-kDa heat shock protein, a component of the *Achlya* steroid receptor complex. *Mol. Cell. Biol.* **10**, 273–281.
Carson-Jurica, M.A., Lee, A.T., Dobson, A.W., Conneely, O.M., Schrader, W.T. & O'Malley, B.W. (1989) Interaction of the chicken progesterone receptor with heat shock (hsp90). *J. Steroid Biochem.* **34**, 1–6.
Catelli, M.G., Binart, N., Jung-Testas, I., Renoir, J.M., Baulieu, E.E., Feramisco, J.R. & Welch, W.J. (1985) The common 90-kDa protein component of non-formed '8S' steroid receptors is a heat shock protein. *EMBO J.* **4**, 3131–3135.

Chappell, T.G., Welch, W.J., Schlossman, D.M., Palter, K.B., Schlesinger, M.J. & Rothman, J.E. (1986) Uncoating-ATPase is a member of the 70-kDa family of stress proteins. *Cell* **45**, 3–13.

Clark, B.D. & Brown, I.R. (1982) Protein synthesis in the mammalian retina, following the intravenous administration of LSD. *Brain Res.* **247**, 97–104.

Clark, B.D. & Brown, I.R. (1985) Axonal transport of a heat shock protein in the rabbit visual system. *Proc. Natl Acad. Sci. USA* **82**, 1281–1285.

Clark, B.D. & Brown, I.R. (1986a) Induction of a heat shock protein in the isolated mammalian retina. *Neurochem. Res.* **11**, 269–279.

Clark, B.D. & Brown, I.R. (1986b) A retinal heat shock protein is associated with elements of the cytoskeleton and binds to calmodulin. *Biochem. Biophys. Res. Commun.* **139**, 974–981.

Clark, B.D. & Brown, I.R. (1987) Altered expression of a heat shock protein in the mammalian nervous system in the presence of agents which affect microtubule stability. *Neurochem. Res.* **12**, 819–823.

Cosgrove, J.W. & Brown, I.R. (1983) Heat shock protein in the mammalian brain and other organs following a physiologically relevant increase in body temperature induced by LSD. *Proc. Natl Acad. Sci. USA* **80**, 569–573.

Cosgrove, J.W. & Brown, I.R. (1984) Effect of intravenous administration of *d*-lysergic acid diethylamide on initiation of protein synthesis in a cell-free system derived from brain. *J. Neurochem.* **42**, 1420–1426.

Cosgrove, J.W., Clark, B.D. & Brown, I.R. (1981) Effect of intravenous administration of *d*-lysergic acid diethylamide on subsequent protein synthesis in a cell-free system derived from brain. *J. Neurochem.* **36**, 1037–1045.

Cruz, T.F., Kandel, R.A. & Brown, I.R. (1991) Interleukin 1 induces the expression of a heat shock gene in chondrocytes. *Biochem. J.* **227**, 327–330.

Currie, R.W. & White, F.P. (1981) Trauma-induced protein in rat tissues: A physiological role for a 'heat shock' protein? *Science* **214**, 72–73.

de Waegh, S. & Brady, S.T. (1989) Axonal transport of a clathrin-uncoating ATPase (hsc70): A role for hsc70 in the modulation of coated vesicle assembly *in vivo*. *J. Neurosci. Res.* **23**, 433–440.

Dechesne, C.J., Kim, H.N., Nowak, T.S. & Wenthold, R.J. (1992) Expression of heat shock protein, hsp72, in the guinea pig and rat cochlea after hyperthermia: Immunochemical and *in situ* hybridization analysis. *Hearing Res.* **59**, 195–204.

Feder, J.J., Rossi, J.M., Solomon, J., Solomon, N. & Lindquist, S. (1992) The consequences of expressing hsp70 in *Drosophila* cells at normal temperatures. *Genes and Devel.* **6**, 1402–1413.

Freedman, M.S., Clark, B.D., Cruz, T.F., Gurd, J.W. & Brown, I.R. (1981) Selective effects of LSD and hyperthermia on the synthesis of synaptic proteins and glycoproteins. *Brain Res.* **207**, 129–145.

Freedman, M.S., Buu, N.N., Ruijs, C.J., Williams, K. & Antel, J.P. (1992) Differential expression of heat shock proteins by human glial cells. *J. Neuroimmunol.* **41**, 231–238.

Georgopouls, C. & Welch, W.J. (1993). Role of the major heat shock proteins as molecular chaperones. *Ann. Rev. Cell Biol.* **9**, 601–634.

Gernold, M., Knauf, U., Gaestal, M., Stahl, J. & Kloetzal, P.M. (1993) Developmental and tissue-specific distribution of mouse small heat shock protein hsp25. *Dev. Genetics* **14**, 103–111.

Gething, M.J., and Sambrook, J. (1992) Protein folding in the cell. *Nature* **355**, 33–45.

Gonzalez, M.F., Shiraishi, K., Hisanga, K., Sagar, S.M., Mandabach, M. & Sharp, F.R. (1989) Heat shock proteins as markers of neural injury. *Mol. Brain Res.* **6**, 93–100.

Gower, D.J., Hollman, C., Lee, K.S. & Tytell, M. (1989) Spinal cord injury and the stress-protein response. *J. Neurosurg.* **70**, 605–611.

Greenberg, S.C. & Lasek, R.J. (1985) Comparison of labelled heat shock proteins in neuronal and non-neuronal cells of *Aplysia californica*. *Neurosci.* **5**, 1239–1245.

Hartl, F.U. & Martin, J. (1992) Protein folding in the cell: The role of molecular chaperones hsp70 and hsp60. *Annu. Rev. Biophys. Bimol. Struct.* **21**, 293–322.
Hickey, E., Brandon, S.E., Sadis, S., Smale, G. & Weber, L.A. (1986) Molecular cloning of sequences encoding the human heat shock proteins and their expansion during hyperthermia. *Gene* **43**, 147–154.
Hickey, E., Brandon, S.E., Smale, G., Lloyd, D. & Weber, L.A. (1989) Sequence and regulation of a gene encoding a human 89-kDa heat shock protein. *Mol. Cell. Biol.* **9**, 2615–2626.
Hightower, L.E. (1991) Heat shock, stress proteins, chaperones, and proteotoxicity, *Cell.* 66, 191–197.
Hightower, L.E. & Guidon, P.T. (1989) Selective release from cultured mammalian cells of heat shock (stress) proteins that resemble glia-axon transfer proteins. *J. Cell. Physiol.* **138**, 257–266.
Inasi, B.S. & Brown, I.R. (1982) Synthesis of a heat shock protein in the microvascular system of the rabbit brain following elevation of body temperature. *Biochem. Biophys. Res. Commun.* **106**, 881–887.
Itoh, H., Toyoshima, I., Mizunuma, H., Kobayashi, R. & Tashima, Y. (1990) Three-step purification method and characterization of the bovine brain 90-kDa heat shock protein. *Arch. Biochem. Biophys.* **282**, 290–296.
Itoh, H., Tashima, Y., Eishi, Y. & Okeda, R. (1993) Localization of hsp90 in rat brain. *Int. J. Biochem.* **25**, 93–99.
Izumoto, S. & Herbert, J. (1993) Widespread constitutive expression of hsp90 messenger RNA in rat brain. *J. Neurosci. Res.* **35**, 20–28.
Johnston, R.N. & Kucey, B.L. (1988) Competitive inhibition of hsp70 gene expression causes thermosensitivity. *Science* **242**, 1551–1554.
Koyashu, S., Nishida, E., Kadowaki, T., Matsuzaki, F., Iida, K., Harada, F., Kasuga, M., Sakai, H. & Yahara, I. (1986). Two mammalian heat shock proteins hsp90 andl hsp100 are actin-binding proteins. *Proc. Natl Acad. Sci. USA* **83**, 8054–8058.
Landry, J., Chretein, P., Lambert, H., Hickey, E. and Weber, L.A. (1989). Heat shock resistance conferred by expression of the human hsp27 gene in rodent cells. *J. Cell Biol.* **109**, 7–15.
Li, G.C., Li, Y-K., Mak, J.Y., Chen, L. & Lee, W.F. (1991) Thermal response of rat fibroblasts stably transfected with the human 70-kDa heat shock protein-coding gene. *Proc. Natl Acad. Sci. USA* **88**, 1681–1685.
Li, Y., Chopp, M., Yoshida, Y. & Levine, S.R. (1992) Distribution of 72-kDa heat shock protein in rat brain after hyperthermia. *Acta Neuropathol.* **84**, 94–99.
Lim, H.H., Jenkins, O.H., Miller, J.M., Myers, M. & Altschuler, R.A. (1992) Hsp72-induction with acoustic overstimulation in the rat cochlea and auditory brainstem. *Soc. Neurosci. Abstr.* **18**, 1192.
Lindquist, S. & Craig, E.A. (1988) The heat shock proteins. *Annu. Rev. Genet.* **22**, 631–677.
Lowenstein, D.H., Chan, P.H. & Miles, M.F. (1991) The stress protein response in cultured neurons: Characterization and evidence for a protective role in excitotoxity. *Neuron* **7**, 1053–1060.
Lowenstein, D.H., Gwinn, R.P. Seven, M.S., Simon, R.P. & McIntosh, T.K. (1994) Increased expression of mRNA encoding calbindin-D28K the glucose-regulated proteins, or the 72kDA heat shock protein in three models of acute CNS injury. *Molec. Brain Res.* (in press).
Manzerra, P. & Brown, I.R. (1990) Time-course induction of a heat shock gene (hsp70) in the rabbit cerebellum after LSD *in vivo*: Involvement of drug-induced hyperthermia. *Neurochem. Res.* **15**, 53–59.
Manzerra, P. & Brown, I.R. (1992a) Distributon of constitutive- and hyperthermia-inducible heat shock mRNA species (hsp70) in the Purkinje layer of the rabbit cerebellum. *Neurochem. Res.* **17**, 559–564.
Manzerra, P. & Brown, I.R. (1992b) Expression of heat shock genes (hsp70) in the rabbit

spinal cord: localization of constitutive- and hyperthermia-inducible mRNA species. *J. Neurosci. Res.* **31**, 606–615.

Manzerra, P., Rush, S.J. & Brown, I.R. (1993) Temporal and spatial distribution of heat shock mRNA and protein (hsp 70) in the rabbit cerebellum in response to hyperthermia. J. *Neurosci. Res.* **36**, 480–490.

Marcuccilli, C.J., Fox, L.E. & Miller, R.J. (1992) Deficient heat shock response in cultured CNS and PNS neurons. *Soc. Neurosci. Abstr.* **18**, 1104.

Marin, R., Valet, J.P. & Tanguay, R.M. (1993) Hsp23 and hsp26 exhibit distinct spatial and temporal patterns of constitutive expression in *Drosophila* adults. *Dev. Genetics* **14**, 69–77.

Marini, A.M., Kozuka, M., Lipsky, R.H. & Nowak, T.S. (1990) 70-kDa heat shock protein induction in cerebellar astrocytes and cerebellar granule cells *in vitro*: comparison with immunocytochemical localization after hyperthermia *in vivo*. *J. Neurochem.* **54**, 1509–1516.

Mathur, S.K., Sistonen, L., Brown, I.R., Murphy, S.P., Sarge, K.D. & Morimoto, R.I. (1993) Differential induction of human heat shock gene transcription in Y79 retinoblastoma cells occurs independent of HSFl activation. *Soc. Neurosci. Abstr.* **19**, 224.

Miller, K., Raese, J.D. & Morrison-Bogorad, M. (1991) Expression of heat shock protein 70 and heat shock cognate 70 messenger RNAs in rat cortex and cerebellum after heat shock or amphetamine treatment. *J. Neurochem.* **56**, 2060–2071.

Moore, S.K., Kozak, C., Robinson, E.A., Ullrich, S.J. & Appella, E. (1989) Murine 86- and 84-kDa heat shock proteins, cDNA sequences, chromosome assignments, and evolutionary origins. *J. Biol. Chem.* **264**, 5343–5351.

Morandi, A., Los, B., Osofsky, L., Autilio-Gambetti, L. & Gambetti, P. (1989) Ubiquitin and heat shock proteins in cultured nerous tissue after different stress conditions. In *Alzheimer's Diseases and Related Disorders* (eds Iqbal, K., Wisniewski, K. & Winblad, H.M.), pp 819–827. New York, Alan R. Liss, Inc.

Morimoto, R.I. (1993) Cells in stress: Transcriptional activation of heat shock genes. *Science* **259**, 1409–1410.

Morimoto, R., Tissieres, A. & Georgopoulos, C. (eds) (1990) *Stress Proteins in Biology and Medicine* New York, Cold Spring Harbor Laboratory Press.

Morimoto, R.I., Sarge, K.D. & Abravaya, K. (1992) Transcriptional regulation of heat shock genes. *J. Biol. Chem.* **267**, 21 987–21 990.

Mosser, D.D., Duchaine, J. & Massie, B. (1993) The DNA-binding activity of the human heat shock transcription factor is regulated *in vivo* by hsp70. *Mol. Cell. Biol.* **13**, 5427–5438.

Neely, J.G., Thompson, A.M. & Gower, D.J. (1991) Detection and localization of heat shock protein 70 in the normal guinea pig cochlea. *Hearing Res.* **52**, 403–406.

New, G.A., Hendrickson, B.R. & Jones, K.J. (1989) Induction of heat shock protein 70 mRNA in adult hamster facial nuclear groups following axotomy of the facial nerve. *Metab. Brain Dis.* **4**, 273–279.

Nishimura, R.N., Dwyer, B.E., Welch, W., Cole, R., de Vellis, J. & Liotta, K. (1988) The induction of the major heat stress protein is purified rat glial cells. *J. Neurosci.* **20**, 12–18.

Nishimura, R.N., Dwyer, B.E., Clegg, K., Cole, R. & de Vellis, J. (1991) Comparison of the heat shock response in cultured cortical neurons and astrocytes. *Mol. Brain Res.* **9**, 39–45.

Nover, L. (1991) *Heat shock response.* Boca Raton, CRC Press Inc.

Nowak, T.S. (1988) Effects of amphetamine in protein synthesis and energy metabolism in mouse brain: Role of drug-induced hyperthermia. *J. Neurochem.* **50**, 285–294.

Nowak, T.S., Bond, U. & Schlesinger, M.J. (1990) Heat shock RNA levels in brain and other tissues after hyperthermia and transient ischaemia. *J. Neurochem.* **54**, 451–458.

Olazabal, U.E., Pfaff, D.W. & Mobbs, C.V. (1992a) Oestrogenic regulation of heat shock protein 90-kDa in the rat ventromedial hypothalamus and uterus. *Mol. Cell. Endocrin.* **84**, 175–183.

Olazabal, U.E., Pfaff, D.W. & Mobbs, C.V. (1992b) Sex differences in the regulation of heat shock protein 70-kDa and 90-kDa in the rat ventromedial hypothalamus by oestrogen. *Brain Res.* **596**, 311–314.

Opperman, H., Levinson, A.D., Levintow, L., Varmus, H.E., Bishop, J.M. & Kawai, S. (1981) A cellular protein that associates with the transforming protein of *Rous sarcoma* virus is also a heat shock protein. *Proc. Natl Acad. Sci. USA* **78**, 1067–1071.

Pardue, M.L., Feramisco, J.R. & Lindquist, S. (eds) (1989) *Stress-induced proteins.* New York, Alan R. Liss Inc.

Pardue, S., Groshan, K., Raese, J.D. & Morrison-Bogorad, M. (1992) Hsp70 mRNA induction is reduced in neurons of aged rat hippocampus after thermal stress. *Neurobiology of Aging* **13**, 661–672.

Pelham, H.R.B (1986) Speculations on the functions of the major heat shock and glucose-regulated proteins. *Cell* **49**, 959–961.

Pratt, W.B. (1987) Transformation of glucocorticoid and progesterone receptors to the DNA-binding state. *J. Cell. Biochem.* **35**, 51–68.

Ramachandran, C., Catelli, M.G., Schneider, W. & Shyamala, G. (1988) Oestrogenic regulation of uterine 90-kDa heat shock protein. *Endocrin.* **123**, 956–961.

Rebbe, N.F., Ware, J., Bertina, R.M., Modrich, P. & Stafford, D.W. (1987) Nucleotide sequence of a cDNA for a member of the human 90-kDa heat shock protein family. *Gene* **53**, 235–245.

Riabowol, K.T., Mizzen, L.A. & Welch, W.J. (1988) Heat shock is lethal to fibroblasts micro-injected with antibodies against hsp70. *Science* **242**, 433–436.

Rordorf, G., Koroshetz, W.J. & Bonventre, J.V. (1991) Heat shock protects cultured neurons from glutamate toxicity. *Neuron* **7**, 1043–1051.

Rose, D.W., Welch, W.J., Kramer, G. & Hardesty, B. (1989) Possible involvement of the 90-kDa heat shock protein in the regulation of protein synthesis. *J. Biol. Chem.* **264**, 6239–6244.

Satoh, J., Yamamura, T., Kunishita, T. & Tabira, T. (1992) Heterogeneous induction of 72-kDa heat shock protein (hsp72) in cultured mouse oligodendrocytes and astrocytes. *Brain Res.* **573**, 37–43.

Schlesinger, M.J. (1990) *Heat Shock Proteins. J. Biol. Chem.* **265**, 12 111–12 114.

Schlossman, D., Schmid, S., Braell, W. & Rothman, J. (1984) An enzyme that removes clathrin: Purification of an uncoating-ATPase. *J. Cell Biol.* **99**, 723–733.

Schuh, S., Yonemoto, W., Brugge, J., Baurer, V.J., Riehl, R.M., Sullivan, W.P. & Toft, D.O. (1985) A 90,000-dalton binding protein common to both steroid receptors and the *Rous scarcoma* virus transforming protein pp60 v-scr. *J. Biol. Chem.* **260**, 14 292–14 296.

Silver, J.C., Andrews, D.R. & Pekkala, D. (1983) Effect of heat shock on synthesis and phosphorylation of nuclear and cytoplasmic proteins in the fungus *Achlya. Can. J. Biochem. Cell. Biol.* **61**, 447–455.

Sloviter, R.S. & Lowenstein, D.H. (1992) Heat shock expression in vulnerable cells of the rat hippocampus as an indicator of excitation-induced neuronal stress. *J. Neurosci.* **12**, 3004–3009.

Solomon, J.M., Rossi, J.M., Golic, K., McGarry, T. & Lindquist, S. (1991) Changes in hsp70 alter thermotolerance and heat shock regulation in *Drosophila. New Biologist* **3**, 1106–1120.

Sprang, G.K. & Brown, I.R. (1987) Selective induction of a heat shock gene in fibre tracts and cerebellar neurons of the rabbit brain detected by *in situ* hybridization. *Mol. Brain Res.* **3**, 89–93.

Stone, G.C. & Dougher, M.M. (1988) Heat stress induces changes in protein synthesis and fast axonal transport in bullfrog sensory neurons. *J. Neurochem.* **51**, 960–966.

Stone, G.C. & Dougher, M.M. (1989) Heat stress increases delivery of a unique sub-population of proteins conveyed by fast axonal transport. *J. Neurochem. Res.* **24**, 477–486.

Swanson, R.A. & Sharp, F.R. (1992) Zinc toxicity and induction of the 72-kDa heat shock protein in primary astrocyte culture. *Glia* **6**, 198–205.

Tanguay, R.M., Wu, Y. & Khandjian, E.W. (1993) Tissue-specific expression of heat shock proteins of the mouse in the absence of stress. *Dev. Genetics* **14**, 112–118.

Tanno, H., Nockels, R.P., Pitts, L.H. & Noble, L.J. (1993) Immunolocalization of heat shock protein after fluid percussive brain injury and relationship to breakdown of the blood-brain barrier. *J. Cereb. Blood Flow and Metab.* **13**, 116–124.

Tytell, M. & Barbe, M.F. (1987) Synthesis and axonal transport of heat shock proteins. In *Axonal Transport* (eds Smith, R.S. & Bisby M.A.), pp 473–492, New York, Alan R. Liss, Inc.

Tytell, M., Greenberg, S.G. & Lasek, R.J. (1986) Heat shock-like protein is transferred from glial to axon. *Brain Res.* **363**, 161–164.

Tytell, M., Barbe, M.F. & Brown, I.R. (1993) Stress (heat shock) protein accumulation in the central nervous system: its relationship to cell stress and damage. In *Advances in Neurology* (ed. Seil, F.J.), pp 292–303. New York, Raven Press Ltd.

Tytell, M., Barbe, M.F. & Brown, I.R. (1994) Induction of heat shock (stress) protein 70 and its mRNA in the normal and light-induced rat retina and after whole body hyperthermia. *J. Neurosci. Res.* (in press).

Ullrich, S.J., Robinson, E.A., Law, L.W. & Willingham, M. (1986) A mouse tumor-specific transplantation antigen is a heat shock related protein. *Proc. Natl Acad. Sci. USA* **83**, 3121–3125.

Uney, J.B., Pickering, S., Staley, K., Johnstone, M.H., Rabbitts, T.H. & Sofroniew, M.V. (1992) A transgenic mouse model to assess the neuroprotective effect of hsp70 in the CA1 and CA2 region of the hippocampus. *Soc. Neurosci. Abstr.* **18**, 1145.

Ungewickell, E. (1985) The 70-kDa mammalian heat shock proteins are structurally and functionally related to the uncoating-protein that releases clathrin triskelia from coated vesicles. *EMBO J* **4**, 3385–3391.

van Bergen en Henegouwen, P.M., Berbers, G.A., Linnemans, W.A. & van Wijk, R. (1987) Subcellular localizaton of 84,000 dalton hsp in mouse neuroblastoma cells; evidence for a cytoplasmic and nuclear location. *Eur. J. Cell. Biol.* **43**, 469–473.

Vass, K., Welch, W.J. & Nowak, T.S. (1988) Localization of 70-kDa stress protein induction in gerbil brain after ischaemia. *Acta Neuropathol. (Berl)* **77**, 128–135.

Vass, K., Berger, M.L., Nowak, T.S., Welch, W.J. & Lassman, H. (1989) Induction of stress protein hsp70 in nerve cells after status epilepticus in the rat. *Neurosci. Lett.* **100**, 259–254.

Weber, L.A. (1992) Relationship of heat shock proteins and induced thermal resistance. *Cell Prolif.* **25**, 101–113.

Welch, W.J. (1990) The mammalian stress response: Cell physiology and biochemistry of stress proteins. In *Stress Proteins in Biology and Medicine* (eds Morimoto, R., Tissieres, A. & Georgopoulos, C.), pp 223–279. New York, Cold Spring Harbor Laboratory Press.

Welch, W.J., Mizzen, L.A. & Arrigo, A.P. (1989) Structure and function of mammalian stress proteins. In *Stress-Induced Proteins* (eds Pardue, M.L., Feramicso, J.R. & Lindquist, S.), pp 187–202. New York, Alan R. Liss. Inc.

White, F.P. (1980) Differences in protein synthesized *in vivo* and *in vitro* by cells associated with the cerebral microvasculature. A protein synthesized in response to trauma? *Neuroscience* **5**, 1793–1799.

White, F.P. (1981) The induction of 'stress' proteins in organ slices from brain, heart and lung as a function of postnatal development. *J. Neurosci.* **1**, 1312–1319.

White, F.P. & Currie, R.W. (1982) A mammalian response to trauma: the synthesis of a 71-kD protein. In *Heat Shock from Bacteria to Man* (eds Schlesinger, M.J., Ashburner, M. & Tissieres, A.), pp 379–386. New York, Cold Spring Harbor Laboratory Press.

Xue, Z. & Grossfield, R.M. (1993) Stress protein synthesis and accumulation after traumatic injury of crayfish CNS. *Neurochem. Res.* **18**, 209–218,

Yamaguchi, K., Barbe, M.F., Brown, I.R. & Tytell, M. (1990) Induction of stress (heat shock) protein 70 and its mRNA in rat corneal epithelium by hyperthermia. *Current Eye Res.* **9**, 913–918.

CHAPTER 3

THE HEAT SHOCK RESPONSE AND GENE EXPRESSION IN BRAIN AFTER ISCHAEMIA

Thaddeus S. Nowak, Jr, Sadao Suga* and Nobuhito Saito†

*Department of Neurology, University of Tennessee College of Medicine, Memphis, Tennessee, *Department of Neurosurgery, Keio University School of Medicine, Tokyo, Japan, †Department of Neurosurgery, Tokyo University School of Medicine, Tokyo, Japan*

Table of Contents

Heat Shock Proteins in the Nervous System
ISBN 0-12-480960-X

T. S. Nowak, S. Suga and N. Saito

3.1 Introduction

Deficits in cerebral blood flow occur in man under conditions of cardiac arrest (global ischaemia) or during a stroke arising from obstruction of flow to a region of brain tissue (focal ischaemia). Characteristic patterns of tissue damage are observed after such insults in both the clinical conditions and in animal models, with selective injury to specific neuron populations after brief global ischaemia (Kirino, 1982; Pulsinelli *et al.*, 1982; Smith *et al.*, 1984; Ross and Duhaime, 1989) and a graded response potentially involving all cell types in the region of a focal insult (Astrup *et al.*, 1981). It is well established that selective neuron loss can evolve with a considerable delay after global insults (Kirino, 1982; Pulsinelli *et al.*, 1982) and there is also some progression of injury with time after focal ischaemia (Selman *et al.*, 1990; Jacewicz *et al.*, 1992). This offers the possibility of intervention during the interval preceding irreversible cell loss, and considerable efforts are directed toward understanding the evolution of cellular pathophysiology after ischaemia.

Changes in gene expression have proved to be sensitive indices of the responses of individual cells to ischaemic insults (Nowak *et al.*, 1993). For a number of years it has been recognized that the brain responds to both global and focal ischaemia with the expression of proteins characteristic of the heat shock stress response (Nowak, 1985; Dienel *et al.*, 1986; Jacewicz *et al.*, 1986; Kiessling *et al.*, 1986). Broadly speaking, the stress response is localized within those cell populations that are susceptible to injury following a given insult (Nowak, 1990), although it appears to be most properly considered as a marker of cells at risk that may either recover or be lost (Sloviter and Lowenstein, 1992). It should be noted that there is a considerable literature on the post-ischaemic stress response in tissues other than brain, including heart (Knowlton *et al.*, 1991), liver (Tacchini *et al.*, 1993) and kidney (Van Why *et al.*, 1992) to mention a few recent studies. Furthermore, it is increasingly evident that the stress response constitutes but one component of a widespread reprogramming of gene expression in post-ischaemic tissues (Nowak *et al.*, 1993; Tacchini *et al.*, 1993), and a primary focus of ongoing work in this area is to define the relationships between specific patterns of altered gene expression and the processes underlying cell loss and survival. Within this larger context, the present overview will summarize available data on stress protein induction in brain after ischaemia.

3.2 Global Ischaemia

3.2.1 The stress response after global ischaemia

3.2.1.1 Transcriptional induction of hsp70 and other heat shock genes

The post-ischaemic stress response was first identified as an increase in the expression of an inducible member of the 70-kDa heat shock protein (hsp70) family, using *in vitro* translation assays to detect the presence of hsp70 mRNA in active polyribosomes during up to 24 h recirculation after global cerebral ischaemia in gerbils (Nowak, 1985) and rats (Dienel *et al.*, 1986). Blot hybridization analyses of extracted RNA also identified a relatively prolonged increase in the levels of hsp70 mRNA in gerbil brain (Nowak *et al.*, 1990), reaching a maximum at 6 h and remaining above control at 24 h, consistent with the *in vitro* translation results. Other studies have yielded generally comparable findings (Abe *et al.*, 1991b). It is apparent that the post-ischaemic stress response evolves more slowly and spans a much longer interval than that observed in brain after hyperthermic insults (Nowak *et al.*, 1990).

These early hybridization studies also demonstrated the importance of using probes that distinguish between inducible sequences of the hsp70 family that are minimally expressed in control tissues and the constitutively expressed hsc70 species that are normally present at significant levels, particularly in brain. For example, only modest two-fold increases in hybridization were detected using a probe that detected constitutively expressed mRNA (Nowak *et al.*, 1990), consistent with recent *in situ* hybridization results detailed below (Kawagoe *et al.*, 1992c), while 8–10-fold increases were observed with an oligonucleotide selective for inducible sequences (Nowak *et al.*, 1990; Miller *et al.*, 1991). Cloned probes with comparable selectivity have been obtained for both rat and gerbil (Brown and Rush, 1990; Sato *et al.*, 1992). The designation 'hsp70' is used here to distinguish both the inducible mRNA and protein from constitutively expressed 'hsc70' species, although the protein induced in rodent brain after ischaemia is apparently equivalent to that more specifically termed 'hsp 72' in other studies involving both rodent and human cells (Watowich and Morimoto, 1988).

In situ hybridization has demonstrated the prominent neuronal localization of hsp70 mRNA expression after brief global ischaemia (Kawagoe *et al.*, 1992c; Nowak, 1991). This is illustrated for gerbil hippocampus in Figure 1, in which the signal is most evident in the densely packed major neuron populations, but hybridization is also readily detected in striatum and throughout ischaemic cortex in this model (Nowak, 1991), and more severe insults result in a more widespread induction of hsp70 (Kawagoe *et al.*, 1992c; Nowak *et al.*, 1992). A neuronal localization of hsp70 hybridization is consistent with the selective involvement of this cell type in the injury that occurs after global ischaemic insults, although the initial distribution of hsp70 mRNA expression does not correlate particularly well with the relative vulnerability of individual neuron populations. At later recirculation intervals,

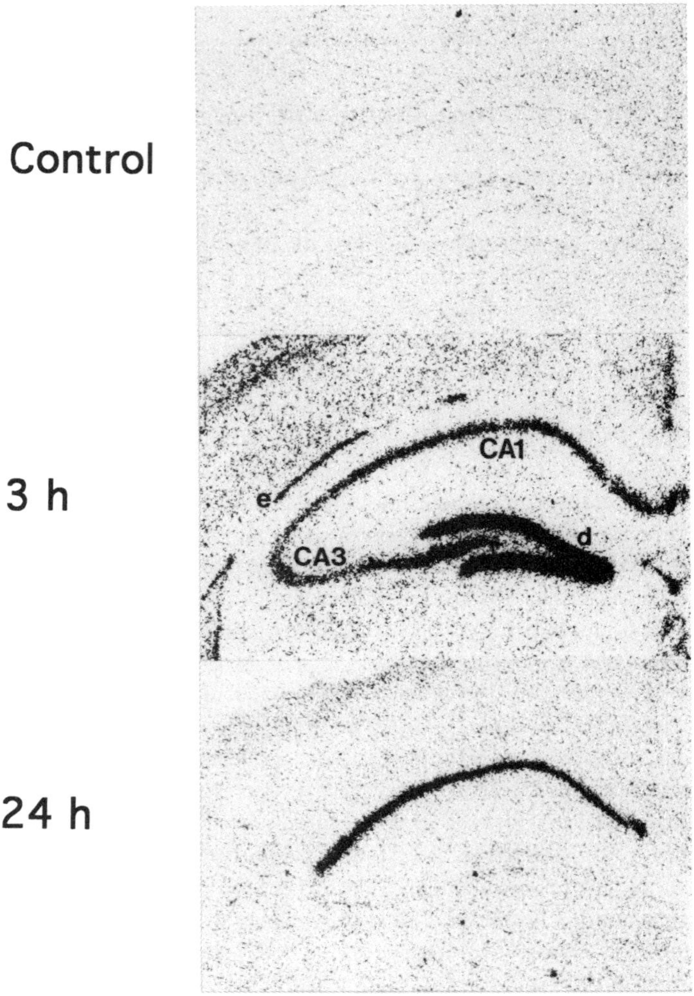

Figure 1 Time-course and distribution of hsp70 mRNA expression in gerbil hippocampus after transient ischaemia. Control tissue shows negligible hybridization. A striking increase in expression is detected in all major hippocampal neuron populations including dentate granule cells (d) at 3 h after 5 min ischaemia. Significant expression is also noted in ependyma (e). Vulnerable neurons in CA1 are distinguished by a sustained increase in hsp70 mRNA levels at 24 h. The photographs illustrate film autoradiography of hybridized [^{35}S]-labelled oligonucleotide as previously described (Nowak, 1991).

however, persistent hsp70-hybridization is restricted to the highly vulnerable neurons in the hippocampal CA1 subfield as well as in dorsolateral cortex (Nowak, 1991), reflecting the distribution of neuronal damage subsequently observed.

While such generalizations inferred from the gerbil results appear to be valid in other ischaemia models, some differences in the time course and distribution of hsp70 mRNA expression have been noted. For example, dentate granule cells that strongly express hsp70 hybridization for a considerable interval after 5 min ischaemia in the gerbil show less striking hybridization even after 10 min ischaemia in the rat (N. Saito and T.S. Nowak, Jr., unpublished observations). It is apparent that the threshold insult required for induction of the stress response will differ among animal models and among cell populations within a species. This issue is addressed below with respect to detailed studies of hsp70 immunoreactivity in several models.

The issue of induction threshold also arises with respect to recent work focused directly on comparisons of hsp70 and cognate hsc70 expression after global ischaemia. It was observed in early blot studies that the normally expressed hsc70 mRNA showed increased expression in gerbil brain after ischaemia, largely in parallel with the changes in hsp70 (Abe *et al.*, 1991b). Cloned cDNA probes have recently been obtained from gerbil brain that include considerable portions of the 3′ untranslated regions of gerbil hsp70 and hsc70 mRNA (Sato *et al.*, 1992), and these have provided the necessary specificity for subsequent *in situ* hybridization studies. In general, modest increases in hsc70 mRNA coincide in distribution and time course with the expression of hsp70 induced after ischaemia of varied duration in the gerbil (Kawagoe *et al.*, 1992c; Aoki *et al.*, 1993). It has been suggested that very brief periods of ischaemia result in selective increases in hsc70 hybridization that are not accompanied by hsp70 induction (Kawagoe *et al.*, 1992a), consistent with observations of a lower temperature threshold for hsc70 induction in brain following hyperthermic insults (Miller *et al.*, 1991), and the modest seizure inducibility of hsc70 but not hsp70 (Wong *et al.*, 1992). Early *in vitro* translation studies did not detect increased expression of constitutive proteins of the hsp70 family in post-ischaemic gerbil brain (Nowak, 1985), and it therefore remains to be determined whether post-ischaemic increases in hsc70 mRNA levels are accompanied by comparable increases in the encoded protein.

3.2.1.2 Translational expression, post-ischaemic protein synthesis deficits, and other factors affecting the detectability of hsp70 and other heat shock genes after ischaemia

As indicated above, the active translation of hsp70 mRNA after ischaemia was initially indicated by the appearance of the characteristic labelled protein among the *in vitro* translation products of gerbil brain extracts (Nowak, 1985). Synthesis was also demonstrated directly *in vivo* after administration of radioactive amino acids during post-ischaemic recirculation in the rat (Kiessling *et al.*, 1986). Hsp70 immunoreactivity has now come to be widely used for the mapping of cell populations affected by ischaemia and other insults. For the most part, hsp70 expression after global ischaemia is restricted to neurons, consistent with the *in situ* hybridization results cited above. There is a widespread distribution of hsp70 staining in hippocampus, cortex and striatum after 10 min ischaemia in the gerbil, but attenuated expression in the vulnerable CA1 neurons (Vass *et al.*, 1988). Neuronal staining is also the predominant feature of global ischaemia in the rat, although more prominent staining of CA1 neurons is detected under most conditions (Chopp *et al.*, 1991; Simon *et al.*, 1991). Detailed studies of the response to varied periods of ischaemia indicate preferential hsp70 expression in CA1 and hilar neurons after very brief ischaemia in both rat and gerbil (Simon *et al.*, 1991; Kirino *et al.*, 1991), with progressive recruitment of less vulnerable populations as the insult duration increases. Reductions in hsp70 expression and changes in distribution consistent with a milder insult occur following interventions that reduce ischaemic injury (Nowak, 1989, 1991; Bergstedt, 1993). Conversely, an additional involvement of glia and vascular elements has been described in hippocampus under conditions of more severe ischaemic injury (Gonzalez *et al.*, 1991), proceeding to focal infarction with islands of hsp70-positive neurons bordered by positive glia (Sharp *et al.*, 1991b). These will be considered more fully below in the setting of focal ischaemic injury.

It is well known that protein synthesis recovers poorly in vulnerable neuron populations after ischaemia (Dienel *et al.*, 1980; Thilmann *et al.*, 1986; Widmann *et al.*, 1991), and it would be expected that such translational deficits would severely limit the expression of hsp70 and other proteins encoded by induced mRNAs. Results in the gerbil are consistent with this prediction, in that immunocytochemical localization of hsp70 after 5 or 10 min ischaemia has consistently failed to detect significant accumulation in CA1 neurons (Vass *et al.*, 1988; Kirino *et al.*, 1991). In contrast, brief 2 min insults allow sufficient translational recovery to support robust accumulation of hsp70 (Kirino *et al.*, 1991), and these results are considered in more detail below in the framework of induced ischaemic tolerance. Findings in the rat are less simply interpreted, since considerable hsp70 immunoreactivity is detected in essentially all CA1 neurons even after 10 min ischaemic insults that result in severe damage to this cell population (Chopp *et al.*, 1991; Simon *et al.*, 1991). From a practical perspective, the successful translation of hsp70 mRNA is consistent with a more modest impairment of CA1 protein synthesis in the rat model (Okawa *et al.*, 1993). While an attenuation of hsp70 accumulation is in fact seen after longer ischaemic insults in the rat (Simon *et al.*, 1991), it is clear from the above studies that the expression of immunoreactive hsp70 in a given cell provides no guarantee of survival after ischaemia.

Other results argue that factors other than translational recovery must also affect the detectability of hsp70 after ischaemia. For example, hsp70 immonocytochemistry fails to reveal detectable protein in any neuron population of gerbil hippocampus at recirculation intervals earlier than 12 h after transient ischaemia (Vass *et al.*, 1988), during which the protein is prominently detected as an *in vitro* translation product (Nowak, 1985), and at which times increases in other proteins such as Fos and Jun (Nowak *et al.*, 1993; Uemura *et al.*, 1991a; Suga and Nowak, 1991) (manuscript in preparation) as well as ornithine decarboxylase (Müller *et al.*, 1991; Paschen *et al.*, 1993) are readily demonstrated. While it is possible that a significant lag-time may be required to reach the limits of immunocytochemical detection, these results are also consistent with findings in cultured cells that some hsp70 epitopes may be masked during the stress response, presumably due to sequestration in functional complexes with other proteins (Milarski *et al.*, 1989). Such a phenomenon is of obvious importance in understanding the physiological chemistry of hsp70 after ischaemia, but also complicates the use of hsp70 staining as a practical marker for the post-ischaemic stress response. Anomalies in hsp72 staining were also suggested in immunoblot studies of neonatal hypoxia-ischaemia (Dwyer *et al.*, 1989).

While most attention has focused on hsp70, a few studies have examined ubiquitin expression after global ischaemic insults. As noted above, only modest post-ischaemic increases in ubiquitin mRNA are detected in gerbil brain (Nowak *et al.*, 1990). Initial immunocytochemical studies in the rat revealed a striking decrease in immunoreactivity considered specific for ubiquitin-conjugates in all major hippocampal neuron populations after ischaemia, with a significantly increased signal in interneurons (Magnusson and Wieloch, 1989). The time-course of recovery was inversely correlated with the relative vulnerability of individual cell populations, paralleling the results for hsp70 expression in gerbil brain (Vass *et al.*, 1988). There is also a correlation between improved ubiquitin recovery and survival of CA1 neurons following hypothermic blunting of ischaemic injury (Yamashita *et al.*, 1991). The failure to recover ubiquitin-conjugate immunoreactivity in CA1 neurons could reflect in part a decreased ubiquitin expression as a result of the protein synthesis deficits noted above. It was also proposed that the observed staining results could arise from specific defects in the ubiquitination pathway (Magnusson and Wieloch, 1989). However, immunoblot studies of micro-dissected CA1 from gerbil hippocampus indicate striking increases in high molecular weight ubiquitin-conjugates at 24 h even after a severe ischaemic insult (Hayashi *et al.*, 1991, 1992), suggesting an increased utilization of the pathway. This latter result is perhaps more consistent with the observed increases in ubiquitin conjugation after hyperthermic stress (Carlson *et al.*, 1987), and raises the possibility that changes in ubiquitin-conjugate detectability in sections may not be strictly interpretable as an index of ubiquitination activity. Increased ubiquitin accumulation within interneurons and non-neuronal cells observed in the immunocytochemistry studies could perhaps contribute to the blot results, and in any case provides one of the few indications that there is an impact of ischaemia on interneurons that are resistant to injury (Magnusson and Wieloch, 1989).

3.2.2 Co-ordinate changes in expression of other genes after global ischaemia

3.2.2.1 Proto-oncogene transcription factors and other early responses

Induction of the proto-oncogene transcription factor, c-*fos*, is perhaps the most studied example of stimulus-induced transcriptional activation in brain and other tissues (Morgan and Curran, 1991), and it is now well-established that c-*fos* and other 'immediate–early' genes are also induced following ischaemia. Blot hybridization studies have demonstrated characteristically rapid and transient increases in c-*fos* mRNA after global ischaemia (Onodera *et al.*, 1989; Kindy *et al.*, 1991). *In situ* hybridization indicates a selective neuronal localization comparable to that of hsp70 (Nowak *et al.*, 1990; Wessel *et al.*, 1991), and the distributions of these mRNAs appear to overlap precisely during early recirculation intervals when both are expressed (J. Ikeda, T. Nakajima, O.C. Osborne, G. Mies and T.S. Nowak, Jr., manuscript in preparation). Ischaemic thresholds are comparable for c-*fos* and hsp70 expression, further suggesting that they may share common induction mechanisms (Nowak *et al.*, 1992). A late component of c-*fos* expression in vulnerable hippocampal regions has also been described in the rat, but it varies in quantitative significance and has not been a consistent observation (Wessel *et al.*, 1991; Jørgensen *et al.*, 1989). Immunocytochemical studies have demonstrated Fos-protein expression, with strikingly attenuated accumulation of immunoreactivity in CA1 (Uemura *et al.*, 1991a), consistent with the lasting translational deficits in these cells. The effect of protein synthesis recovery on expression of immediate early genes is considered in more detail below in the context of induced ischaemic tolerance.

Global ischaemia triggers changes in the expression of a diverse and rapidly growing list of mRNAs and proteins, and the observed results are generally consistent with the principles established from above studies of hsp70 and c-*fos*. Superoxide dismutase (SOD) mRNA is induced in neurons after ischaemia, with attenuated accumulation of the protein in the vulnerable CA1 region (Matsuyama *et al.*, 1993). Ornithine decarboxylase (ODC) has long been known to increase after ischaemia (Dienel *et al.*, 1985), and appears to be less limited by the translational impairment in CA1, showing significant increases in ODC immunoreactivity in these vulnerable neurons even under conditions of relatively severe ischaemia (Müller *et al.*, 1991; Paschen *et al.*, 1993). This is presumably due, in part, to a relatively prolonged expression of its mRNA (Dempsey *et al.*, 1991) accompanied by partial and transient recovery of amino acid incorporation (Thilmann *et al.*, 1986), further supporting the suggestion that factors other than protein synthesis recovery may be responsible for the limited detectability of hsp70 in the gerbil noted above. Several growth factor mRNAs are known to be induced in rat hippocampus after ischaemia, predominantly in dentate granule cells (Lindvall *et al.*, 1992), while expression of the dynorphin peptide localized in these neurons has been shown to decrease (Fried and Nowak, 1987). There undoubtedly are post-ischaemic changes in many other neuronal mRNAs and proteins that have been already noted or remain to be described, as well as additional responses in brain contributed by glia (Kindy *et al.*, 1992) and other reactive cell populations (Saito *et al.*, 1993).

3.2.2.2 Microtubule-associated protein 2 and other late responses

Like the stress response, most of the above noted changes in gene expression are at least as pronounced in resistant as in vulnerable neuron populations after ischaemia.

The search continues for responses that may be early and specific predictors of post-ischaemic neuron loss. Decreases in cytoskeletal proteins have proven to be sensitive indicators of cellular pathology after both global and focal ischaemia (Kitawaga *et al.*, 1989; Seubert *et al.*, 1989; Yanagihara *et al.*, 1990; Inuzuka *et al.*, 1990). Among the several proteins examined, microtubule-associated protein 2 (MAP2) appears to be relatively vulnerable to both lysosomal (Matus and Green, 1987) and calcium-dependent proteolysis (Billger *et al.*, 1988). The neuronal, and more specifically dendritic, localization of MAP2 make it particularly useful as a selective marker of injury to neurons, with very early changes noted under some conditions (Kitagawa *et al.*, 1989; Yoshimi *et al.*, 1991).

Beyond providing a passive index of neuron loss, however, recent studies at the mRNA level demonstrate dynamic changes in MAP2 expression in response to ischaemia that appear to be closely associated with the progression of neuronal injury and recovery (Saito and Nowak, 1993). It has been appreciated for some years that the MAP2 gene gives rise to two mRNAs derived by alternative splicing, MAP2b that encodes a 200-kDa protein normally expressed in mature neurons, and a truncated MAP2c that is translated to form a 70-kDa protein expressed in the developing nervous system and in regions of sustained neuron turnover in the adult brain (Garner and Matus, 1988; Viereck *et al.*, 1989; Kindler *et al.*, 1990). After ischaemia (produced by 10 min cardiac arrest in the rat), the normal MAP2b transcript shows transiently increased expression in cortex and major hippocampal populations that is followed by a return to pre-ischaemic values in surviving cells, but by a progressive decline in CA1 and vulnerable cortical neurons coincident with cell loss. In addition, however, there is a striking accumulation of MAP2c mRNA at 12–48 h after, that is restricted to potentially vulnerable neurons of CA1 and outer layers of dorsolateral cortex (Saito and Nowak, 1993). The time-course and distribution of MAP2c expression coincide precisely with the persistent phase of hsp70 hybridization detailed above, as illustrated in Figure 2. It should be noted that a significant proportion of CA1 neurons survive under the conditions of this model, and it remains to be determined whether MAP2c hybridization distinguishes between dying and surviving neurons. Nevertheless, this shift in MAP2 splicing provides a strikingly selective correlate to neuronal vulnerability after ischaemia. It is well known that severe heat shock is associated with defects in RNA processing (Yost and Lindquist, 1986), but any possible mechanistic relationship between this shift in MAP2 splicing and the post-ischaemic stress response remains to be identified.

Finally, an increase in clathrin-immunoreactivity restricted in CA1 neurons has been described during intervals of 3 h to 2 days after ischaemia in the gerbil (Yoshimi *et al.*, 1991). In view of the role of clathrin as a substrate of the uncoating-ATPase activity of hsc70 (Ungewickel, 1985; Chappell *et al.*, 1986), this observation opens the way for future studies of changes in hsc70 function that may be uniquely associated with the post-ischaemic stress response in vulnerable neurons.

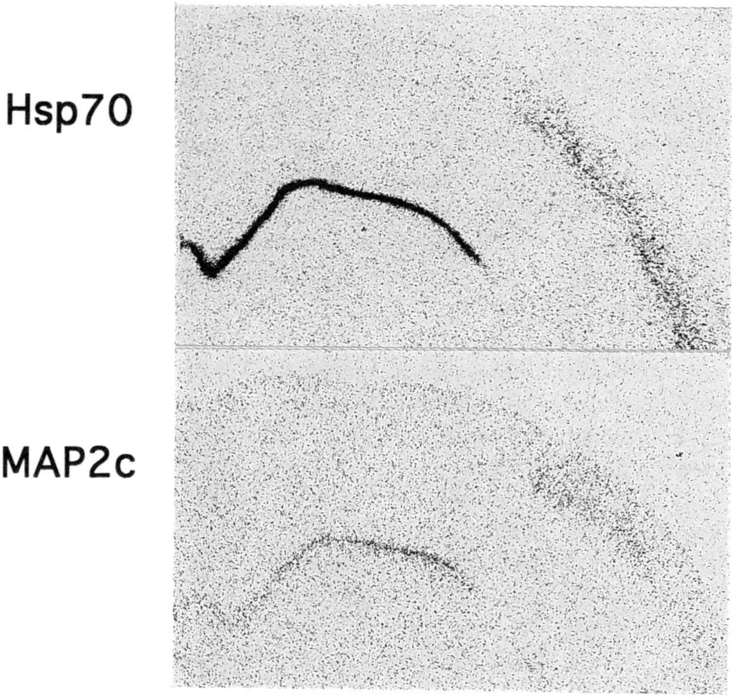

Figure 2 Localization of hsp70 and MAP2c mRNAs in rat brain at 24 h recirculation after transient ischaemia. Expression of MAP2c occurs only late in the recirculation period (12–48 h) and is restricted to vulnerable neurons of CA1 and outer cortical layers that also show prolonged hsp70 hybridization.

3.3 Focal ischaemia

3.3.1 Heat shock gene expression within an infarct

Induction of hsp70 and other changes in gene expression have also been documented after focal ischaemia. Blot hybridization studies after middle cerebral-artery occlusions in adult rats, or in hypoxia-ischaemia models in young animals, have documented significant increases in mRNAs encoding hsp70 (Abe *et al.*, 1991a; Gubits *et al.*, 1993), as well as c-*fos*, c-*jun*, *zif*/268 and a number of other immediate–early genes (Abe *et al.*, 1991a; Gubits *et al.*, 1993; An *et al.*, 1993), for the most part restricted to the ipsilateral hemisphere. However, even the earliest studies employing *in vivo* labelling and electrophoretic analysis of micro-dissected brain tissue identified gradients of hsp70 expression, presumably associated with variations in transcriptional and translational activity according to the level of residual cerebral blood flow (Jacewicz *et al.*, 1986). Subsequent *in situ* hybridization studies have demonstrated that transcription is predictably attenuated in the severely ischaemic territory after permanent focal ischaemia, with levels of hsp70 and c-*fos* mRNAs greatly reduced relative to those observed in the surrounding rim of tissue that receives collateral flow (Welsh *et al.*, 1992). Similar results have been obtained for c-*fos* and *jun*B after long periods of transient focal ischaemia (An *et al.*, 1993). With the higher resolution provided by immunocytochemistry, hsp70 expression within severely ischaemic regions appears to be restricted to cells of the vascular endothelium, while astrocytes are labelled at the infarct rim (Sharp *et al.*, 1991b; Kinouchi *et al.*, 1993). Vascular expression has also been noted bilaterally after severe neonatal hypoxia-ischaemia (Ferriero *et al.*, 1990).

In addition to ischaemic severity, duration of flow deficit is a major variable affecting the extent of injury after focal ischaemia. The ischaemic core suffers severe metabolic disruption, leading eventually to necrosis and loss of essentially all cellular elements unless flow is relatively rapidly restored. Final infarct volume is usually established with occlusions lasting 2–3 h, while much of the severely ischaemic tissue will survive if reperfusion is initiated within 30–60 min (Jacewicz *et al.*, 1992). In such reperfusion models there is striking expression of hsp70 (Kinouchi *et al.*, 1993), c-*fos* and *jun*B (An *et al.*, 1993), as well as increased hsc70 hybridization (Kawagoe *et al.*, 1992b), within the ischaemic but reperfused tissue. Immunocytochemical studies under such conditions confirm an initial neuronal localization within the ischaemic territory, with a transition to glial and vascular expression as the ischaemic duration increases (Kinouchi *et al.*, 1993). Hsp70 immunoreactivity therefore appears to define affected but potentially surviving cell populations after focal ischaemia.

3.2.2 Immediate–early genes and remote responses to focal ischaemia

While hsp70 expression remains localized to the region immediately surrounding a focal injury, c-*fos* and a number of immediate early genes are expressed throughout the ipsilateral hemisphere (An *et al.*, 1993; Welsh *et al.*, 1992; Uemura *et al.*, 1991b; Gass *et al.*, 1992; Lindsberg *et al.*, 1993). Expression of Fos and several other transcription factor proteins in regions remote from the lesion is blocked by treatment with antagonists of the *N*-methyl-D-aspartate (NMDA) receptor, while a component at the border of the lesion is resistant to inhibition (Uemura *et al.*, 1991b; Gass *et al.*, 1992). These findings support the interpretation that such remote responses are the result of transient depolarizations associated with spreading depression that occurs following focal damage, which is well known to induce c-*fos* (Herrara and Robertson, 1990), and both spreading depression and its associated c-*fos* expression are sensitive to NMDA antagonists (Herrara and Robertson, 1990; Gill *et al.*, 1992). A wide range of other changes in gene expression, still to be defined, may be expected to occur after focal ischaemia as a consequence of the cascades of transcriptional activation and repression initiated following spreading depression.

3.3.3 Gene expression and neuronal vulnerability in the infarct periphery

The intermediate zone between a severely ischaemic core and surrounding well-perfused tissue constitutes an ischaemic 'penumbra' (Astrup *et al.*, 1981). Many cells survive in this region, but selective neuronal injury is observed. Hsp70 is selectively localized in neurons in this territory (Kinouchi *et al.*, 1993; Gonzalez *et al.*, 1989), and this also constitutes the zone of immediate–early gene expression that is insensitive to NMDA antagonists (Uemura *et al.*, 1991b; Gass *et al.*, 1992). It remains to be determined whether expression of hsp70 or other induced proteins correlates with neuronal survival in this region (Kinouchi *et al.*, 1993).

3.4 The stress response following seizures and neurotoxic insults

Selective neuronal inductions of the stress response, but with patterns of expression distinct from those observed following ischaemic insults, have been observed following administration of several compounds that injure or destroy neurons. These agents span a wide range of pharmacological actions, including both a known excitotoxin, kainic acid, that induces seizures, and an excitotoxin antagonist, dizocilpine (MK-801), that is a potent anticonvulsant, suggesting that there must be diverse triggers to this final common response. In view of proposed excitotoxic mechanisms of post-ischaemic neuronal injury, these results also may be of relevance to understanding the mechanisms underlying the stress response after ischaemia.

Kainic acid was initially observed to induce hsp70 mRNA in blot hybridizations of rat hippocampus following stereotaxic injections (Uney *et al.*, 1988). Immunocytochemical localization after intraventricular or systemic administration demonstrated diffuse neuronal expression (Gonzalez *et al.*, 1989; Vass *et al.*, 1989), with parallels to the distribution of c-*fos* induction in response to this compound (Le Gal La Salle, 1988; Popovici *et al.*, 1990; Jørgensen *et al.*, 1991). Hsp70 induction required conditions of sustained status epilepticus associated with neuronal injury after kainic acid administration (Vass *et al.*, 1989), and similar results have been obtained following flurothyl-induced seizures (Lowenstein *et al.*, 1990). The pattern of kainate-induced hsp70 immunoreactivity is in some respects reciprocal to that seen after global ischaemia, in that CA1 neurons show preferential staining, while the response is attenuated in CA3 neurons that are more vulnerable to the excitotoxin (Vass *et al.*, 1989). Detailed studies examining the relationship between stimulus intensity and hsp70 expression in a model of repetitive electrical activation concluded that hsp70 was not induced by neuronal activation *per se*, but was expressed only under stimulus conditions that approached the threshold for neuronal injury (Sloviter and Lowenstein, 1992). This is consistent with the failure to demonstrate hsp70 expression in response to brief depolarization as a result of spreading depression (Nowak *et al.*, 1991) or following brief seizures induced by pentylene tetrazole (S. Nadi and T.S. Nowak, Jr., unpublished observations) electroconvulsive shock, or kindling (Wong *et al.*, 1992).

It has been appreciated for some time that NMDA receptor antagonists such as MK-801 can result in transient morphological changes in specific neuron populations, notably in cingulate cortex (Olney *et al.*, 1991). Hsp70 immunoreactivity is strikingly induced in the same neurons under these conditions (Olney *et al.*, 1991; Sharp *et al.*, 1991a), and parallel inductions of c-*fos* expression have also been observed (Dragunow and Faull, 1990). Treatment with a number of other compounds can attenuate both hsp70 expression and the morphological changes (Olney *et al.*, 1991; Sharp *et al.*, 1992). While the mechanisms of induction and inhibition remain to be established, these studies strengthen the link between hsp70 induction and the potential for injury in response to diverse challenges, and also provide further support for an association between cell survival and a successful stress response.

3.5 Signal transduction mechanisms mediating post-ischaemic changes in gene expression

Much is known regarding the molecular mechanisms underlying the heat shock response, with defined roles of upstream heat shock sequence elements (HSE) and the several recognized variants of heat shock factors (HSF) that bind to the regulatory sequences to activate transcription (Pelham, 1982; Scharf *et al.*, 1990; Rabindran *et al.*, 1991; Schuetz *et al.*, 1991). Less well characterized are the signal transduction cascades by which HSF may be activated in response to diverse stimuli, and the potential interactions between these and factors acting via other regulatory elements to control transcription of specific heat shock genes under the complex conditions associated with ischaemia *in vivo*. As noted above, post-ischaemic hsp70-induction occurs in concert with a range of other changes in gene expression that are not necessarily components of a classical heat shock response, suggesting that additional mechanisms may contribute to its expression. There is support for this suggestion in sequence and functional activity data for both rodent and human hsp70 promoters, demonstrating the presence of multiple regulatory elements responsive to calcium/cAMP, phorbol esters, and other signals (Wu *et al.*, 1986; Choi *et al.*, 1991; Voellmy *et al.*, 1985; Leung *et al.*, 1990; Hunt and Calderwood, 1990). While the characterized genes include that encoding a human hsp72-like protein (Leung *et al.*, 1990), no data is yet available for the specific rodent gene that is transcribed following ischaemia. A consideration of the mechanisms regulating the post-ischaemic stress response must at present be considered largely speculative.

As shown in Figure 3, transient ischaemia results in the activation of HSF in gerbil hippocampus with a delayed time-course relative to that observed after

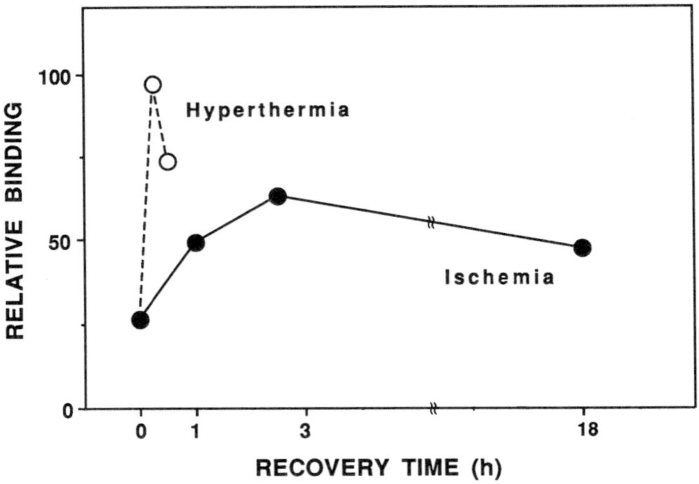

Figure 3 Time-course of heat shock factor (HSF) activation in gerbil hippocampus after hyperthermic and ischaemic insults. Delayed activation of HSF is observed after ischaemia relative to the rapid increase in response to temperature elevation, consistent with the respective time courses of hsp70 mRNA expression observed after these insults. Ischaemia was produced by 5 min bilateral carotid artery occlusion, while body temperature was elevated to 42.5–43.0°C in the hyperthermic insult. Whole cell extracts of dissected hippocampi were prepared and analysed for binding of a synthetic heat shock element (HSE) sequence by a gel-shift assay, essentially as previously described (Blake *et al.*, 1991). Symbols indicate the mean values obtained by densitometry of duplicate samples, expressed as a proportion of peak levels observed in hyperthermic tissue.

hyperthermic insults (W. Valentine, H. Abe and T.S. Nowak, Jr., unpublished observations). This finding is consistent with the differing time course of hsp70 mRNA expression established for ischaemia and heat shock in brain (Nowak *et al.*, 1990), and confirms a role of HSF activation in the post-ischaemic stress response *in vivo*. Increases in HSF activity have also been observed *in vitro* following hypoxia-ischaemia in cultured mouse myogenic cells, and presence of an intact HSE was shown to be required for transcriptional induction via a human hsp70 promoter in this model (Benjamin *et al.*, 1990). It remains to be determined whether HSF activation is alone sufficient for post-ischaemic hsp70 expression after hypoxia-ischaemia. In addition, the signal transduction mechanisms that mediate post-ischaemic HSF activation have yet to be established. Mechanisms of HSF activation even under conditions of a classical heat shock response are not fully elucidated. Suggested roles of denatured proteins (Hightower, 1980; Ananthan *et al.*, 1986) or oxidative stress (Courgeon *et al.*, 1988) still require a determination of the details of signal transduction. Recent studies in liver specifically argue against a role of oxidative stress mechanisms in hsp70 expression during post-ischaemic recirculation in that tissue, since hsp70 fails to be expressed following oxidative stress sufficient to induce heme oxygenase, although heme oxygenase is co-expressed with hsp70 after ischaemia (Tacchini *et al.*, 1993). While ATP depletion may coincide with the threshold for activating ischaemic or hypoxic hsp70 induction (Benjamin *et al.*, 1992), the resultant depolarization and associated ion fluxes and transmitter release may be of greater mechanistic importance in brain *in vivo.*

The post-ischaemic coinduction of hsp70 with c-*fos* and other gene products detailed above provides one means of identifying potential signal transduction pathways that may be involved in the post-ischaemic stress response. For example, c-*fos*, c-*jun* and other immediate–early genes are induced by NMDA receptor activation in cultured neurons (Szekely *et al.*, 1989, 1990), and NMDA antagonists have been reported to attenuate both c-*fos* and hsp70 expression after global ischaemia (Nowak, 1991; Uemura *et al.*, 1991a). However, hypothermia is a potential confounding artifact in the *in vivo* pharmacological studies (Nowak, 1991; Buchan and Pulsinelli, 1990), and hsp70 expression does not overlap with the NMDA receptor- and depolarization-mediated responses observed in remote regions of hemisphere after a focal insult (Welsh *et al.*, 1992; Uemura *et al.*, 1991b; Gass *et al.*, 1992). Furthermore, stimulus-induced potentiation in hippocampal slices, that can also be blocked by NMDA antagonists, is associated with co-ordinate changes in expression of a broad range of mRNAs, but hsp70 is induced inconsistently in this model (Mackler *et al.*, 1992). Given the range of pharmacological agents that appear to act via distinct receptors to induce hsp70, it may be proposed that these several pathways converge under sustained stimulus conditions to activate common mechanisms that are closely linked to cellular injury (Sloviter and Lowenstein, 1992).

The role of calcium as both physiological and pathological mediator has received considerable attention in the context of ischaemic injury (Siesjö and Bengtsson, 1989). Of particular relevance are recent studies indicating the post-ischaemic sequestration of calmodulin in complexes with calcium, with the most persistent effect observed in vulnerable CA1 neurons (Picone *et al.*, 1989; DeGraba *et al.*, 1992). The potential involvement of intracellular calcium in the stress response has been controversial. Hyperthermic insults can increase intracellular Ca^{2+}, but this response has been dissociated from effects on transcription in some experiments (Drummond *et al.*, 1986, 1988). HSF activation by calcium has been reported in cell extracts (Mosser *et al.*, 1990), and recent studies also suggest its role in the stress response of cultured cells (Price and Calderwood, 1991). In addition, there is evidence for functional interaction between the heat shock and Ca^{2+}/cAMP response elements in the human hsp70 gene in which this has been studied (Choi *et al.*, 1991). These results suggest that calcium is involved in several signal transduction pathways that could contribute to increases in hsp70 transcription. We have recently observed that modest post-ischaemic hyperthermia, which can occur spontaneously during recirculation in the gerbil (Kuroiwa *et al.*, 1990), is required for maximal expression of hsp70 mRNA after threshold 2 min ischaemic insults (Suga and Nowak, 1993) (S. Suga and T.S. Nowak, Jr., manuscript in preparation). It remains to be determined whether this reflects a temperature dependence of HSF activation or indirect effects via amplification of other post-ischaemic responses mediating hsp70 expression or injury.

3.6 Ischaemic tolerance

3.6.1. Models of induced ischaemic tolerance and expression of hsp70

Numerous reports have now described phenomena of induced ischaemic tolerance *in vivo*, with largely parallel developments in brain and in other tissues. Brief periods of global ischaemia in both rat and gerbil brain result in an attenuation of injury after subsequent otherwise damaging insults (Kirino *et al.*, 1991; Kitagawa *et al.*, 1990; Kato *et al.*, 1991; Liu *et al.*,. 1992). The main features of all such protection paradigms include priming ischaemic insults of approximately 2 min duration, with protective effects observed at intervals of 1–7 days, depending on the severity of the subsequent insult. Moderate hyperthermic challenges have also been reported to confer ischaemic tolerance in brain (Kitagawa *et al.*, 1991; Chopp *et al.*, 1989). Tolerance has also been described in heart following brief ischaemia (Sharma *et al.*, 1992). Considerable caution is necessary in interpreting such observations *in vivo*, including in brain, for example, an assessment of physiological parameters such as temperature and metabolic activity that could affect the impact of the second test insult. *In vitro* studies in cultured neurons, somewhat closer to classical thermotolerance experiments, suggest that brief hyperthermic stress can induce cross-tolerance to glutamate toxicity (Rordorf *et al.*, 1991; Lowenstein *et al.*, 1991), and that prior excitotoxin exposure can itself protect against subsequent challenges. The available data would indicate that much can be learned regarding post-ischaemic pathophysiology from an identification of those factors that contribute to tolerant states.

In this regard, much of the work in the above tolerance studies has centered on hsp70 as an index of the stress response. As noted previously, brief global ischaemia in brain is associated with more rapid recovery of translation in CA1 neurons, and increased hsp70 protein expression in comparison with longer insults (Simon *et al.*, 1991; Kirino *et al.*, 1991), providing a rationale for its involvement in protection mechanisms. A strikingly similar result is obtained in liver, with a 1 h insult allowing expression of both hsp70 mRNA and protein, but an apparent failure of the translational component of the response after 2 h ischaemia (Tacchini *et al.*, 1993). Hyperthermic induction of ischaemic tolerance *in vivo* is less readily understood in terms of a direct relationship to the stress response, since such insults give rise to preferential hsp70 expression in glia and cells of the vasculature (Marini *et al.*, 1990; Sprang and Brown, 1987). Hsp70 induction was observed in heat-stressed neurons under conditions that induced tolerance to glutamate (Rordorf *et al.*, 1991; Lowenstein *et al.*, 1991).

3.6.2 Other changes in gene expression associated with tolerance

It is difficult to envision an isolated role for hsp70 or other stress proteins in the induction of ischaemic tolerance in view of the demonstrated and growing complexity of changes in gene expression that occur after ischaemia, both in brain and in other tissues (e.g. Tacchini *et al.*, 1993; Sharma *et al.*, 1992). To appreciate the full range of such changes that could contribute to the development of tolerance it is necessary to consider variations in both transcriptional and translational responses that occur as insults are reduced to threshold levels (Nowak *et al.*, 1993; Suga and Nowak, 1991). The importance of varied translation is illustrated in Figure 4, in which expression of hsp70, Fos and Jun proteins are compared after ischaemic insults of 5 min or 2–2.5 min in the gerbil. The shorter insult gives rise to preferential hsp70 expression in CA1 neurons as previously described, at least in part as a result of a more rapid recovery of protein synthesis activity, while expression in dentate and CA3 is attenuated. Fos immunoreactivity remains undetectable in CA1 even after 2 min ischaemia, presumably because the transient expression of c-*fos* mRNA is too rapid to allow translation even after the shorter insult. Jun-like immunoreactivity, however, is robustly expressed in CA1 after 2 min but not after 5 min ischaemia, suggesting overlap of translational recovery with a more prolonged expression of the encoding mRNA. Since available data indicates relatively transient c-*jun* mRNA expression after ischaemia (Kindy *et al.*, 1991; Wessel *et al.*, 1991), it remains to be determined whether this staining pattern reflects translation of c-*jun* or a related mRNA. These results illustrate the practical considerations that will apply to any such responses after brief ischaemia, and indicate the complex reprogramming of gene expression that occurs in CA1 neurons in association with the induction of tolerance.

3.7 Summary and Conclusions

From the above observations, it is evident that the stress response can be a valuable indicator of pathophysiology involving diverse cell types in brain after global and focal ischaemia. Since hsp70 is expressed in many cell populations that will survive an ischaemic insult it is clearly not a marker of irreversible injury. On the contrary, there is frequently an attenuated expression of the immunoreactive protein in the most vulnerable neuron populations, reflecting effects of translational impairment as well as potential masking of immunoreactivity during an ongoing stress. For these reasons, detection of induced hsp70 mRNA is the most reliable marker of the post-ischaemic stress response.

Increased hsp70 immunoreactivity is observed in vulnerable neurons following brief periods of global ischaemia that induce tolerance to subsequent more severe insults. These results strongly suggest that heat shock stress protein expression is a

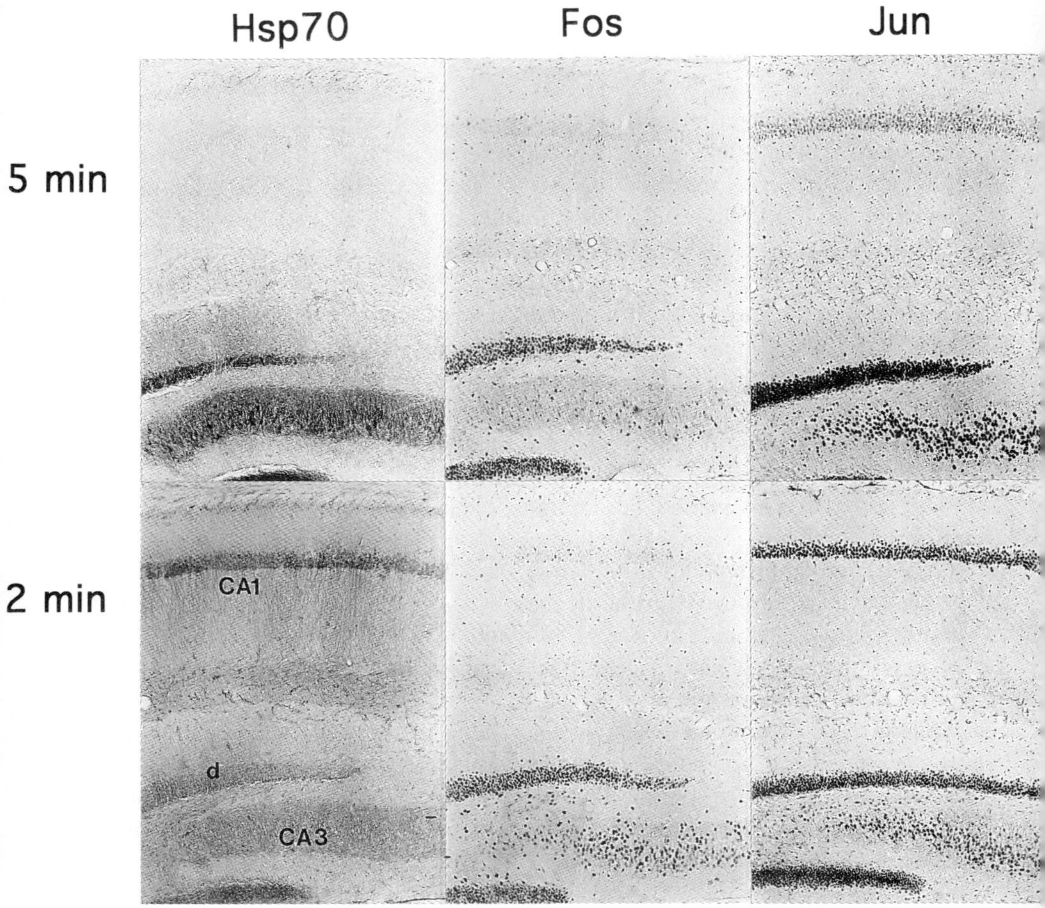

Figure 4 Distributions of hsp70, Fos and Jun immunoreactivities after ischaemic insults of varied duration. Upper panels illustrate the negligible expression of these proteins in CA1 neurons following 5 min ischaemia that results in cell loss, consistent with profound translational impairment demonstrated to persist in this region. Increases in both hsp70 and Jun staining are observed after 2 min insults that produce negligible injury and that may induce ischaemic tolerance. Fos expression is also attenuated in CA3 after the longer insult, reflecting the impact of delayed protein synthesis recovery on translation of the very transiently expressed c-*fos* mRNA. Both hsp70 and Jun show more striking accumulation in dentate (d) and CA3 after 5 min ischaemia, perhaps as a result of an increased impact of ischaemia on these cell populations after the longer insult. Animals were perfused at recirculation intervals resulting in maximal expression of the several immunoreactivities as follows: hsp70, 24 h; Fos, 3 h; Jun, 6 h.

component of the adaptive response of neurons to ischaemia. Given the importance of temperature as a critical variable affecting both neuronal injury and the post-ischaemic stress response, potential contributions of changes in this and other physiological parameters to the tolerant state remain to be fully investigated. Furthermore, it is evident that diverse changes in gene expression occur following brief ischaemia, and it is a considerable challenge for future studies to better characterize the range of these responses and the integrated mechanisms underlying their combined impact on ischaemic vulnerability.

Fundamental questions remain regarding the mechanisms regulating the induction of hsp70 and other responsive genes after ischaemia. The evidence for prolonged post-ischaemic heat shock factor (HSF) activation in hippocampus is consistent with the sustained hsp70 mRNA expression observed in this region, indicating that induction, at least in part, involves this component of classical heat shock regulation. The mechanism by which HSF is activated after ischaemia, and its overlap with signal transduction pathways that respond to hyperthermia and other stresses, remain to be determined. The co-ordinate expression of c-*fos* and hsp70 observed after global ischaemia as well as in response to heat shock and other stimuli, and the presence of additional regulatory elements upstream of known hsp70 genes, suggest that other mechanisms may also contribute to the observed responses. Delayed appearance of the alternatively spliced MAP2c transcript in neurons that show prolonged hsp70 mRNA expression suggests its selective association with neuronal vulnerability. Further knowledge of the mechanisms involved in these and other responses is critical to an assessment of their respective values as markers of pathology, as well as to understanding the physiology that underlies ischaemic injury.

Finally, while the majority of studies have focused on the highly inducible member of the hsp70 family that is generally not detected in control tissues, recent results suggest that increased expression of constitutively expressed hsc70 is also an important component of the response to ischaemic as well as hyperthermic stresses in brain. Available evidence indicates that hsc70 may be induced with lower stimulus thresholds than hsp70, raising the possibility that it may be particularly important under conditions of tolerance induction. The pattern of ubiquitin-conjugate expression after ischaemia indicates a widespread cellular distribution, with increased interneuron expression that has not been observed for hsp70, and future investigations of other heat shock proteins will undoubtedly identify further complexity in the post-ischaemic stress response. Nevertheless, the principles identified from an early focus on hsp70 expression should continue to provide a useful foundation on which to build a more general understanding of the relationship between changes in gene expression and brain pathology after ischaemia and other insults.

References

Abe, K., Kawagoe, J., Sato, S., Sahara, M. & Kogure, K. (1991a) Induction of the 'zinc finger' gene after transient focal ischaemia in rat cerebral cortex. *Neurosci. Lett.* **123**, 248–250.

Abe, K., Tanzi, R.E. & Kogure, K. (1991b) Induction of HSP70 mRNA after transient ischaemia in gerbil brain. *Neurosci. Lett.* **125**, 166–168.

An, G., Lin, T.-N. Lui, J.-S., Xue, J.-J., He, Y.-Y. & Hsu, C.Y. (1993) Expression of c-*fos* and c-*jun* family genes after focal cerebral ischaemia. *Ann. Neurol.* **33**, 457–464.

Ananthan, J., Goldberg, A.L. & Voellmy, R. (1986) Abnormal proteins serve as eukaryotic stress signals and trigger the activation of heat shock genes. *Science* **232**, 522–524.

Aoki, M., Abe, K., Kawagoe, J., Sato, S., Nakamura, S. & Kogure, K. (1993) Temporal profile of the induction of heat shock protein 70 and heat shock cognate protein 70 mRNAs, after transient ischaemia in gerbil brain. *Brain Res.* **601**, 185–192.

Astrup, J., Siesjö, B.K. & Symon, L. (1981) Thresholds in cerebral ischaemia. The ischaemic penumbra. *Stroke* **12**, 723–725.

Benjamin, I.J., Horie, S., Greenberg, M.L., Alpern, R.J. & Williams, R.S. (1992) Induction of stress proteins in cultured myogenic cells. Molecular signals for the activation of heat shock transcription factor during ischaemia. *J. Clin. Invest.* **89**, 1685–1689.

Benjamin, I.J., Kröger, B. & Williams, R.S. (1990) Activation of the heat shock transcription factor by hypoxia in mammalian cells. *Proc. Natl Acad. Sci. USA* **87**, 6263–6267.

Bergstedt, K. (1993) Ischaemic and hypoglycemic brain damage. Studies on protein synthesis and heat shock protein expression in the rat brain. *Thesis*, Lund University, Sweden.

Billger, M., Wallin, M. & Karlsson, J.-O. (1988) Proteolysis of tubulin and microtubule-associated proteins1 and 2 by calpain I and II. Difference in sensitivity of assembled and disassembled microtubules. *Cell Calcium* **9**, 33–44.

Blake, M.J., Udelsman, R., Feulner, G.J., Norton, D.D. & Holbrook, N.J. (1991) Stress-induced heat shock protein 70 expression in adrenal cortex: an adrenocorticotropic hormone-sensitive, age-dependent response. *Proc. Natl Acad. Sci. USA* **88**, 9873–9877.

Brown, I.R. & Rush, S.J. (1990) Expression of heat shock genes (hsp70) in the mammalian brain: Distinguishing constitutively expressed and hyperthermia-inducible mRNA species. *J. Neurosci. Res.* **25**, 14–19.

Buchan, A. & Pulsinelli, W.A. (1990) Hypothermia but not the *N*-methyl-D-aspartate antagonist, MK-801, attenuates neuronal damage in gerbils subjected to transient global ischaemia. *J. Neurosci.* **10**, 311–316.

Carlson, N., Rogers, S. & Rechsteiner, M. (1987) Micro-injection of ubiquitin: Changes in protein degradation in HeLa cells subjected to heat shock. *J. Cell Biol.* **104**, 547–555.

Chappell, T.G., Welch, W.J., Schlossman, D.M., Palter, K.B., Schlesinger, M.J. & Rothman, J.E. (1986) Uncoating-ATPase is a member of the 70-kilodalton family of stress proteins. *Cell* **45**, 3–13.

Choi, H.-K., Li, B., Lin, Z., Huang, L.E. & Liu, A.Y.-C. (1991) cAMP and cAMP-dependent protein kinase regulate the human heat shock protein 70 gene promoter activity. *J. Biol. Chem.* **266**, 11 858–11 865.

Chopp, M., Chen, H., Ho, K.-L., Dereski, M.O., Brown, E., Hetzel, F.W. & Welch, K.M.A. (1989) Transient hyperthermia protects against subsequent forebrain ischaemic cell damage in the rat. *Neurology* **39**, 1396–1398.

Chopp, M., Li, Y., Dereski, M.O., Levine, S.R., Yoshida, Y. & Garcia, J.H. (1991) Neuronal injury and expression of 72-kDa heat shock protein after forebrain ischaemia in the rat. *Acta Neuropathol.* **83**, 66–71.

Courgeon, A.-M., Rollet, E., Becker, J., Maisonhaute, C. & Best-Belpomme, M. (1988) Hydrogen peroxide (H_2O_2) induces actin and some heat shock proteins in *Drosophila* cells. *Eur. J. Biochem.* **171**, 163–170.

DeGraba, T.J., Ostrow, P.T., Strong, R.A., Earls, R.M., Zong, Z.J. & Grotta, J.C. (1992) Temporal relation of calcium-calmodulin binding and neuronal damage after global ischaemia in rats. *Stroke* **23**, 876–882.

Dempsey, R.J., Carney, J.M. & Kindy, M.S. (1991) Modulation of ornithine decarboxylase mRNA following transient ischaemia in the gerbil. *J. Cereb. Blood Flow Metab.* **11**, 979–985.

Dienel, G.A., Cruz, N.F. & Rosenfeld, S.J. (1985) Temporal profiles of proteins responsive to transient ischaemia. *J. Neurochem.* **44**, 600–610.

Dienel, G.A., Kiessling, M., Jacewicz, M. & Pulsinelli, W.A. (1986) Synthesis of heat shock proteins in rat brain cortex after transient ischaemia. *J. Cereb. Blood Flow Metab.* **6**, 505–510.

Dienel, G.A., Pulsinelli, W.A. & Duffy, T.E. (1980) Regional protein synthesis in rat brain following acute hemispheric ischaemia. *J. Neurochem.* **35**, 1216–1226.

Dragunow, M. & Faull, R.L.M. (1990) MK-801 induces c-*fos* protein in thalamic and neocortical neurons of rat brain. *Neurosci. Lett.* **111**, 39–45.

Drummond, I.A.S., Livingstone, D. & Steinhardt, R.A. (1988) Heat shock protein synthesis and cytoskeletal rearrangements occur independently of intracellular-free calcium increases in *Drosophila* cells and tissues. *Radiat. Res.* **113**, 402–413.

Drummond, I.A.S., McClure, S.A., Poenie, M., Tsien, R.Y. & Steinhardt, R.A. (1986) Large changes in intracellular pH and calcium observed during heat shock are not responsible for the induction of heat shock proteins in *Drosophila melanogaster. Mol. Cell. Biol.* **6**, 1767–1775.

Dwyer, B.E., Nishimura, R.N. & Brown, I.R. (1989) Synthesis of the major inducible heat shock protein in rat hippocampus after neonatal hypoxia-ischaemia. *Exp. Neurol.* **104**, 28–31.

Ferriero, D.M., Soberano, H.Q., Simon, R.P. & Sharp, F.R. (1990) Hypoxia-ischaemia induces heat shock protein-like (hsp72) immunoreactivity in neonatal rat brain. *Dev. Brain Res.* **53**, 145–150.

Fried, R.L. & Nowak. T.S.J. (1987) Opioid peptide levels in gerbil brain after transient ischaemia: Lasting depletion of hippocampal dynorphin. *Stroke* **18**, 765–770.

Garner, C.C. & Matus, A. (1988) Different forms of microtubule-associated protein 2 are encoded by separate mRNA transcripts. *J. Cell Biol.* **106**, 779–783.

Gass, P., Spranger, M., Herdegen, T., Bravo, R., Köck, P., Hacke, W. & Kiessling, M. (1992) Induction of Fos and Jun proteins following focal ischaemia in the rat cortex: differential effect of MK-801. *Acta Neuropathol.* **84**, 545–553.

Gill, R., Andiné, P., Hillered, L., Persson, L. & Hagberg, H. (1992) The effect of MK-801 on cortical spreading depression in the penumbral zone following focal ischaemia in the rat. *J. Cereb. Blood Flow Metab.* **12**, 371–379.

Gonzalez, M.F., Lowenstein, D., Fernyak, S., Hisanaga, K., Simon R. & Sharp, F.R. (1991) Induction of heat shock protein 72-like immunoreactivity in the hippocampal formation following transient global ischaemia. *Brain Res. Bull.* **26**, 241–250.

Gonzalez, M.F., Shiraishi, K., Hisanaga, K., Sagar, S.M., Mandabach, M. & Sharp, F.R. (1989) Heat shock proteins as markers of neural injury. *Mol. Brain Res.* **6**, 93–100.

Gubits, R.M., Burke, R.E., Casey-McIntosh, G., Bandele, A. & Munell, F. (1993) Immediate early gene induction after neonatal hypoxia-ischaemia. *Mol. Brain Res.* **18**, 228–238.

Hayashi, T., Takada, K. & Matsuda, M. (1991) Changes in ubiquitin and ubiquitin-protein conjugates in the CA1 neurons after transient sublethal ischaemia. *Mol. Chem. Neuropathol.* **15**, 75–82.

Hayashi, T., Takada, K. & Matsuda, M. (1992) Subcellular distribution of ubiquitin-protein conjugates in the hippocampus following transient ischaemia. *J. Neurosci. Res.* **31**, 561–564.

Herrera, D.G. & Robertson, H.A. (1990) Application of potassium chloride to the brain surface induces the c-*fos* proto-oncogene: Reversal by MK-801. *Brain Res.* **510**, 166–170.

Hightower, L.E. (1980) Cultured animal cells exposed to amino acid analogues or puromycin rapidly synthesize several polypeptides. *J. Cell Physiol.* **102**, 407–427.

Hunt, C. & Calderwood, S. (1990) Characterization and sequence of a mouse *hsp70* gene and its expression in mouse cell lines. *Gene* **87**, 199–204.

Inuzuka, T., Tamura, A., Sato, S., Kirino, T., Yanagisawa, K., Toyoshima, I. & Miyatake, T. (1990) Changes in the concentrations of cerebral proteins following occlusion of the middle cerebral artery in rats. *Stroke* **21**, 917–922.

Jacewicz, M., Kiessling, M. & Pulsinelli, W.A. (1986) Selective gene expression in focal cerebral ischaemia. *J. Cereb. Blood Flow Metab.* **6**, 263–272.

Jacewicz, M., Tanabe, J. & Pulsinelli, W.A. (1992) The CBF threshold and dynamics for focal cerebral infarction in spontaneously hypertensive rats. *J. Cereb. Blood Flow Metab.* **12**, 359–370.

Jørgensen, M.B., Deckert, J., Wright, D.C. & Gehlert, D.R. (1989) Delayed c-*fos* proto-oncogene expression in the rat hippocampus induced by transient global cerebral ischaemia: An *in situ* hybridization study. *Brain Res.* **484**, 393–398.

Jørgensen, M.B., Johansen, F.F. & Diemer, N.H. (1991) Post-ischaemic and kainic acid-induced c-*fos* protein expression in the rat hippocampus. *Acta Neurol. Scand.* **84**, 352–356.

Kato, H., Liu, Y., Araki, T. & Kogure, K. (1991) Temporal profile of the effects of pretreatment with brief cerebral ischaemia on the neuronal damage following secondary ischaemic insult in the gerbil: Cumulative damage and protective effects. *Brain Res.* **553**, 238–242.

Kawagoe, J., Abe, K. & Kogure, K. (1992a) Different thresholds of HSP70 and HSC70 heat shock mRNA induction in post-ischaemic gerbil brain. *Brain Res.* **599**, 197–203.

Kawagoe, J., Abe, K., Sato, S., Nagano, I., Nakamura, S. & Kogure, K. (1992b) Distributions of heat shock protein (HSP) 70 and heat shock cognate protein (HSC) 70 mRNAs after transient focal ischaemia in rat brain. *Brain Res.* **587**, 195–202.

Kawagoe, J., Abe, K., Sato, S., Nagano, I., Nakamura, S. & Kogure, K. (1992c) Distributions of heat shock protein 70 mRNAs and heat shock cognate protein 70 mRNAs after transient global ischaemia in gerbil brain. *J. Cereb. Blood Flow Metab.* **12**, 794–801.

Kiessling, M., Dienel, G.A., Jacewicz, M. & Pulsinelli, W.A. (1986) Protein synthesis in post-ischaemic rat brain: A two-dimensional electrophoretic analysis. *J. Cereb. Blood Flow Metab.* **6**, 642–649.

Kindler, S., Schulz, B., Goedert, M. & Garner, C.C. (1990) Molecular structure of microtubule associated protein 2b and 2c from rat brain. *J. Biol. Chem.* **265** 19679–19684.

Kindy, M., Bhat, A.N. & Bhat, N.R. (1992) Transient ischaemia stimulates glial fibrillary acid protein and vimentin gene expression in the gerbil neocortex, striatum and hippocampus. *Mol. Brain Res.* **13**, 199–206.

Kindy, M.S., Carney, J.P., Dempsey, R.J. & Carney, J.M. (1991) Ischaemic induction of proto-oncogene expression in gerbil brain. *J. Mol. Neurosci.* **2**, 217–228.

Kinouchi, H., Sharp, F.R., Hill, M.P., Koistinaho, J., Sagar, S.M. & Chan, P.H. (1993) Induction of 70-kDa heat shock protein and hsp70 mRNA following transient focal cerebral ischaemia in the rat. *J. Cereb. Blood Flow Metab.* **13**, 105–115.

Kirino, T. (1982) Delayed neuronal death in the gerbil hippocampus following ischaemia. *Brain Res.* **239**, 57–79.

Kirino, T., Tsujita, Y. & Tamura, A. (1991) Induced tolerance in ischaemia in gerbil hippocampal neurons. *J. Cereb. Blood Flow Metab.* **11**, 299–307.

Kitagawa, K., Matsumoto, M., Niinobe, M., Mikoshiba, K., Hata, R., Ueda, H., Handa, N., Fukunaga, R., Isaka, Y., Kimura, K. & Kodama, T. (1989) Microtubule-associated protein 2 as a sensitive marker for cerebral ischaemic damage. Immunohistochemical investigation of dendritic damage. *Neuroscience* **31**, 401–411.

Kitagawa, K., Matsumoto, M., Tagaya, M., Hata, R., Ueda, H., Niinobe, M., Handa, N., Fukunaga, R., Kimura, K., Mikoshiba, K. & Kamada, T. (1990) "Ischaemic tolerance" phenomenon found in brain. *Brain Res.* **528**, 21–24.

Kitagawa, K., Matsumoto, M., Tagaya, M., Kuwabara, K., Hata, R., Handa, N., Fukunaga, R., Kimura, K. & Kamada, T. (1991) Hyperthermia-induced neuronal protection against ischaemic injury in gerbils. *J. Cereb. Blood Flow Metab.* **11**, 449–452.

Knowlton, A.A., Brecher, P. & Apstein, C.S. (1991) Rapid expression of heat shock protein in the rabbit after brief cardiac ischaemia. *J. Clin. Invest.* **87**, 139–147.

Kuroiwa, T., Bonnekoh, P. & Hossmann, K.-A. (1990) Prevention of post-ischaemic hyperthermia prevents ischaemic injury of CA1 neurons in gerbils. *J. Cereb. Blood Flow Metab.* **10**, 550–556.

Le Gal La Salle, G. (1988) Long-lasting and sequential increase of c-*fos* oncoprotein expression in kainic acid-induced status epilepticus. *Neurosci. Lett.* **88**, 127–130.

Leung, T.K.C., Rajendran, M.Y., Monfries, C., Hall, C. & Lim, L. (1990) The human heat shock protein family. Expression of a novel heat-inducible HSP70 (HSP70B′) and isolation of its cDNA and genomic DNA. *Biochem. J.* **267**, 125–132.

Lindsberg, P.J., Nowak, T.S., Jr., Frerichs, K.U., Siren, A.-L. & Hallenbeck, J.M. (1993) Dissociated heat shock protein and c-*fos* gene expression in focal microvascular brain damage (Abstract). *J. Cereb. Blood Flow Metab.* **13**, S485.

Lindvall, O., Ernfors, P., Bengzon, J., Kokaia, Z., Smith, M.-L., Siesjö, B.K. & Persson, H. (1992) Differential regulation of mRNAs for nerve growth factor, brain-derived neurotrophic factor, and neurotrophin 3 in the adult rat brain, following cerebral ischaemia and hypoglycemic coma. *Proc. Natl Acad. Sci. USA* **89**, 648–652.

Liu, Y., Kato, H., Nakata, N. & Kogure, K. (1992) Protection of rat hippocampus against ischaemic neuronal damage by pretreatment with sublethal ischaemia. *Brain Res.* **586**, 121–124.

Lowenstein, D.H., Chan, P.H. & Miles, M.F. (1991) The stress-protein response in cultured neurons: Characterization and evidence for a protective role in excitotoxicity. *Neuron* **7**, 1053–1060.

Lowenstein, D.H., Simon, R.P. and Sharp, F.R. (1990) The pattern of 72-Kda heat shock protein-like immunoreactivity in the rat brain following flurothyl-induced status epilepticus. *Brain Res.* **531**, 173–182.

Mackler, S.A., Brooks, B.P. & Eberwine, J.H. (1992) Stimulus-induced coordinate changes in mRNA abundance in single postsynaptic hippocampal CA1 neurons. *Neuron* **9**, 539–548.

Magnusson, K. & Wieloch, T. (1989) Impairment of protein ubiquitination may cause delayed neuronal death. *Neurosci Lett.* **96**, 264–270.

Marini, A.M., Kozuka, M., Lipsky, R.L. & Nowak, T.S., Jr. (1990) 70-kilodalton heat shock-protein induction in cerebellar astrocytes and cerebellar granule cells *in vitro*: Comparison with immunocytochemical localization after hyperthermia *in vivo*. *J. Neurochem.* **54**, 1509–1516.

Matsuyama, T., Michishita, H., Nakamura, H., Tsuchiyama, M., Shimizu, S., Watanabe, K. & Sugita, M. (1993) Induction of copper-zinc superoxide dismutase in gerbil hippocampus after ischaemia. *J. Cereb. Blood Flow Metab.* **13**, 135–144.

Matus, A. & Green, G.D.J. (1987) Age-related increase in a cathepsin D-like protease that degrades brain microtubule-associated proteins. *Biochemistry* **26**, 8083–8086.

Milarski, K.L., Welch, W.J. & Morimoto, R.I. (1989) Cell cycle-dependent association of HSP70 with specific cellular proteins. *J. Cell Biol.* **108**, 413–423.

Miller, E.K., Raese, J.D. & Morrison-Bogorad, M. (1991) Expression of heat shock protein 70 and heat shock cognate 70 messenger RNAs in rat cortex and cerebellum after heat shock or amphetamine treatment. *J. Neurochem.* **56**, 2060–2071.

Morgan, J.I. & Curran, T. (1991) Stimulus-transcription coupling in the nervous system: Involvement of the inducible proto-oncogenes *fos* and *jun*. *Annu. Rev. Neurosci.* **14**, 421–451.

Mosser, D.D., Kotzbauer, P.T., Sarge, K.D. & Morimoto, R.I. (1990) *In vitro* activation of heat shock transcription factor DNA-binding by calcium and biochemical conditions that affect protein confirmation. *Proc. Natl Acad. Sci. USA* **87**, 3748–3752.

Müller, M., Cleef, M., Röhn, G., Bonnekoh, P., Pajunen, A.E.I., Bernstein, H.-G. & Paschen, W. (1991) Ornithine decarboxylase in reversible cerebral ischaemia: An immunohistochemical study. *Acta Neuropathol.* **83**, 39–45.

Nowak, T.S., Jr (1985) Synthesis of a stress protein following transient ischaemia in the gerbil. *J. Neurochem.* **45**, 1635–1641.
Nowak, T.S., Jr (1989) MK-801 prevents 70-kDa stress protein-induction in gerbil brain after ischemia: The heat shock response as a marker for excitotoxic pathology. In *Pharmacology of Cerebral Ischaemia 1988: Proceedings of the Second International Symposium on the Pharmacology of Cerebral Ischaemia* (ed Krieglstein, J,), pp 229–234. Boca Raton, CRC Press.
Nowak, T.S., Jr (1990) Protein synthesis and the heat shock stress response after ischemia. *Cerebrovasc. Brain Metab. Rev.* **2**, 345–366.
Nowak, T.S., Jr (1991) Localization of 70-kDa stress protein mRNA induction in gerbil brain after ischaemia. *J. Cereb. Blood Flow Metab.* **11**, 432–439.
Nowak, T.S., Jr, Bond, U. & Schlesinger, M.J. (1990a) Heat shock RNA levels in brain and other tissues after hyperthermia and transient ischaemia. *J. Neurochem.* **54**, 451–458.
Nowak, T.S., Jr, Ikeda, J. & Nakajima, T. (1990b) 70-kilodalton heat shock protein and c-*fos* gene expression following transient ischaemia. *Stroke* **21** (Suppl. III), 107–111.
Nowak, T.S., Jr, Osborne, O.C. & Ikeda, J. (1992) Role of altered gene expression in development of neuronal changes after ischaemia. In *Maturation Phenomenon in Cerebral Ischaemia* (ed Ito, U., Kirino, T., Kuroiwa, T. & Klatzo, I.), pp 121–128. Berlin, Springer-Verlag.
Nowak, T.S., Jr, Osborne, O.C. & Mies, G. (1991) Cortical spreading depression induces the proto-oncogene, c-*fos*, but not the stress protein, hsp70 (Abstract). *J. Cereb. Blood Flow Metab.* **11**, S215.
Nowak, T.S., Jr, Osborne, O.C. & Suga, S. (1993) Stress protein and proto-oncogene expression as indicators of neuronal pathophysiology after ischaemia. In *Neurobiology of Ischaemic Brain Damage – Progress in Brain Research Vol. 96* (ed. Kogure, K., Hossmann, K.-A. & Siesjö, B.K.), pp 195–208. Amsterdam, Elsevier Science Publishers.
Okawa, M., Halaby, I., Nowak, T.S., Jr. & Pulsinelli, W.A. (1993) Time-course of protein synthesis recovery in hippocampal neuron populations after brief ischaemia in the rat (Abstract). *J. Cereb. Blood Flow Metab.* **13**, S586.
Olney, J.W., Labruyere, J., Wang, G., Wozniak, D.F., Price, M.T. & Sesma, M.A. (1991) NMDA antagonist neurotoxicity: Mechanism and prevention. *Science* **254**, 1515–1518.
Onodera, H., Kogure, K., Ono, Y., Igarashi, K., Kiyota, Y. & Nagaoka, A. (1989) Proto-oncogene c-*fos* is transiently induced in the rat cerebral cortex after forebrain ischaemia. *Neurosci. Lett.* **98**, 101–104.
Paschen, W., Cleef, M., Röhn, G., Müller, M. & Pajunen, A.E.I. (1993) Ischaemia-induced disturbances of polyamine synthesis. In *Neurobiology of Ischemic Brain Damage – Progress in Brain Research Vol. 96* (eds Kogure, K., Hossmann, K.-A. & Siesjö, B.K.), pp 147–160. Amsterdam, Elsevier.
Pelham, H.R.B. (1982) A regulatory upstream promoter element in the *Drosophila* hsp70 heat shock gene. *Cell* **30**, 517–528.
Picone, C.M., Grotta, J.C., Earls, R., Strong, R. & Dedman, J. (1989) Immunohistochemical determination of calcium–calmodulin binding predicts neuronal damage after global ischaemia. *J. Cereb. Blood Flow Metab.* **9**, 805–811.
Popovici, T., Represa, A., Crépel, V., Barbin, G., Beaudoin, M. & Ben-Ari, Y. (1990) Effects of kainic acid-induced seizures and ischaemia on c-*fos*-like proteins in rat brain. *Brain Res.* **536**, 183–194.
Price, B.D. & Calderwood, S.K. (1991) Ca^{2+} is essential for multistep activation of the heat shock factor in permeabilized cells. *Mol. Cell. Biol.* **11**, 3365–3368.
Pulsinelli, W.A., Brierley, J.B. & Plum, F. (1982) Temporal profile of neuronal damage in a model of transient forebrain ischaemia. *Ann. Neurol.* **11**, 491–498.
Rabindran, S.K., Giorgi, G., Clos, J. & Wu, C. (1991) Molecular cloning and expression of a human heat shock factor, HSF1. *Proc. Natl Acad. Sci. USA* **88**, 6906–6910.
Rordorf, G., Koroshetz, W.J. & Bonventre, J.V. (1991) Heat shock protects cultured neurons from glutamate toxicity. *Neuron* **7**, 1043–1051.

Ross, D.T. & Duhaime, A.C. (1989) Degeneration of neurons in the thalamic reticular nucleus following transient ischaemia due to raised intracranial pressure: Excitotoxic degeneration mediated via non-NMDA receptors? *Brain Res.* **501**, 129–143.

Saito, K., Nowak, T.S., Jr., Markey, S.P. & Heyes, M.P. (1993) Mechanisms of delayed increases in kyrnurenine pathway metabolism in damaged brain regions following transient cerebral ischaemia. *J. Neurochem.* **60**, 180–192.

Saito, N. & Nowak, T.S., Jr. (1993) Delayed expression of mRNA encoding microtubule-associated protein 2c (MAP2c) in vulnerable neuron populations after ischemia (Abstract). *J. Cereb. Blood Flow Metab.* **13**, S467.

Sato, S., Abe, K., Kawagoe, J., Aoki, M. & Kogure, K. (1992) Isolation of complementary DNAs for heat shock protein (HSP) 70 and heat shock cognate protein (HSC) 70 genes and the expressions in post-ischaemic gerbil brain. *Neurol. Res.* **14**, 375–380.

Scharf, K.-D., Rose, S., Zott, W., Schöff, F. & Nover, L. (1990) Three tomato genes code for heat stress transcription factors with a region of remarkable homology to the DNA-binding domain of the yeast HSF. *EMBO J.* **9**, 4495–4501.

Schuetz, T.J., Gallo, G.J., Sheldon, L., Tempst, P. & Kingston, R.E. (1991) Isolation of a cDNA for HSF2: Evidence for two heat shock factor genes in humans. *Proc. Natl Acad. Sci. USA* **88**, 6911–6915.

Selman, W.R., Ricci, A.J., Crumrine, R.C., LaManna, J.C., Ratcheson, R.A. & Lust, W.D. (1990) The evolution of focal ischaemic damage: A metabolic analysis. *Metab. Brain Dis.* **5**, 33–44.

Seubert, P., Lee, K. & Lynch, G. (1989) Ischemia triggers NMDA receptor-linked cytoskeletal proteolysis in hippocampus. *Brain Res.* **492**, 366–370.

Sharma, H.S., Wünsch, M., Brand, T., Verdouw, P.D. & Schaper, W. (1992) Molecular biology of the coronary vascular and myocardial responses to ischemia. *J. Cardiovasc. Pharmacol.* **20**, S23–S31.

Sharp, F.R., Butman, M., Wang, S., Koistinaho, J., Graham, S.H., Sagar, S.M., Noble, L., Berger, P. & Longo, F.M. (1992) Haloperidol prevents induction of the hsp70 heat shock gene in neurons injured by phencyclidine (PCP), MK801, and ketamine. *J. Neurosci. Res.* **33**, 605–616.

Sharp, F.R., Jasper, P., Hall, J., Noble, L. & Sagar, S.M. (1991a) MK-801 and ketamine induce heat shock protein HSP70 in injured neurons in posterior cingulate and retrosplenial cortex. *Ann. Neurol.* **30**, 801–809.

Sharp, F.R., Lowenstein, D., Simon, R. & Hisanaga, K. (1991b) Heat shock protein hsp72-induction in cortical and striatal astrocytes and neurons following infarction. *J. Cereb. Blood Flow Metab.* **11**, 621–627.

Siesjö, B.K. & Bengtsson, F. (1989) Calcium fluxes, calcium antagonists, and calcium-related pathology in brain ischaemia, hypoglycemia, and spreading depression: A unifying hypothesis. *J. Cereb. Blood Flow Metab.* **9**, 127–140.

Simon, R.P., Cho, H., Gwinn, R. & Lowenstein, D.H. (1991) The temporal profile of 72-kDa heat shock protein expression following global ischaemia. *J. Neurosci.* **11**, 881–889.

Sloviter, R.S. & Lowenstein, D.H. (1992) Heat shock protein expression in vulnerable cells of the rat hippocampus as an indicator of excitation induced neuronal stress. *J. Neurosci.* **12**, 3004–3009.

Smith, M.-L., Auer, R.N. & Siesjö, B.K. (1984) The density and distribution of ischaemic brain injury in the rat following 2–10 min of forebrain ischaemia. *Acta Neuropathol. (Berl.)* **64**, 319–332.

Sprang, G.K. & Brown, I.R. (1987) Selective induction of a heat shock gene in fibre tracts and cerebellar neurons of the rabbit brain detected by *in situ* hybridization. *Mol. Brain. Res.* **3**, 89–93.

Suga, S. & Nowak, T.S., Jr. (1991) Localization of immunoreactive Fos and Jun proteins in gerbil brain: Effect of transient ischaemia (Abstract). *J. Cereb. Blood Flow Metab.* **11**, S352.

Suga, S. & Nowak, T.S., Jr. (1993) Pharmacology of the post-ischemic stress response: Effects of temperature on hsp70 expression after transient ischaemia and hypothermic action of NBQX in the gerbil. In *Pharmacology of Cerebral Ischemia 1992* (ed Krieglstein, J. & Oberpichler-Schwenk, H.), pp 279–286. Stuttgart, Wissenschaftliche Verlag.

Szekely, A.M., Barbaccia, M.L., Alho, H. & Costa, E. (1989) In primary cultures of cerebellar granules cells the activation of *N*-methyl-D-aspartate-sensitive glutamate receptors induces c-*fos* mRNA expression. *Mol. Pharmacol.* **35**, 401–408.

Szekely, A.M., Costa, E. & Grayson, D.R. (1990) Transcriptional program coordination by *N*-methyl-D-aspartate-sensitive glutamate receptor stimulation in primary cultures of cerebellar neurons. *Mol. Pharmacol.* **38**, 624–633.

Tacchini, L., Schiaffonati, L., Pappalardo, C., Gatti, S. & Bernelli-Zazzera, A. (1993) Expression of HSP70, immediate–early response and heme oxygenase genes in ischaemic-reperfused rat liver. *Lab. Invest.* **68**, 465–471.

Thilmann, R., Xie, Y., Kleihues, P. & Kiessling, M. (1986) Persistent inhibition of protein synthesis precedes delayed neuronal death in post-ischaemic gerbil hippocampus. *Acta Neuropathol.* **71**, 88–93.

Uemura, Y., Kowall, N.W. & Beal, M.F. (1991a) Global ischaemia induces NMDA receptor-mediated c-*fos* expression in neurons resistant to injury in gerbil hippocampus. *Brain Res.* **542**, 343–347.

Uemura, Y., Kowall, N.W. & Moskowitz, M.A. (1991b) Focal ischaemia in rats causes time-dependent expression of c-*fos* protein immunoreactivity in widespread regions of ipsilateral cortex. *Brain Res.* **552**, 99–105.

Uney, J.B., Leigh, P.N., Marsden, C.D., Lees, A. & Anderton, B.H. (1988) Stereotaxic injection of kainic acid into the striatum of rats induces synthesis of mRNA for heat shock protein 70. *FEBS Lett.* **235**, 215–218.

Ungewickel, E. (1985) The 70-kd mammalian heat shock proteins are structurally and functionally related to the uncoating protein that releases triskelion from coated vesicles. *EMBO J.* **4**, 3385–3391.

Van Why, S.K., Hildebrandt, F., Ardito, T., Mann, A.S., Siegel, N.J. & Kashgarian, M. (1992) Induction and intracellular localization of HSP-72 after renal ischaemia. *Am. J. Physiol.* **263**, F769–F775.

Vass, K., Berger, M.L., Nowak, T.S., Jr., Welch, W.J. & Lassmann, H. (1989) Induction of stress protein HSP-70 in nerve cells after status epilepticus in the rat. *Neurosci. Lett.* **100**, 259–264.

Vass, K., Welch, W.J. & Nowak, T.S., Jr. (1988) Localization of 70kDa stress-protein induction in gerbil brain after ischaemia. *Acta Neuropathol.* **77**, 128–135.

Viereck, C., Tucker, R.P. & Matus, A. (1989) The adult rat olfactory system expresses microtubule-associated proteins found in the developing brain. *J. Neurosci.* **9**, 3547–3557.

Voellmy, R., Ahmed, A., Schiller, P., Bromley, P. & Rungger, D. (1985) Isolation and functional analysis of a human 70 000 dalton heat shock protein gene segment. *Proc. Natl Acad. Sci. USA* **82**, 4949–4953.

Watowich, S.S. & Morimoto, R.I. (1988) Complex regulation of heat shock- and glucose-responsive genes in human cells. *Mol. Cell. Biol.* **8**, 393–405.

Welsh, F.A., Moyer, D.J. & Harris, V.A. (1992) Regional expression of heat shock protein 70 mRNA and c-*fos* mRNA, following focal ischaemia in rat brain. *J. Cereb. Blood Flow Metab.* **12**, 204–212.

Wessel, T.C., Joh, T.H. & Volpe, B.T. (1991) *In situ* hybridization analysis of c-*fos* and c-*jun* expression in the rat brain following transient forebrain ischaemia. *Brain Res.* **567**, 231–240.

Widmann, R., Kuroiwa, T., Bonnekoh, P. & Hossmann, K.-A. (1991) [^{14}C]Leucine incorporation into brain proteins in gerbils after transient ischaemia: Relationship to selective vulnerability of hippocampus. *J. Neurochem.* **56**, 789–796.

Wong, M.-L., Weiss, S.R.B., Gold, P.W., Doi, S.Q., Banerjee, S., Licinio, J., Lad, R., Post, R.M. & Smith, M.A. (1992) Induction of constitutive heat shock protein 73 mRNA in the dentate gyrus by seizures. *Mol. Brain Res.* **13**, 19–25.

Wu, B.J., Kingston, R.E. & Morimoto, R.I. (1986) Human hsp70 promoter contains at least two distinct regulatory domains. *Proc. Natl Acad. Sci. USA* **83**, 629–633.

Yamashita, K., Eguchi, Y., Kajiwara, K. & Ito, H. (1991) Mild hypothermia ameliorates ubiquitin synthesis and prevents delayed neuronal death in the gerbil hippocampus. *Stroke* **22**, 1574–1581.

Yanagihara, T., Brengman, J.M. & Mushynski, W.E. (1990) Differential vulnerability of microtubule components of cerebral ischaemia. *Acta Neuropathol.* **80**, 499–505.

Yoshimi, K., Takeda, M., Nishimura, T., Kudo, T., Nakamura, Y., Tada, K. & Iwata, N. (1991) An immunohistochemical study of MAP2 and clathrin in gerbil hippocampus after cerebral ischaemia. *Brain Res.* **560**, 149–158.

Yost, H.J. & Lindquist, S. (1986) RNA splicing is interrupted by heat shock and is rescued by heat shock protein synthesis. *Cell* **45**, 185–193.

CHAPTER 4

HEAT SHOCK PROTEINS IN THE RETINA AND OPTIC NERVE

Michael Tytell

Department of Neurobiology and Anatomy, and Department of Ophthalmology, Programme in Neuroscience,
Bowman Gray School of Medicine of Wake Forest University, Winston-Salem, North Carolina 27157, USA

Table of Contents

4.1 Introduction

Several of the chapters in this book make it clear that a dramatic elevation in the transcription of heat shock genes and accumulation of the encoded proteins are fundamental features of the response of nerve tissue to a wide variety of insults. Clearly, heat shock proteins (hsps) must have many important functions in the way in which glia and neurons attempt to cope with injury. However, work on the significance of the increased hsps for the survival of injured glia and neurons is at an early stage and many questions remain to be answered. The retina, as 'an approachable part of the brain' (Dowling, 1987), offers some advantages for pursuing such questions on the functional significance of both normal and stress-induced levels of hsps.

The uniform layered structure of the retina facilitates morphological analysis of

Heat Shock Proteins in the Nervous System
ISBN 0-12-480960-X

the normal and stress-altered distribution of hsps in relation to various cell types and their functions. An example of this regularity of structure is provided in Figure 1, a micrograph of a haematoxylin and eosin (H&E) stained section of a rat retina. It contains many of the same assortment of cellular components as other parts of the central nervous system, as well as sensory elements, the photoreceptors. The various parts of the photoreceptors make up four of the outermost layers of the retina: (a) the outer segments, which consist mainly of stacks of membrane disks containing the photoreceptive pigment (Figure 1, layer 2); (b) the inner segments, which can be considered the powerhouses of the photoreceptors, containing much of their protein synthetic apparatus as well as densely packed arrays of mitochondria, providing the energy required for maintenance of the membrane potential (Figure 1, layer 3); (c) the outer nuclear layer comprised of mainly the photoreceptor nuclei with just a thin rim of relatively organelle-free cytoplasm (Figure 1, layer 4); and (d) the outer plexiform layer, which includes the short axon-like extensions of the photoreceptors that synapse on the dendrites of neurons of the inner retina (Figure 1, layer 5).

The small neurons of the retina make up the inner nuclear layer (Figure 1, layer 6). This layer includes their cell bodies and some of their interconnecting dendrites and axons. These neurons, classified as amacrine, bipolar, and horizontal cells, based on their dendritic and synaptic morphologies, are similar to small interneurons found in the brain because their processes are relatively short, remaining within the neighbourhood of their respective cell bodies. This layer also contains the cell bodies of a specialized type of glial cell unique to the retina, the Müller cell. Müller cells may be related to astrocytes, since they often contain intermediate filaments composed of glial fibrillary acidic protein (Newman, 1986). These cells also resemble in their structure, the radial glia found in the developing central nervous system because they have a bipolar organization, with cytoplasmic processes extending from the inner segment layer to the vitreal surface of the inner retina.

The retinal equivalent of large neurons with long axons are the retinal ganglion cells (RGCs), whose cell bodies make up the innermost layer of the retina, which is named after them (Figure 1, layer 8). The extensive dendritic arbors of the RGCs in the inner plexiform layer (Figure 1, layer 7) receive synaptic input from the small neurons of the INL and relay the information to the visual centres of the brain via their axons, comprising the optic nerve, which are a centimetre or more in length even in a small mammal like the rat. Scattered among the RGCs are typical astrocytes.

The one major CNS cell-type and structure not found in the retina is the oligodendrocyte and the myelin sheaths which they form. These cells are present in the optic nerve only after it exits the rear of the globe.

4.2 Hsps in the normal retina

Even without experimentally applied metabolic stress, several of the hsps have been detected in the retina using a variety of approaches. A low level of the mRNA

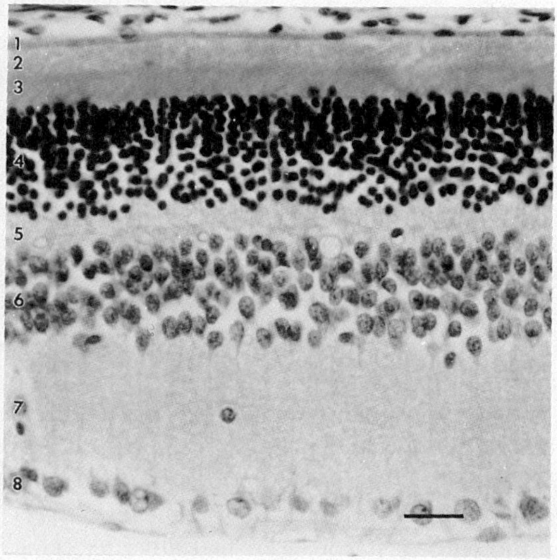

Figure 1 Microscopic structure of the rat retina. An example of a region in the superior part of the retina stained with haematoxylin and eosin to illustrate the nuclear and cytoplasmic structures. Each of the distinct layers are numbered along the left edge as follows: (1) pigment epithelium (no pigment is present in the albino rat); (2) outer segments of photoreceptors; (3) inner segments of photoreceptors; (4) outer nuclear layer; (5) outer plexiform layer; (6) inner nuclear layer; (7) inner plexiform layer; (8) retinal ganglion cell layer. The eye was fixed in Carnoy's solution, embedded in paraffin and sectioned along the superior-inferior plane at 8 μm. Details of the procedure can be found in Yamaguchi *et al.* (1991). The scale bar at the bottom right is 20 μm.

corresponding to hsc70 has been found in the rabbit retina by Northern blot analysis (Masing *et al.*, 1990) and, by *in situ* hybridization, that mRNA was shown to be present in all the nuclear layers of the retina, the ONL (photoreceptors) and the INL (amacrine, bipolar, and horizontal cells) of the rat (Tytell *et al.*, 1994) and the RGC of the rabbit (Masing *et al.*, 1990). When retinal protein synthesis was examined one or more hours after an intravitreal injection of ^{35}S-methionine, hsc70 was shown to be one of the more prominently labelled retinal proteins in the rat (Tytell and Barbe, 1987; Tytell *et al.*, 1989; Bizzi *et al.*, 1991). This fact appeared to hold true for the normal rabbit retina as well, if one compares the two-dimensional electrophoresis gel patterns presented by Clark and Brown (1985, 1986) with those of the rat (Bizzi *et al.*, 1991), even though those authors did not point out hsc70 in their gel patterns. Low levels of several other hsps were also synthesized in the unstressed rat retina, including hsp70, the 80, 90, and 100-kDa hsps, and ubiquitin (8-kDa) (Bizzi *et al.*, 1991).

The relative abundance of the inducible and constitutive forms of hsp in the retina is particularly apparent in Western blot and immunocytochemical analyses. Barbe *et al.* (1988) first demonstrated that there was detectable hsp70 in a Western blot of whole retina of an unstressed rat probed with the C92F3A-5 monoclonal antibody (MAb), shown by Welch and Suhan (1986) to be specific for the inducible form of the 70-kDa hsp family. Subsequent immunohistochemical analyses of retinal tissue sections from unstressed rats using the MAb designated N27F3-4 (both this MAb and the C92 MAb were obtained from StressGen Biotechnologies Corp., Victoria, British Columbia, Canada), which binds to both the constitutive and inducible members of the 70-kDa hsp family (hereafter referred to as 70-kDa hsps; Welch and Suhan, 1986) has revealed an intriguing heterogeneous distribution of immunoreactive protein (Figure 2). In general, the highest levels of immunoreactive 70-kDa hsps were found in those retinal layers containing the cytoplasmic portions of the various retinal cells, the inner segment layer, outer plexiform layer, and inner plexiform layer, whereas the outer and inner nuclear layers had lower amounts of immunoreactive material (Figure 2 and Tytell *et al.*, 1989; Yamaguchi *et al.*, 1991). Additionally, the cytoplasm of the large retinal ganglion cells usually demonstrated abundant immunoreactive 70-kDa hsps (Figure 2 rat and human retinas and Tytell *et al.*, 1989; Yamaguchi *et al.*, 1991). Only one part of the retina, the outer segment layer, had little detectable immunoreactive material. This distribution of 70-kDa hsps in the unstressed retina appears to be a general characteristic of that tissue regardless of the species, as illustrated in Figure 2, which presents the patterns of retinal immunoreactivity for 70-kDa hsps in the frog, rat, and human. When other retinal tissue sections from these organisms were probed with the C92 MAb specific for hsp70, the distribution of

---

Figure 2 Distribution of immunoreactive 70-kDa hsps in unstressed vertebrate retinas. Examples are shown of the distribution of immunoperoxidase staining for 70-kDa hsps in retinas from *Xenopus laevis* (FROG), a three-month old Fischer-344 rat (RAT), and a 78-year old Caucasian male (HUMAN). The layers of each retina are numbered as in Figure 1. For each of the immunostained sections shown on the left, a corresponding non-specific immunoperoxidase control section is shown on the right in which the anti-70-kDa hsps antibody was omitted (N27F3–4, StressGen Biotechnologies, Victoria, British Columbia, Canada). Both the frog and human retinas were pigmented, so the dark material seen in the pigment epithelium (layer 1) is primarily melanin, although a low level of immunoreactive 70-kDa hsps is present there. The frog photoreceptors have very long outer segments; thus, in that retina, layer 2 is more than twice as wide as in the rat and human. On the other hand, in the frog there is only a single layer of photoreceptor nuclei forming the outer nuclear layer (layer 4), compared to the multilayered outer nuclear layers of the rat and human. Despite the structural and species differences, the overall distribution of immunoreactive 70-kDa hsps is similar. In particular, the inner segment and outer plexiform layers (3 and 5, respectively) have the most intense immunoreactivity. Additionally, in the rat and human, scattered cells in the bottom portion of the inner nuclear layer (6) and many retinal ganglion cells (8) have intense perinuclear staining. Also intensely stained are occasional cells in layer 6 of unknown identity. Scale bars in the lower right corners of the frog and rat micrographs are 20 μm; magnification of the human retina micrograph is identical to that of the rat. The frog and rat eyes were fixed in Carnoy's solution and the human eye in a modified formulation of Carnoy's solution called methacarn (Mitchell *et al.*, 1985).

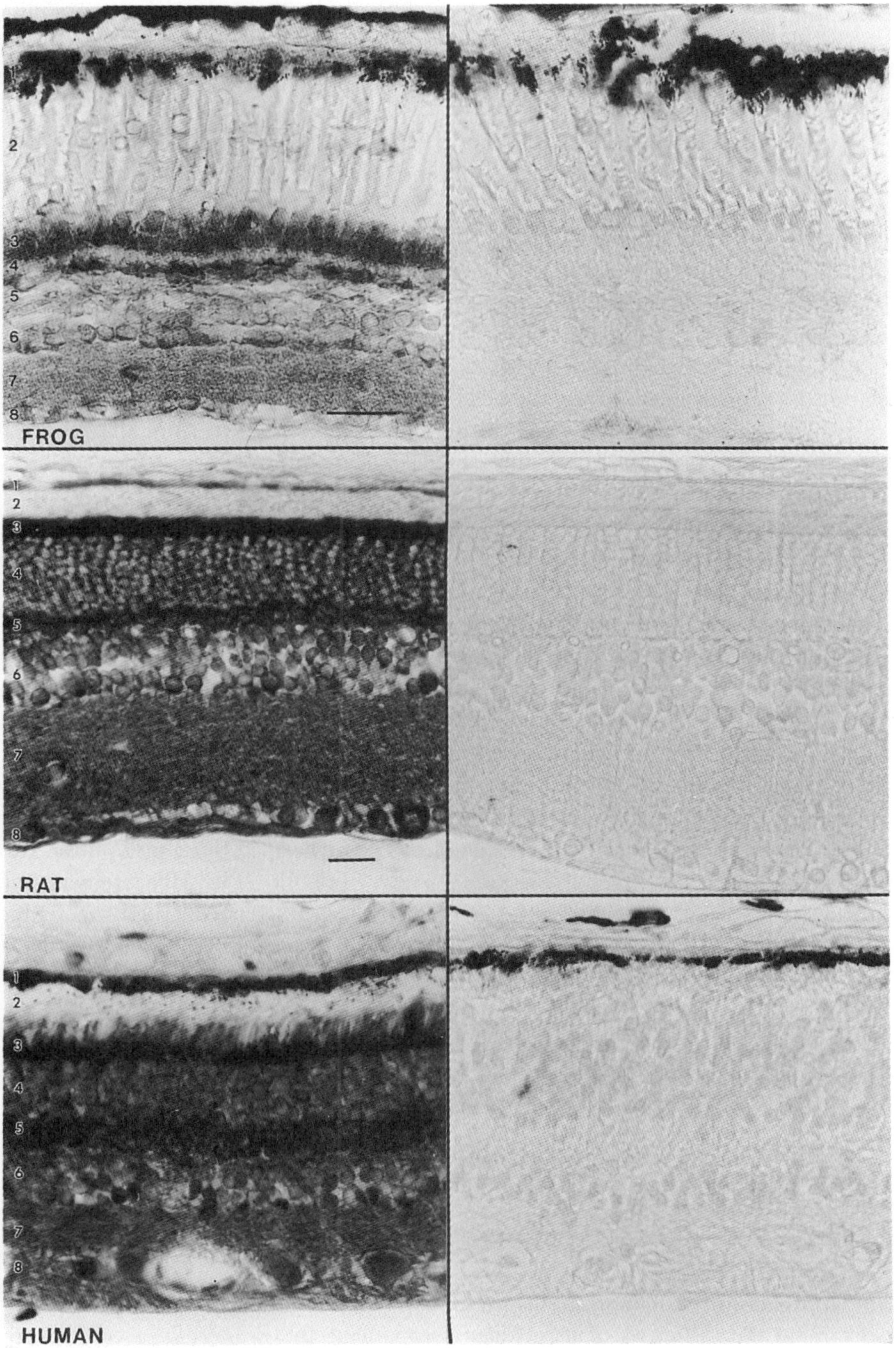
1
2
3
4
5
6
7
8
FROG
1
2
3
4
5
6
7
8
RAT
1
2
3
4
5
6
7
8
HUMAN

immunoreactivity was similar to that seen with the N27 MAb, but the amount was much less, being detectable primarily in the inner segment and outer plexiform layers, the same regions that had the greatest N27 immunoreactive material (not shown).

The fact that immunoreactivity for the 70-kDa hsps is abundant in those retinal layers containing most of the cytoplasmic portions of the retinal cells makes sense in relation to what we know about the intracellular distribution of hsc70 in unstressed cells. Hsc70 is a relatively abundant cytoplasmic protein that participates in the maturation of nascent polypeptides being synthesized on free ribosomes (Beckmann *et al.*, 1990), the importation of certain new proteins into mitochondria and endoplasmic reticulum (Chirico *et al.*, 1988; Deshaies *et al.*, 1988), and the cycling of coated vesicles. It would not be expected to be as abundant in the retinal nuclei unless the cells were experiencing acute metabolic stress, nor would one expect to find much in the outer segment layer, since that portion of the photoreceptors consists mainly of stacks of membrane discs with very little cytoplasm.

The presence in the unstressed retina of a low level of hsp70 is not without precedent, as Beckmann *et al.* (1990) have reported this to be true for normal HeLa cells. Whether the synthesis of hsp70 under non-stressful conditions has functional significance or merely represents some looseness in the regulation of the gene remains to be determined. However, a rationale can be provided to support the hypothesis that the presence of immunoreactive hsp70 in the retina is a reflection of the fact that even the apparently normal retina is chronically in a state of mild metabolic stress. This stress may arise from one or both of the following properties of the retina. Firstly, the middle layers of the retina (INL, OPL, ONL, IS, and OS) are avascular, depending on diffusion of oxygen and other nutrients from capillaries located in the choroidal layer outside of the pigment epithelium and on the vitreal surface, some of which penetrate partway into the IPL. Measurements of oxygen tension at the level of the ONL and IS in the dark-adapted cat retina showed it to be 5 mmHg or less, a state of borderline hypoxia (Linsenmeier, 1986). Secondly, the absorption of light energy by the photoreceptor OS produces potentially damaging reactive by-products, such as lipid peroxides (Organisciak *et al.*, 1983; Wiegand *et al.*, 1983; De La Paz and Anderson, 1992), in part because of the higher than average content in photoreceptors of long-chain polyunsaturated fatty acids (Anderson and Andrews, 1982).

4.3 Induction of hsps in cultured retinal cells

One approach to evaluating the hsp response of a complex tissue like the retina is to examine the responses of its cellular components individually in a cell culture system. Such studies have been done with two types of retinal cells that are easier than the others to maintain in long-term cultures, the Müller cell and the pigment epithelial cell. The Müller cell is the predominant glial cell of the retina and resembles an astrocyte in that it plays an important role in maintaining proper extracellular ion concentrations and produces intermediate filaments composed of glial fibrillary acidic protein in response to injury (review: Newman, 1986). The retinal pigment epithelial cell is a specialized simple epithelium which forms the outermost layer of the retina and plays a critical role in the metabolic support of the photoreceptors, in part by phagocytizing and recycling the tips of the photoreceptors as they are diurnally shed.

In cultures of rabbit Müller cells, the production of three hsps with molecular weights of about 60-, 68-, and 90-kDa were monitored with three monoclonal antibodies (Wakakura *et al.*, 1987; for antibody characterization, see: LaThangue, 1984; LaThangue and Chan, 1984; LaThangue *et al.*, 1984). Generally, a low level of these hsps was detected in about 10–20% of unstressed cells. When the cells were infected with *Herpes simplex* virus, the production of the three hsps increased dramatically, with 40–60% of the cells showing strongly positive immunoreactivity in their nuclei and cytoplasm (Wakakura *et al.*, 1987). Similar increases in immunoreactivity and intracellular distributions of the three hsps also were detected 1 h after a 42–43°C heat shock of 1.5 or 4 h. The low level of hsp immunoreactivity seen in unstressed cultured Müller cells is consistent with the relatively lighter immunostaining of the inner nuclear layer of the unstressed intact retina (Tytell *et al.*, 1987, 1994), the layer which contains the Müller cell bodies. Additionally, this work clearly shows that the Müller cells contribute to the overall increase in hsps seen in the intact retina after heat stress (Clark and Brown, 1982, 1986; Tytell and Barbe, 1987; Tytell *et al.*, 1989, 1994).

In primary cultures of human retinal pigment epithelial cells, the induction of 70-, 84-, and 110-kDa hsps was observed after incubation in the analogue of proline called L-azetidine-2-carboxylic acid (5 mM), but not after 30–60 min of exposure to as much as 10^{-3} M hydrogen peroxide (Wong and Lin, 1989). The results using the proline analogue showed that the retinal pigment epithelial cells were capable of an hsp response similar to that of the rest of the retina (see above). The hydrogen peroxide experiment was an attempt to mimic the presence of reactive lipid peroxides known to form in the retina upon light exposure (Organisciak *et al.*, 1983; Wiegand *et al.*, 1983; De La Paz and Anderson, 1992), but may have been ineffective because it could not gain access to the same sites as those lipid peroxides. This speculation is supported by recent work on an SV-40 transformed immortal cell line derived from human fetal RPE by Kerendian *et al.* (1992), which showed that exposure to organic hydroperoxides did elevate hsp70 levels (Kerendian *et al.*, 1992). Furthermore, those researchers found that unstressed fresh bovine RPE contained low levels of hsp70 like the rest of the retina (see above).

Recently, the hsp response of cultured rat retinal ganglion cells was studied (Kitano and Caprioli, 1993). They were found to accumulate increased levels of hsp70 in response to a 42°C heat stress or sublethal hypoxia (10% O_2). This increase in hsp70 content correlated with significantly greater resistance of the cells to potentially lethal hypoxic injury (0% O_2, 3 h).

The observations on cultured Müller, RPE, and retinal ganglion cells illustrate that the hsp repertoire of those individual cells is similar to that of the whole retina. The continuation of studies of these individual retinal cells in culture in combination with the retinal organ culture and *in vivo* retinal studies will help to determine how these separate cell responses contribute to the response of the intact tissue, and the capacity of the stress response to protect the retina from various injuries.

4.4 Heat stress or chemical induction of hsps in the organ cultured retina

Although the cell culture approach allows great control of the conditions used to examine the hsp response, it suffers from the potential disadvantage that the isolated cell-type may behave differently than it does when it is together with the other cells that comprise the tissue. Additionally, in the case of the retina, it may be technically very difficult to obtain cultures of the neuronal cell-types and the photoreceptors. As a compromise between cell culture and *in vivo* studies, one can use organ culture of the intact tissue to obtain better control over experimental conditions than achievable *in vivo*, yet still evaluate cellular interactions. Two studies show that the retinal stress response can be studied in organ culture. The adult rabbit retina can be dissected from the eye placed in organ culture for nearly two hours without eliciting a stress response (Clark and Brown, 1986). Yet, the retinal tissue remained responsive and a brief 43°C heat stress caused a marked increase in the synthesis of hsp70 (*ibid*). Further study of this response in the isolated rabbit retina showed that it was influenced by the state of the microtubule cytoskeleton (Clark and Brown, 1987). When retinal microtubules were stabilized in the presence of taxol, the increase in hsp70 synthesis following a 5 min, 43°C heat stress was partly suppressed compared to a normal heat-stressed retina. Depolymerization of the microtubules with colchicine, on the other hand, increased the amount of hsp70 synthesized after the heat stress, although colchicine alone had no effect on hsp70 production.

Very recently, the cultured eyecup of *Xenopus laevis* has also been employed to examine the retinal stress response (Beasley *et al.*, 1993). In this preparation, the anterior portion of the eye is removed and posterior portion is left intact and can be maintained in culture for at least 24 h. Although the eyecup showed a stress-like reaction to placement in culture, as evidenced by an increase in perinuclear 70-kDa hsps staining in the photoreceptors within 10 min, it returned to the normal pattern shown in Figure 2 after about 1 h. From that point, exposure of the eyecup to 100 μM $NaAsO_2$ for 2 h or to a 35°C heat stress for 2 h elicited a prominent rise in the synthesis of both hsc70 and hsp70, as detected in Western blots probed with the N27 MAb (*ibid.*). This model is being used currently to examine the potential for the stress-induced increase in hsps to protect the retina against glutamate excitotoxicity.

4.5 Retinal induction of hsps after whole body hyperthermia

By raising an animal's body temperature, essentially giving it an artificial fever, hsp production can be increased in many tissues throughout the body. Although such a whole animal approach is complicated by the fact that it also elicits many other physiological changes whose interactions with the hsp response are largely unknown, it has the advantage of allowing direct study of the functional significance of elevated hsps on the survival of cells in a damaged tissue. Two procedures have been used to stimulate retinal hsp production: drug-induced hyperthermia and hyperthermia caused by elevated ambient temperature. In the rabbit, rectal temperature was increased via an intravenous injection of D-lysergic acid diethylamide (LSD), which caused about a 3°C rise in rectal temperature in one hour (Brown, Heikkila, and Cosgrove, 1982). In the rat, pentobarbital anaesthesia was induced and then the animal was placed in a chamber maintained at 40–42°C, which led to a rise of about 4–5°C in rectal temperature (from 37°C to 41–42°C) after about 20–30 min (Tytell and Barbe, 1987; Barbe *et al.*, 1988). Additionally, 12-day chicken embryos were heat stressed by placing the eggs in a plastic bag immersed in water at 45°C for 2 h. Changes in retinal hsp accumulation, as well as induction of mRNA for heat shock transcription factors (a set of proteins responsible for turning on the hsp70 gene) were then examined by immunohistochemistry and *in situ* hybridization, respectively (Nakai and Morimoto, 1993).

The hsp response of the retina *in vivo* was first documented in the rabbit retina by Clark and Brown (1982). After labelling newly synthesized retinal proteins via an intravitreal injection of ^{35}S-methionine, they found that proteins of 74-kDa and 95-kDa hsp were both synthesized in markedly greater amounts in retinas from rabbits with LSD-induced hyperthermia than in the retinas of normothermic rabbits. Although in that report, no further analysis was performed on those proteins to confirm that they were members of the hsp family, the authors noted the similarity of their molecular weights to hsps reported in tissue culture systems, and subsequent work showed the suggested correspondence to be correct (Brown, 1983; Cosgrove and Brown, 1983; Brown, Lowe, and Moran, 1985; Clark and Brown, 1985). In particular, using high-resolution *in situ* hybridization, LSD-induced hyperthermia has been shown to induce the hsp70 gene in the retinal ganglion cells of the rabbit retina (Masing *et al.*, 1990).

In the rat retina model used by this author and coworkers, environmentally induced hyperthermia caused significant increases within four hours in the synthesis of three ^{35}S-methionine-labelled retinal proteins with relative molecular weights of 64-, 74-, 110-kDa (Barbe, 1987). The elevated synthesis of those proteins persisted through six hours after hyperthermia (Barbe, 1987; Tytell and Barbe, 1987; Barbe *et al.*, 1988). Immunochemical analysis showed that the 64-kDa protein was equivalent to hsp70; the lower relative molecular weight was at least partially attributable to the sample buffer used for gel electrophoresis (Barbe *et al.*, 1988). The 74- and 110-kDa hsps have not been further characterized, but similar observations were made by Bizzi *et al.* (1991), who detected at 6 h after a 41°C heat stress in the rat retina the increased synthesis of hsps of 80-, 90-, and 100-kDa, as well as hsc70 and hsp70, in fluorographs of two-dimensional polyacrylamide gels. Analysis of the duration of the response revealed that protein synthesis returned to control levels by 12 h post-hyperthermia (Barbe, 1987). However, the increased hsp70 synthesized within the first few hours of the heat stress was reasonably long-lived. Immunoblot analysis of total hsp70 content of the retina (Barbe *et al.*, 1988), as well as densitometric analyses of anti-hsp70 immunoreaction product in retinal tissue sections showed that hsp70 continued to accumulate through at least 18 h after hyperthermia. Only by 24 h after hyperthermia did the level of hsp70 begin to decline, although it still had not returned to control levels even by 48–50 h after heat stress (Barbe *et al.*, 1988; Tytell *et al.*, 1994).

Detailed analysis of the contributions of the various cell types in the rat retina to the 70-kDa hsp-response after heat stress was conducted by *in situ* hybridization and immunohistochemistry, followed by densitometry to measure relative changes in the levels of 70-kDa hsp mRNAs and immunoreactive protein (Tytell *et al.*, 1994). The greatest increase in mRNA was detected in the outer nuclear layer, which contains the photoreceptor nuclei, and the greatest increase in protein immunoreactivity was observed in the inner segment layer, which is largely comprised of portions of the photoreceptors containing a high density of mitochondria and protein synthetic machinery. These results are consistent with the hypothesis that the 70-kDa hsps would concentrate in regions of a cell most vulnerable to metabolic stress. Protein immunoreactivity was greatly increased also in the outer and inner plexiform layers and in retinal ganglion cell cytoplasm. The outer plexiform layer includes the synaptic terminals of the photoreceptors and dendrites of intra-retinal neurons and the inner plexiform layer includes axons of those same neurons as well as their synaptic terminals with dendrites of retinal ganglion cells. Additionally, both layers contain cytoplasmic processes of the Müller cells.

It is notable that by both mRNA and immunoreactive protein localization, the photoreceptors seem to have the most vigorous hsp response of all the retinal cells. This prominent role of the photoreceptors in the retinal stress response is supported further by the observations that destruction of the photoreceptors by light damage suppressed the post-hyperthermia hsp response (Tytell *et al.*, 1994) and that loss of photoreceptors because of a genetic defect reduced overall retinal 70-kDa hsps immunoreactivity (Yamaguchi *et al.*, 1991).

A somewhat different picture of the relative contributions of retinal cells to the heat shock response was observed in the chicken embryo. Nakai and Morimoto (1993) found that the greatest increase in hsp70 mRNA after heat stress occurred in the retinal pigment epithelium of this model and that this result correlated with a strong induction of mRNA for two heat shock transcription factors. Additionally, they showed that mRNA for a 47-kDa hsp, known to be associated with collagen (Nakai *et al.*, 1992; Takechi *et al.*, 1992), was also induced in those cells.

4.6 Using the retina to study intracellular transport of 70-kDa hsps

The retina with its ganglion cell axons projecting into the brain has been used extensively to study axonal transport of protein in a wide variety of species. Its utility in such work derives from the structure of the optic system. The eye acts as a natural container for injected radioactive precursors, which are then incorporated into proteins by the retinal ganglion cells. The movement of those labelled proteins within the axons of the retinal ganglion cells in the optic nerve and tract can then be analysed relatively easily by removing segments of the nerve and tract at various times after the intravitreal injection of labelled precursor. Such work has identified five rate components of anterograde transport and one rate component of retrograde transport (review: Grafstein and Forman, 1980). These rate components correspond to functionally distinct classes of intracellular organelles and molecules. Three of the anterograde transport components supply most of the constituents of the axon and they are: (1) the fast component (FC) with a rate of about 250–400 mm/day, which includes most of the membranous organelles of the axon except for mitochondria; (2) slow component b (SCb) with a rate of about 2–5 mm/day, which includes many of the enzymes of intermediary metabolism and actin and clathrin with their associated proteins; (3) slow component a (SCa) with a rate of about 0.2–1.5 mm/day, which corresponds mainly to microtubules and neurofilaments (reviews: Lasek, 1982; Brady and Lasek, 1982; Lasek *et al.*, 1984).

This technique has been applied to evaluate the transport of hsps in optic axons and thereby determine with which class of intra-axonal constituents they associate. Initially, Clark and Brown (1985) presented evidence in the rabbit optic system that hsp70 induced following hyperthermia, was part of the slow component of transport, not the fast. However, they did not distinguish between SCa and SCb. A later study further specified that 70-kDa hsps, as well as two of the higher molecular weight hsps induced by hyperthermia, were all included in SCb only (Tytell and Barbe, 1987). These observations provided the first results suggesting that, at least within retinal ganglion cell axons, hsps are primarily associated with the axoplasmic structures and proteins comprising SCb, which includes cytoskeletal constituents such as actin microfilaments, clathrin, and spectrin, the regulatory molecule, calmodulin, and metabolic enzymes such as nerve specific enolase, creatine phosphokinase, aldolase, and pyruvate kinase (reviews: Brady and Lasek, 1982, and Lasek *et al.*, 1984, provide background on the constituents of FC, SCb, and SCa). Subsequent more detailed studies of hsc70 axonal transport in unstressed rats further refined the earlier observations. de Waegh and Brady (1989) confirmed that the 70-kDa axonally transported protein was, in fact, immunoreactive hsc70, and additionally suggested that its abundance in SCb along with clathrin supported the hypothesis that it was playing a role in the regulation of axonal clathrin. This hypothesis was substantiated by Black *et al.* (1991) who demonstrated in a series of carefully conducted immunoprecipitation studies that hsc70 and clathrin exist as a complex in intact optic axons. They proposed that this complex of hsc70 and clathrin, which includes other cytoplasmic poplypeptides as well, probably represents a unit with distinctive, although as yet, uncharacterized functions in the axon.

Additional support for special functional properties of the molecular complex of proteins in SCb related to cell stress can be found in the observations of Bizzi *et al.* (1991) on the transport of several hsps in normal and heat shocked rat optic axons. Besides replicating the observations that hsc70 and hsp70 are constituents of SCb, they showed that ubiquitin and ubiquitin carboxyl-terminal hydrolase, PGP 9.5 (Wilkinson *et al.*, 1989), an enzyme involved in processing poly-ubiquitin conjugates, are also part of that complex of axonal proteins. However, they did not detect in SCb or the other two main axonal transport components any of the >70-kDa hsps as reported by Tytell and Barbe (1987), so the presence of the higher molecular weight hsps in SCb remains controversial. None the less, the co-transport of hsc70, hsp70, ubiquitin and PGP 9.5 in axons makes it seem likely that these and perhaps other stress proteins, contribute jointly to the stabilization and or processing of many of the proteins of the axon.

There is one more particularly noteworthy implication of the segregation of hsp transport to SCb which relates to the potential contribution of the neuron soma to the stress response of its axon. If an axon that is more than a few mm in length is injured, the slow rate of transport of hsps means that it would take several days for increased amounts of hsps synthesized in the soma to reach the site of injury. That delay would be too long for the hsps to promote the survival of the damaged axon.

Therefore, for the hsp response to be of more than limited use in the protection of the nervous system from potentially lethal damage, it would be most efficient if additional hsps were supplied to an injured axon by the glial cells at the site of injury. In fact, glia to axon transfer of hsps has been shown to occur in the squid giant axon (Tytell *et al.*, 1986) and the medial giant axon of the crayfish (Sheller *et al.*, 1991). Thus, the possibility of cell-to-cell transfer of hsps must be considered in studies of the role of hsps in protection of the nervous system from damage.

4.7 Retinal heat shock protein changes with age and in a model of retinitis pigmentosa

In the retina *in vivo*, only a few studies have been conducted to evaluate how 70-kDa hsps distribution and content may be altered during degenerative changes. Age-related degeneration occurs in the retina as it does in other parts of the central nervous system. In the albino rat, there is a gradual decline in the number of photoreceptors with age (O'Steen *et al.*, 1987). Since in tissue culture models of cellular ageing, the capacity of the cells to elevate hsps in response to heat stress becomes impaired (Liu *et al.*, 1989, 1991), it is possible that the age-related photoreceptor loss could be, at least in part, a result of some disturbance to the hsp response system. To begin to examine that question, this author and coworkers compared the normal- and post-heat stress 70-kDa hsps distribution in three-month and 20-month Fischer-344 rats (Tytell *et al.*, 1993). Although no difference was detected in retinal 70-kDa hsps distribution in normal young and old rats, after they were made hyperthermic using the same procedure as described earlier, the old rats demonstrated an anomalous response. Relative to heat stressed young rats, the old rats showed small, but significant decreases in levels of 70-kDa hsps in the retinal layers that included the various parts of the photoreceptor cells, the outer segments, inner segments, outer nuclear layer, and outer plexiform layer. If this response is a product of age-related defects in the hsp response system, it may leave the photoreceptors in the aged retina more vulnerable to metabolic stresses that arise as a course of normal function (see section 4.1) and could contribute to the age-related loss of rods known to occur in rats (O'Steen *et al.*, 1974; Lai *et al.*, 1978), and humans (Gartner and Henkind, 1981; Gao and Hollyfield, 1992), and to the increased extent of photoreceptor degeneration in aged Fischer-344 rats subjected to chronic psychological stress (O'Steen *et al.*, 1987).

The relationship between the level of immunoreactive 70-kDa hsps and genetically based retinal degeneration has been investigated in a strain of albino rat known as the Royal College of Surgeons (RCS) rat. That animal has a genetic defect resembling human retinitis pigmentosa, in which there is progressive degeneration of the photoreceptors beginning in childhood and leading to blindness in early adulthood. The progress of the disease in the rat begins prior to sexual maturity and continues until most of the photoreceptors are lost by three months of age. The levels of 70-kDa hsps detected by immunohistochemistry in all retinal layers of these rats, except for the photoreceptor outer segments, were increased dramatically at 22 days of age, a time during which photoreceptor cell death was occurring at a maximal rate (Yamaguchi *et al.*, 1991). By 90 days of age, after the disease had run its course and most of the photoreceptors were gone, immunoreactive 70-kDa hsps in the remaining retinal cells were actually below that seen in age-matched Sprague-Dawley rats with normal retinas (*ibid*). A similar time course of increased immunoreactive hsp70 detected by Western blots of whole retinas from RCS rats was reported as well by Clarke *et al.* (1991).

4.8 Evidence for the protective effect of heat shock protein *in vivo*

In order to test in a straightforward manner whether an elevation in cellular 70-kDa hsp content can reduce the vulnerability of the cells to a potentially lethal metabolic stress, one needs a method to stimulate 70-kDa hsp synthesis without causing significant cell death. That criterion is best met by controlled hyperthermia, induced by placing the animal to be tested in an elevated temperature environment. Although this approach has been used in the unanaesthetized rat (Blake *et al.*, 1990), this author considers the pentobarbital-anaesthetized rat preferable because the animal does not experience the discomfort of the hot environment and the barbiturate anaesthesia suppresses the normal thermoregulatory responses of the rat so that its body temperature rises quickly in response to the heat treatment. It is also an advantage in a study of the potential protectiveness of elevated 70-kDa hsps to work with a tissue with cells in regular layers so that the loss of cells following injury can be related to the thinning of the layers. The retina has such a structural organization and provides one more advantage: in the albino, the photoreceptors are very sensitive to light damage, so that one can, in a noninvasive manner, produce a wide range of photoreceptor damage by simply exposing the animal to a known level of illuminance for a specific period of time.

This author and coworkers took advantage of the above assets of the Sprague–Dawley strain of albino rats to determine if prior hyperthermic induction of hsp synthesis and accumulation made photoreceptor cells more resistant to light damage (Barbe *et al.*, 1988). That study demonstrated that hyperthermia dramatically increased the number of photoreceptors that survived after the light exposure. In recent work (Tytell *et al.*, 1994), it was shown that the time of maximal protection of the photoreceptors against light damage, 18 h post-hyperthermia, correlated with the period during which both hsp70 and its mRNA were maximally elevated. This result strongly suggests that hsp70 can be cytoprotective in the retina, but it remains circumstantial. Also, the experiment cannot separate the contribution of hsp70 to cytoprotection from that of other hsps induced by hyperthermia or from the possibility that other hormonal and physiological changes that occurred during the treatment also were cytoprotective.

To show directly that 70-kDa hsps can protect photoreceptors from light damage *in vivo*, it is necessary to raise the levels of those proteins selectively in the retina. Initial work in this author's laboratory has focused on the possibility that exogenously administered 70-kDa hsps might be taken up by the retina in a functionally useful way. The reason for considering that possibility came from the observation that hsp70 is transferred from glia cells to the giant axon in the squid (Tytell *et al.*, 1986). A mixture of about 95% hsc70 and 5% hsp70 was purified from bovine brain [prepared by D.J. Gower (Department of Neurobiology and Anatomy, Bowman Gray School of Medicine) according to Schlossman *et al.*, 1984; Schmid and Rothman, 1985]. The mixture, referred to as 70-kDa hsps, was labelled with the fluorescent molecule, 7-amino-4-methyl coumarin 3-acetate succinimidyl ester (Molecular Probes, Inc., Eugene, OR). Upon injection into the vitreous of the rat eye, it appeared to be taken up by the retina (Tytell *et al.*, 1993). Additional investigations are planned to confirm that the fluorescent material detected in the retina was intact 70-kDa hsps.

The above encouraging, initial result, prompted other experiments to evaluate the sensitivity of photoreceptors to light damage after intravitreal injections of 2, 10, 15, or 20 μg of the purified 70-kDa hsps. Those doses of 70-kDa hsps, if taken up completely by the retina would significantly raise retinal levels of the proteins, since densitometric measurements of immunostained Western blots of retinal samples suggest that the adult rat retina contains about 4 μg of 70-kDa hsps (Tytell, unpublished observation). However, the efficiency of retinal uptake of the injected protein has not yet been determined. None the less, the eyes that received intravitreal injections of all but the lowest dose of the 70-kDa hsps four hours prior to the start of a 24-h exposure to 150–175 foot-candles of fluorescent light had significantly more surviving photoreceptors than control eyes that had received injections of saline alone or saline containing 20 μg of bovine serum albumin (Tytell *et al.*, 1993; Tytell, unpublished observations). The extent of the protective effect of intravitreally injected 70-kDa hsps was highly variable between eyes, probably as a consequence of the natural variation in the degree of light damage between eyes and the potential for the injected protein to be unevenly distributed throughout the eye (Tytell, unpublished observations). Initially, it appeared that only the 10 μg dose of the proteins was consistently protective and that the effect declined when 20 μg was administered. However, in repetitions of these experiments with more detailed analyses of the numbers of surviving photoreceptors, it was confirmed that all the doses of the 70-kDa hsps above 2 μg had significant cytoprotective effects, but there was no clear dose-response relationship (Tytell, unpublished observations). Work is in progress to determine the source of variability in the cytoprotective effect, the distribution within the various cells of the retina of the injected 70-kDa hsps, and whether the exogenously administered proteins function in the same manner as endogenously synthesized 70-kDa hsps.

Despite the variability in the protective effect of the injected 70-kDa hsps against photoreceptor light damage, it is clear that the proteins have a positive impact on cell stress tolerance. Furthermore, the cytoprotective activity of exogenous 70-kDa hsps is not unique to the retina, as other work has shown it to occur in cultured smooth muscle cells stressed by nutrient deprivation (Johnson *et al.*, 1990). These observations raise the intriguing possibility that 70-kDa hsps could be used therapeutically to reduce the extent of cell death in damaged neural and other tissues.

4.9 Acknowledgements

The author thanks T. C. Beasley (Department of Neurobiology and Anatomy, Bowman Gray School of Medicine) and A. J. Sweatt (Departments of Ophthalmology and Neurobiology and Anatomy, Bowman Gray School of Medicine) for providing the micrographs of 70-kDa hsps immunostained *Xenopus laevis* retina. Support for the author's work on the rat retina was provided by National Eye Institute grant EY-07616 and for that on the human retina by a grant from the Claude D. Pepper Older Americans Independence Centre.

References

Anderson, R.E. and Andrews, L.D. (1982) Biochemistry of photoreceptor membranes in vertebrates and invertebrates. In *Visual Cells in Evolution* (ed. Westfall, J.), pp. 1–22. New York, Raven Press.

Barbe, M.F. (1987) Heat shock protein production correlates with protection of the retina against light damage. *Ph.D. Thesis*, 1–126.
Barbe, M.F., Tytell, M., Gower, D.J. & Welch, W.J. (1988) Hyperthermia protects against light damage in the rat retina. *Science* **214**, 1817–1820.
Beasley, T. C., Sweatt, A.J. and Tytell, M. (1993) Heat shock protein 70 (hsp70) in *Xenopus laevis* retina. *Soc. Neurosci. Abst.* **19**, in press.
Beckmann, R.P., Mizzen, L.A. & Welch, W.J. (1990) Interaction of hsp70 with newly synthesized proteins: Implications for protein folding and assembly. *Science* **248**, 850–854.
Beckmann, R.P., Lovett, M. & Welch, W.J. (1992) Examining the function and regulation of hsp70 in cells subjected to metabolic stress. *J. Cell Biol.* **117**, 1137–1150.
Bizzi, A., Schaetzle, B., Patton, A., Gambetti, P. & Autilio-Gambetti, L. (1991) Axonal transport of two major components of the ubiquitin system: free ubiquitin and ubiquitin carboxyl-terminal hydrolase PGP 9.5. *Brain Res.* **548**, 292–299.
Black, M.M., Chestnut, M.H., Pleasure, I.T. & Keen, J.H. (1991) Stable clathrin: uncoating protein (hsc70) complexes in intact neurons and their axonal transport. *J. Neurosci.* **11**, 1163–1172.
Blake, M.J., Nowak, T.S., Jr & Holbrook, N.J. (1990) *In vivo* hyperthermia induces expression of hsp70 mRNA in brain regions controlling the neuroendocrine response to stress. *Mol. Brain Res.* **8**, 89–92.
Brady, S.T. & Lasek, R.J. (1982) Axonal transport: a cell biological method for studying proteins that associate with the cytoskeleton. In *Methods in Cell Biology, vol. 25B*, pp. 365–398.
Brown, I.R., Heikkila, J.J. & Cosgrove, J.W. (1982) Analysis of protein synthesis in the mammalian brain using LSD and hyperthermia as experimental probes. In *Molecular Approaches to Neurobiology* (ed. Brown, I.R.), pp 221–253. New York. Academic Press.
Brown, I.R. (1983) Hyperthermia induces the synthesis of a heat shock protein by polysomes isolated from faetal and neonatal mammalian brain. *J. Neurochem.* **40**, 1490–1493.
Brown, I.R., Lowe, D.G. & Moran, L.A. (1985) Expression of heat shock genes in faetal and maternal rabbit brain. *Neurochem. Res.* **10**, 1277–1284.
Chirico, W.J., Waters, M.G. & Blobel, G. (1988) 70K heat shock related proteins stimulate protein translocation into microsomes. *Nature* **332**, 805–810.
Clark, B.D. and Brown, I.R. (1982) Protein synthesis in the mammalian retina following the intravenous administration of LSD. *Brain Res.* **247**, 97–104.
Clark, B.D. & Brown, I.R. (1985) Axonal transport of a heat shock protein in the rabbit visual system. *Proc. Natl Acad. Sci. USA* **82**, 1281–1285.
Clark, B.D. & Brown, I.R. (1986) Induction of a heat shock protein in the isolated mammalian retina. *Neurochem. Res.* **11**, 269–279.
Clark, B.D. & Brown, I.R. (1987) Altered expression of a heat shock protein in the mammalian nervous system in the presence of agents which affect microtubule stability. *Neurochem. Res.* **12**, 819–823.
Clarke, I.S., Dzialoszynski, T., Sanford, S.E. & Trevithick, J.R. (1991) A possible relationship between cataract, increased levels of the major heat shock protein hsp70 and decreased levels of S-antigen in the retina of the RCS rat. *Exp. Eye Res.* **53**, 545–548.
Cosgrove, J.W. and Brown, I.R. (1983) Heat shock protein in mammalian brain and other organs after a physiologically relevant increase in body temperature induced by *D*-lysergic acid diethylamide. *Proc. Natl Acad. Sci. USA* **80**, 569–573.
De La Paz, M.A. & Anderson, R.E. (1992) Lipid peroxidation in rod outer segments. *Invest. Ophthalmol. Vis. Sci.* **33**, 2091–2096.
de Waegh, S. & Brady, S.T. (1989) Axonal transport of a clathrin uncoating ATPase (HSC70): a role for HSC70 in the modulation of coated vesicle assembly *in vivo*. *J. Neurosci. Res.* **23**, 433–44.
Deshaies, R.J., Koch, B.D., Werner-Washburne, M., Craig, E.A. & Schekman, R. (1988) A subfamily of stress proteins facilitates translocation of secretory and mitochondrial precursor polypeptides. *Nature* **332**, 800–805.

Dowling, J.E. (1987) *The Retina – an Approachable Part of the Brain*, Cambridge, MA, Belknap Press of Harvard University Press.

Gao, H. & Hollyfield, J.G. (1992) Aging of the human retina. *Invest. Ophthalmol. Vis. Sci.* **33**, 1–17.

Gartner, S. & Henkind, P. (1981) Aging and degeneration of the human macula. 1. Outer nuclear layer and photoreceptors. *Br. J. Ophthalmol.* **65**, 23–28.

Grafstein, B. & Forman, D.S. (1980) Intracellular transport in neurons. *Physiol. Rev.* **60**, 1167–1283.

Johnson, A.D., Berberian, P.A. & Bond, M.G. (1990) Effect of heat shock proteins on survival of isolated aortic cells from normal and atherosclerotic cynomolgus macaques. *Atheroscl.* **84**, 111–119.

Kerendian, J., Enomoto, H. & Wong, C.G. (1992) Induction of stress proteins in SV-40 transformed human RPE-derived cells by organic oxidants. *Curr. Eye Res.* **11**, 385–396.

Kitano, S. & Caprioli, J. (1993) Expression of 72-kDa heat shock protein in cultured rat retinal ganglion cells. Ischemic tolerance and protective effect of hyperthermia. *Invest. Ophthalmol. Vis. Sci.* **34**, 1430

Lai, Y.-L., Jacoby, R.O. & Jonas, A.M. (1978) Age-related and light-associated retinal changes in Fischer rats. *Invest. Ophthalmol. Vis. Sci.* **17**, 634–638.

Lasek, R.J. (1982) Translocation of the neuronal cytoskeleton and axonal locomotion. *Phil. Trans. R. Soc. Lond.* **B299**, 313–327.

Lasek, R.J., Garner, J.A. & Brady, S.T. (1984) Axonal transport of the cytoplasmic matrix. *J. Cell Biol.* **99**, 212s–221s.

LaThangue, N.B. (1984) A major heat shock protein defined by monoclonal antibodies. *EMBO J.* **3**, 1871–1879.

LaThangue, N.B., Shriver, K., Dawson, C. & Chan, W.L. (1984) *Herpes simplex* virus infection causes the accumulation of a heat shock protein. *EMBO J.* **3**, 267–277.

LaThangue, N.B. & Chan, W.L. (1984) The characterization and purification of DNA-binding proteins present within *Herpes simplex* virus infected cells using monoclonal antibodies. *Arch. Virol.* **79**, 13–33.

Linsenmeier, R.A. (1986) Effects of light and darkness on oxygen distribution and consumption in the cat retina. *J. Gen. Physiol.* **88**, 521–542.

Liu, A.Y.-C., Lin, Z., Choi, H.-S., Sorhage, F. & Li, B. (1989) Attenuated induction of heat shock gene expression in aging diploid fibroblasts. *J. Biol. Chem.* **264**, 12 037–12 045.

Liu, A.Y.-C., Choi, H.S., Lee, Y.K. & Chen, K.Y. (1991) Molecular events involved in transcriptional activation of heat shock genes become progressively refractory to heat stimulation during aging of human diploid fibroblasts. *J. Cell. Physiol.* **149**, 560–566.

Masing, T.E., Rush, S.J. & Brown, I.R. (1990) Induction of heat shock gene (hsp 70) in rabbit retinal ganglion cells detected by *in situ* hybridization with plastic-embedded tissue. *Neurochem. Res.* **15**, 1229–1235.

Mitchell, D., Ibrahim, S. & Gusterson, B.A. (1985) Improved immunohistochemical localization of tissue antigens using modified Methacarn fixation. *J. Histochem. Cytochem.* **33**, 491–495.

Nakai, A., Satoh, M., Hirayoshi, K. & Nagata, K. (1992) Involvement of the stress protein HSP47 in procollagen processing in the endoplasmic reticulum. *J. Cell Biol.* **117**, 903–914.

Nakai, A. & Morimoto, R.I. (1993) Characterization of a novel chicken heat shock transcription factor, heat shock factor 3, suggest a new regulatory pathway. *Mol. Cell. Biol.* **13**, 1983–1997.

Newman, E.A. (1986) The Müller Cell. In *Astrocytes* (eds. Federoff, S. and Vernadakis, A.), vol. 1, pp 149–171. New York, Academic Press.

O'Steen, W.K., Anderson, K.V. & Shear, C.R. (1974) Photoreceptor degeneration in albino rats: dependency on age. *Invest. Ophthalmol.* **13**, 334–339.

O'Steen, W.K., Sweatt, A.J. & Brodish, A. (1987) Effects of acute and chronic stresss on the neuronal retina of young, mid-age and aged Fischer-344 rats. *Brain Res.* **426**, 37–46.

Organisciak, D.T., Favreau, P. & Wang, H.M. (1983) The enzymatic estimation of organic hydroperoxides in the rat retina. *Exp. Eye Res.* **36**, 337–349.

Schlossman, D.M., Schmid, S.L., Braell, W.A. & Rothman, J.E. (1984) An enzyme that removes clathrin coats: purification of an uncoating-ATPase. *J. Cell Biol.* **99**, 723–733.

Schmid, S.L. & Rothman, J.E. (1985) Two classes of binding sites for uncoating protein in clathrin triskelions. *J. Biol. Chem.* **260**, 10050–10056.

Sheller, R.A., Grossfeld, R.M. & Bittner, G.D. (1991) A possible role for glia and stress proteins in the maintenance of axoplasmic proteins. *Soc. Neurosci. Abstr.* **17**, 376.

Takechi, H., Hirayoshi, K., Nakai, A., Kudo, H., Saga, S. & Nagata, K. (1992) Molecular cloning of a mouse 47-kDa heat shock protein (HSP47), a collagen-binding stress protein, and its expression during the differentiation of F9 teratocarcinoma cells. *Eur. J. Biochem.* **206**, 323–329.

Tytell, M., Greenberg, S.G. & Lasek, R.J. (1986) Heat shock-like protein is transferred from glia to axon. *Brain Res.* **363**, 161–164.

Tytell, M. & Barbe, M.F. (1987) Synthesis and axonal transport of heat shock proteins. In *Neurology and Neurobiology vol 25 – Axonal Transport* (eds Smith, R.S. & Bisby, M.A.), pp 473–492. New York, Alan R. Liss, Inc.

Tytell, M., Barbe, M.F. & Gower, D.J. (1989) Photoreceptor protection from light damage by hyperthermia. In *Inherited and Enviromentally Induced Retinal Degenerations* (eds LaVail, M.M., Anderson, R.E. & Hollyfield, J.G.), pp 523–538. New York, Alan R. Liss, Inc.

Tytell, M., Barbe, M.F. & Brown, I.R. (1993) Stress (heat shock) protein accumulation in the central nervous system – Its relationship to cell stress and damage. In *Advances in Neurology* (ed. Seil, F.J.), pp 293–303. New York, Raven Press, Ltd.

Tytell, M., Barbe, M.F. & Brown, I.R. (1994) Induction of heat shock (stress) protein 70 and its mRNA in the normal and light-damaged rat retina after whole body hyperthermia. *J. Neurosci. Res.* (in press).

Tytell, M., Yamaguchi, K. & Yamaguchi, K. (1993) Role of heat shock protein 70 (hsp70) in photoreceptor cell survival in the aged rat. In *Retinal Degeneration: Clinical and Laboratory Applications* (eds. Hollyfield, J.G., LaVail, M.M. and Anderson, R.E.), New York, Plenum Publishing Corp. (in press).

Wakakura, M., Kennedy, P.G.E., Foulds, W.S. & Clements, G.B (1987) Stress proteins accumulate in cultured retinal glial cells during herpes simplex viral infection. *Exp. Eye Res.* **45**, 557–567.

Welch, W.J. & Suhan, J.P. (1986) Cellular and biochemical events in mammalian cells during and after recovery from physiological stress. *J. Cell Biol.* **103**, 2035–2052.

Wiegand, R.D., Giusto, N.M., Rapp, L.M. & Anderson, R.E. (1983) Evidence for rod outer segment lipid peroxidation following constant illumination of the rat retina. *Invest. Ophthalmol. Vis. Sci.* **24**, 1433–1435.

Wilkinson, K.D., Lee, K., Deshpande, S., Duerksen-Hughes, P., Boss, J.M. & Pohl, J. (1989) The neuron specific protein PGP 9.5 is a ubiquitin carboxyl-terminal hydrolase. *Science* **246**, 670–673.

Wong, C.G. & Lin, N. (1989) Induction of stress proteins in cultured human RPE-derived cells. *Curr. Eye Res.* **8**, 537–545.

Yamaguchi, K., Gaur, V.P., Tytell, M., Hollman, C.R. & Turner, J.E. (1991) Ocular distribution of 70-kDa heat shock protein in rats with normal and dystrophic retinas. *Cell Tissue Res.* **264**, 497–506.

CHAPTER 5

HEAT SHOCK PROTEINS AND NEUROPROTECTION IN CNS CULTURE

Barney E. Dwyer,* and Robert N. Nishimura†

The Molecular Neurobiology Laboratory and the* in vitro *Remyelination Laboratory,† Veterans Affairs Medical Center, Sepulveda, CA 91343 and the Department of Neurology*† and the Brain Research Institute*, UCLA School of Medicine, Los Angeles, California*

Table of Contents

5.1 Introduction

The phenomenon of acquired thermotolerance and the association of heat shock protein synthesis with this phenomenon raises the possibility that a cell's response to injury might be manipulated experimentally. Several intriguing studies in central nervous system (CNS) tissue showed that a 'priming' stress and/or synthesis of hsp70 was associated with neuroprotection *in vivo* and *in vitro* (Barbe *et al.*, 1988; Chopp *et al.*, 1989; Khan and Sotelo, 1989; Kitagawa *et al.*, 1990, 1991; Kirino, 1991; Lowenstein *et al.*, 1991; Nishimura *et al.*, 1991b; Rordorf *et al.*, 1991). The focus of this chapter is the 70-kDa family of heat shock proteins: constitutively synthesized hsc70 and stress-inducible hsp70. Several questions which we consider particularly germane for establishing a specific role for heat shock proteins in neuroprotection were raised in a previous review (Dwyer and Nishimura, 1992a). In the course of this chapter we will show how we have attempted to address these questions in our laboratory and comment on efforts to demonstrate specific effects of heat shock proteins in cell injury.

Heat Shock Proteins in the Nervous System
ISBN 0-12-480960-X

5.2 Inducible hsp70 proteins

It is imperative that efforts to identify a specific role for 70-kDa heat shock proteins in CNS injury must start with identification of the proteins involved. This is not necessarily straightforward and species differences may exist. A single inducible member of the 70-kDa heat shock protein family, dnaK, has been described in *E. coli* (Georgopoulos *et al.*, 1990). At least two closely related hsp70 genes have been found in *Drosophila melanogaster* as well as several non-inducible hsc70 'cognates' (Pauli and Tissieres, 1990). At least nine members of the 70-kDa heat shock protein family have been characterized in yeast cells, several of which are inducible (Craig, 1990). These observations suggest a very relevant question to those using cell culture or *in vivo* animal models to study the mechanism of neuronal injury and neuroprotection: How many inducible hsp70 proteins are expressed? At least three inducible members of the 70-kDa heat shock protein family have been described in human cells (Wu *et al.*, 1985; Voellmy *et al.*, 1985; Leung *et al.*, 1990). In addition, three human MHC-linked hsp70 genes have been identified (Milner and Campbell, 1990). The expression of three closely related mouse hsp70 genes was reported by Lowe and Moran (1986), as well as several weakly inducible mouse 70-kDa family members (Zakeri *et al.*, 1988; Matsumoto *et al.*, 1990; Anderson *et al.*, 1989). The localization of the latter in germ cells or in tumor cells raises the possibility that cell specific differences in hsp70 expression within a species may be expected. A new member of the mouse 70-kDa heat shock protein family, a 66-kDa protein designated $p66^{mot\text{-}1}$, which possibly contributes to regulation of cell division, was recently described (Wadhwa *et al.*, 1993). In rat, less is known about the number of members of the 70-kDa heat shock protein family. This constitutes an important gap in knowledge because rat brain cell cultures and rat models of neuronal injury *in vivo* are frequently used to investigate mechanisms of brain injury including the protective effects of 'priming' stress and heat shock proteins. The question posed earlier should perhaps be rephrased: If there are several inducible hsp70 proteins in rat brain cells, how many are there? What cells are they expressed in? Can they be distinguished? And, what criteria can be used to identify them?

5.3 Identification of hsp70 in brain cells

Hsp70 mRNA. We detected two heat-inducible transcripts in primary cultures of rat astrocytes but only one in a transformed rat astroglial cell line (Nishimura *et al.*, 1988b, 1992a). This is consistent with two genes or with a single gene containing different lengths of 3′-non-coding sequences. This pattern of hybridization is reproduced here using probes based on rat and human sequences, suggesting that species-specific sequence differences in the probes used do not explain our observation (Figure 1).

Direct sequence information is not yet available in our laboratory to resolve this question but there is evidence that more than one inducible rat hsp70 gene exists. Mestril *et al.* (submitted manuscript) reported that two inducible hsp70 transcripts in heat shocked rat heart could be distinguished in Northern blots. We were unable to confirm this in CNS cells. A probe containing mostly 3′-untranslated sequences from their rat inducible gene which was reported to detect only the smaller transcript in heat shocked rat hearts (Mestril *et al.*) hybridized to both heat-inducible transcripts in rat astrocytes (Figure 1). However, evidence in favour of multiple inducible hsp70 proteins has been obtained from ongoing PCR studies (Frank Sharp, personal communication). In addition, identification of a rat testes specific hsp70 (Wisniewski *et al.*, 1990) raises the possibility that several, tissue or cell specific hsp70 proteins may exist.

Detecting hsp70 mRNA, even if several genes are expressed, should pose little problem since sequence specific probes can be made; however, probe specificity should always be verified. Figure 1e clearly shows that under reduced stringency a probe containing human hsp70 sequence detects several bands in addition to hsp70 mRNA in both neurons and astroglia. Probe specificity is a particularly important consideration for *in situ* hybridization. It should be verified under conditions approximating the stringency of the *in situ* hybridization. Lastly, it is important to remember that for the question of hsp70 induction and neuroprotection the protein end-product is of most interest. Although sequence specific probes appear capable of detecting hsp70 mRNA under appropriate conditions it is possible that hsp70 mRNA can be induced with little or no synthesis of hsp70 protein (Dwyer *et al.*, 1991; Nowak, 1991; Kawagoe *et al.*, 1992).

Hsp70 protein: Evidence for synthesis of stress-induced hsp70 has generally been obtained from one or more of the following techniques: migration of radiolabelled proteins on 1- and 2-dimensional electrophoretic gels followed by autoradiography and Western blotting or immunocytochemistry. Identification of hsp70 solely by the appearance of an inducible protein on gel electrophoresis is inadequate. Estimated molecular size for hsp70 has ranged from 64-kDa (Barbe *et al.*, 1988) to 68-kDa in our hands, to 74-kDa (Cosgrove and Brown, 1983). Many other proteins have been reported to migrate with a molecular size around 68-kDa including p68, a Ca^{2+}-binding protein (Moss *et al.*, 1988), nuclear protein p68, an RNA-dependent ATPase (Iggo and Lane, 1989), and a 68-kDa inhibitor of calcium-activated neutral proteinase (Banik *et al.*, 1990). In our opinion Western blot analysis with specific antibodies must be used to verify the identity of hsp70. Figure 2 shows the pattern of antibody staining of heat shocked rat cortical neurons, rat forebrain astrocytes, and spontaneously transformed rat astroglial cells, using several commercially available monoclonal antibodies.

Resolution of hsp70 and hsc70 on one-dimensional gels should be sufficient to allow detection of hsp70 and hsc70 on Western blots. Hsp70 in astrocytes and transformed astroglial cells was detected by all the antibodies tested but only C-92 was selective for hsp70. Weak staining of neuronal hsp70 by C-92 confirmed our

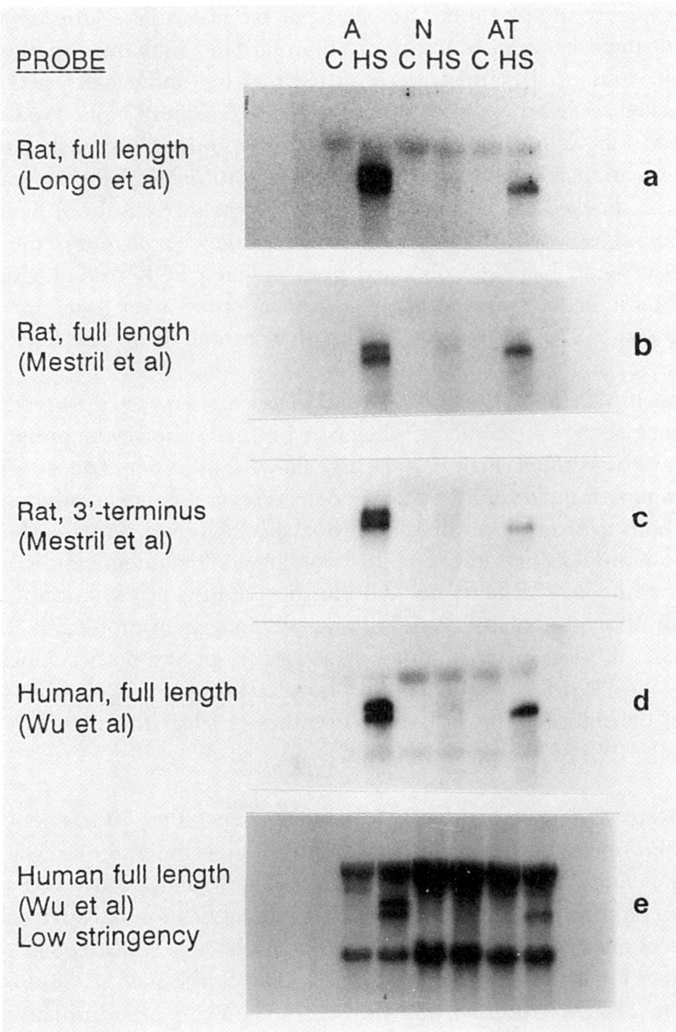

Figure 1 Northern blot of control (C) and heat shocked (HS) cultured cortical neurons, forebrain astrocytes and spontaneously transformed rat astroglial cells. Culture conditions for neurons (N), astrocytes (A), and transformed astrocytes (AT) were previously described (Nishimura *et al.*, 1991a, 1992a). The transformed rat astroglial cell line was derived from a subpopulation of primary rat-forebrain astrocytes which spontaneously transformed in culture (Bressler and de Vellis, 1985). Heat shock was at 45°C, 20 min followed by 3 h of recovery at 37°C. Total RNA from astrocytes (3 μg), ATs (3 μg), and neurons (10 μg) was extracted and Northern blots were run as previously described (Nishimura *et al.*, 1991a). From the top, the probe sequences used were: (a) a full length rat hsp70 cDNA sequence in pBluescript SK (Longo *et al.*, 1993); (b) full length rat cDNA sequence in pBluescript SK$^+$ (pG12 A/D Bluescript SK – Mestril *et al.*); (c) a probe containing largely 3′ untranslated region of the rat inducible hsp70 cDNA – Mestril *et al.*; (d) and (e) full-length human hsp70 cDNA sequence subcloned from plasmid pH 2.3

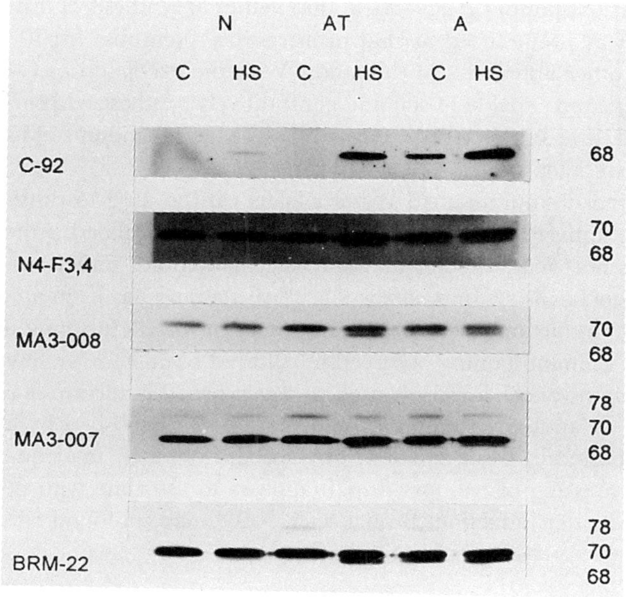

Figure 2 Immunoblot analysis of 70-kDa heat shock proteins control (C) and heat shocked (HS) cultured cortical neurons (N), forebrain astrocytes (A), and spontaneously transformed rat astroglial cells (AT). Protein electrophoresis and Western blotting was performed as previously described (Nishimura *et al.*, 1988b) except that the highly sensitive ECL detection system (Amersham) was used. Each lane contains 20 µg of total cellular protein. Cells were heat shocked at 45°C for 20 min and allowed to recover at 37°C for 4 h. Inducible hsp70 (68-kDa), constitutively synthesized hsc70 (70-kDa), and glucose-regulated protein GRP-78 (78-kDa) are indicated in the right margin. Monoclonal antibodies used in this study are indicated in the left margin: C-92 (W. Welch, UCSF, San Francisco, CA), N4-F3,4 (W. Welch), MA3-008 and MA3-007 (Affinity Bioreagents) and BRM-22 (Sigma Chemical Company). Differential immunostaining of the 70-kDa heat shock proteins in neurons, astrocytes, and transformed astroglial cells with additional antibodies was previously reported (Nishimura *et al.*, 1991a, 1992a). In those studies we experienced difficulty in detecting hsp70 with C-92 on immunoblots of neurons and transformed astroglial cell using the Vectastain ABC kit. Our ability to stain hsp70 with C-92 in this study probably reflects the increased sensitivity (estimated at 100-times) of the ECL system.

(Wu *et al.*, 1985) into pT7/T3–19. ^{32}P-labelled riboprobes were prepared from plasmids and Northern blots (a–d) were hybridized at 65°C and washed as previously described (Nishimura *et al.*, 1991a). Blot (e) was hybridized at low stringency (50°C) and RNase A treatment was omitted in the wash step. Typically, two heat-inducible transcripts have been found in rat. Size estimates ranging from 2.5 and 2.8 kb in size to 2.8 and 3.1 for the doublet probably reflect conditions of electrophoresis. Species differences may exist; only a single stress-inducible transcript has been reported in rabbit (Brown and Rush, 1990) and gerbil (Kawagoe *et al.*, 1992). Hybridization to any of the probes was barely detectable in heat shocked neurons, consistent with our previous conclusion that cultured cortical neurons mount a very weak heat shock response under the conditions used here. Lastly only the smaller of the two transcripts is detected irrespective of probe in transformed rat astroglial cells.

previous report (Nishimura *et al.*, 1991a) that neuronal synthesis of hsp70 after heat shock was not as pronounced as that of astrocytes. Neuronal hsp70 was not detected by the other antibodies in this study. With the exception of C-92 all of the antibodies appeared capable of staining constitutively synthesized hsc70. It is noteworthy that all three brain-cell types have relatively large amounts of hsc70, a point to which we will later refer.

It is clear that immunostained Western blots can be used to distinguish hsp70 from the constitutively synthesized hsc70 and glucose-regulated protein, GRP78. However, it is not clear whether antibody specificity holds under conditions used for immunocytochemistry. Immunocytochemical staining has been a common way to detect hsp70 induction *in vivo* after brain injury or after a 'priming' stress. Based on patterns of immunostaining observed in cultured brain cells we have concluded that when immunocytochemistry is used to detect hsp70 induction *in vitro* or *in vivo* the specificity of antibody staining should always be established by Western blot analysis. Furthermore, tissue fixation, non-specific staining, inadequate blocking, and potential masking of epitopes (hsp70 appears to associate with other proteins including hsc70 after induction: Brown *et al.*, 1993) may confound efforts to establish the presence or absence of hsp70 in tissue.

5.4 Hsp70 induction in cultured neuronal and glial cells

Many types of stress induce hsp70: Patterns of hsp70 induction in cultured CNS cells are briefly summarized here.

Glial cells: Supraphysiological temperature is a strong inducer of hsp70 in cultured forebrain astrocytes (Nishimura *et al.*, 1988b; Satoh *et al.*, 1992), cerebellar astrocytes (Marini *et al.*, 1990) and oligodendroglia (Nishimura *et al.*, 1988b; Satoh *et al.*, 1992). Hsp70 can also be induced in astrocytes by reduced extracellular pH (Nishimura *et al.*, 1989) and oxidative stress (micromolar concentrations of hydrogen peroxide) although less dramatically (Nishimura *et al.*, 1988a). The physiologically important heavy metal zinc induced hsp70 in cultured astrocytes (Swanson and Sharp, 1992; Nishimura *et al.*, unpublished) as did micromolar amounts of the toxic heavy metal lead (results not shown) although hsp70 induction was weak. In contrast, heme oxygenase (recently christened hsp32 – Ewing *et al.*, 1992) was more strongly induced than hsp70 in astrocytes exposed to hydrogen peroxide (Nishimura *et al.*, 1988a; Dwyer *et al.*, 1992) zinc and lead (unpublished) suggesting that part of the pathophysiology of zinc and lead toxicity may be oxidative stress.

Some treatments can modulate the heat shock response in brain cells. We have found that 1,10-phenanthroline (1,10-PA), an iron chelator and DNA intercalating agent, could promote the induction of hsp70 synthesis in astrocytes exposed to an extracellular pH (pH 6.0) which normally does not result in hsp70 induction (Nishimura and Dwyer, 1993). The use of a second, chemically different iron chelator, deferoximine did not result in hsp70 induction suggesting that the action of 1,10-PA may be related to its DNA intercalating properties.

Neurons: Cultured cortical neurons did not synthesize hsp70 as vigorously as astrocytes within 3 h after heat shock by the criteria of gel electrophoresis of labelled proteins, immunostained Western blots, and Northern blot analysis (Nishimura *et al.*, 1991a). In a recent report, Marcuccilli *et al.* (1992) failed to detect heat-induced hsp70 synthesis in cultured hippocampal neurons, acutely dissociated superior cervical ganglion neurons and cultured cerebellar granule cells by immunocytochemistry and Western blotting. This is consistent with the observation that hyperthermia *in vivo* did not result in marked expression of hsp70 transcripts in rabbit cortical and hippocampal neurons as assessed by *in situ* hybridization (Sprang and Brown, 1987) [hsc70 transcript was detectable in neuronal populations]. Furthermore, Li *et al.* (1992b) were unable to detect hsp70 in cerebral cortex after *in vivo* hyperthermia by immunocytochemistry. [Note: forebrain neurons in brain structures associated with the neuroendocrine response to stress including the paraventricular and dorsomedial hypothalamic nuclei, the median eminence, and the medial habenula appear to be an exception (Blake *et al.*, 1990, Li *et al.*, 1992b).]

The predominantly glial pattern of hsp70 induction in forebrain after *in vivo* hyperthermia has recently been extended to rabbit spinal cord (Manzerra and Brown, 1992a). The response of cerebellar neurons to hyperthermia is different in some respects. It is also noteworthy because of *in vivo/in vitro* differences. Marini *et al.* (1990) reported strong induction of hsp70 *in vivo* in the cerebellar granule cell layer. The implication was that granule cell neurons synthesized hsp70, however, abundant astrocyte cell bodies and their processes were shown to be present in this region by glial fibrillary acidic protein (GFAP) immunostaining. Hsp70 mRNA induction was detected by *in situ* hybridization in rabbit cerebellar granule cells and Bergmann glia after hyperthermia (Sprang and Brown, 1987, Manzerra and Brown, 1992b). Li *et al.* (1992b) also found strong induction of hsp70 in rat cerebellar granule cells by immunocytochemistry. In contrast, Marini *et al.* (1990) were unable to detect hsp70 synthesis after heat shock in cultured cerebellar granule neurons, although hsp70 was clearly detected in cultured cerebellar astrocytes. More recently, cultured cerebellar granule cells were shown to be capable of hsp70 synthesis (Lowenstein *et al.*, 1991). Astrocyte contamination of granule cell cultures reported at about 5% (Marini *et al.*, 1990) and 3% (Lowenstein *et al.*, 1991) was not likely the cause for the difference, but results suggest that culture conditions can influence heat shock protein induction.

In contrast to hyperthermia, a predominantly neuronal pattern of hsp70 induction, *in vivo*, has been reported in forebrain after global and focal ischaemia (Chopp *et al.*, 1991; Gonzalez *et al.*, 1989; Nowak, 1991; Sharp *et al.*, 1991b; Simon *et al.*, 1991; Vass *et al.*, 1988; Kawagoe *et al.*, 1992; Li *et al.*, 1992a; Kinouchi *et al.*, 1993), neonatal hypoxia-ischaemia (Ferriero *et al.*, 1990; Blumenfeld *et al.*, 1992), seizures (Gonzalez *et al.*, 1989; Lowenstein *et al.*, 1990; Vass *et al.*, 1989), afferent stimulation (Sloviter and Lowenstein, 1992), exposure to NMDA receptor antagonists MK-

801 and ketamine (Olney *et al.*, 1991; Sharp *et al.*, 1991a), and spinal cord compression injury (Gower *et al.*, 1989). After penetrating brain injury (Brown *et al.*, 1989), and percussive brain injury (Tanno *et al.*, 1993), hsp70 induction was prominent in both neurons and glia.

We have failed to see significant *in vitro* neuronal induction of hsp70 within three hours of exposure to non-thermal stress including low pH, low pH combined with 1–10 phenanthroline, and oxidative stress induced by micromolar amounts of hydrogen peroxide (unpublished). These conditions were selected to approximate biochemical changes associated with ischaemic, traumatic, and toxic neuronal injury *in vivo*. For example, cerebral ischaemia may result in oxidative injury, cellular acidosis, and calcium overload from membrane injury or excitotoxic amino acid release (recent review: Schurr and Rigor, 1992). The *in vitro/in vivo* difference is a significant discrepancy and requires explanation. To address this point we hypothesized that conditions found *in vivo* during injury could promote neuronal hsp70 synthesis. One of these potential factors is locally released prostaglandins in injured tissue. However, preliminary studies failed to establish altered neuronal hsp70 synthesis in the presence of prostaglandins (Nishimura *et al.*, 1992b).

Regulation of the heat shock response may be fundamentally different in neurons and astrocytes. Preliminary results by Marcuccilli *et al.* (1993, in press) shows a difference in the distribution of heat shock transcription factors between neurons and glia in culture. Culture conditions also may influence the heat shock response. Possibly, the signal transduction pathway activating the heat shock response is less active or impaired in our cultured cortical neurons although the results shown in Figure 3 suggest this is unlikely. In this experiment, neurons transiently transfected with a plasmid containing human hsp70 with an intact promoter were able to induce significant amounts of hsp70 mRNA in comparison with nontransfected neurons.

Perhaps, cultured cortical neurons are a good model of hyperthermia *in vivo* and less useful for studying the neuronal heat shock response after ischaemia or trauma. In this regard Lowenstein *et al.* (1991) failed to detect hsp70 synthesis in cultured cerebellar neurons exposed to the 'excitotoxic' amino acid L-glutamate. [L-glutamate was not tested as an inducer of heat shock proteins on our cortical neurons since we have found that glutamate (1 mM, 24 h) did not result in significant neurotoxicity]. However, it should be kept in mind that reports of striking inductions of hsp70 in many of the neuronal injury models cited above were solely immunocytochemistry studies and the specificity of the antibody staining was not verified by Western blot analysis.

5.5 Heat shock proteins and cell protection

At the beginning we listed several papers which showed that a 'priming' stress was associated with mitigated brain injury when intact animals were subject to a

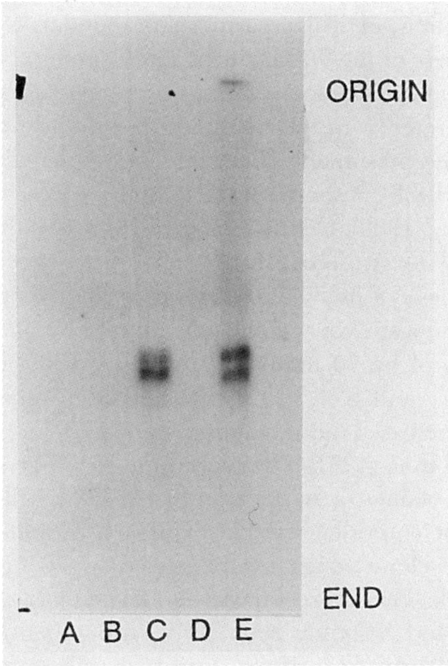

Figure 3 Expression of hsp70 in Lipofectin-transfected neurons. Cultured cortical neurons were transfected with plasmid pHhsp70 containing the 5′-promoter and entire coding region of human hsp70 (kindly provided by Dr R. I. Morimoto) using the Lipofectin procedure (Gibco-BRL). Neurons were switched back to their original medium after 5 h and returned to their incubator for 16 h. Neurons were switched to serum-free DMEM/F-12 medium and heat shocked (43°C, 15 min). After a three-hour recovery at 37°C total cellular RNA was extracted and Northern blots were prepared and hybridized with a human hsp70 riboprobe as previously described (Nishimura *et al.*, 1991a); five μg total RNA were applied to each lane. (A) control neurons; (B) heat shocked neurons; (C) heat shocked, transfected neurons. For comparison: (D) control astrocytes; (E) heat shocked astrocytes.

subsequent damaging injury. The implication from many of these studies was that the 'priming' stress induced heat shock proteins and that their presence contributed to neuroprotection. In practice it is extremely difficult to dissociate non-specific protective effects of hyperthermia (or ischaemia) which are potentially very important in their own right from a specific protective effect of heat shock proteins *in vivo*. For example, 'priming' hyperthermia was reported to reduce ischaemia-induced acidosis in brain (Chopp *et al.*, 1989). A recent commentary suggests the importance of protein kinase C-linked receptors, like the A_1 adenosine receptor, in ischaemic preconditioning in heart muscle (Cohen and Downey, 1993).

Even in cell culture systems it has been difficult to isolate the effects of individual heat shock proteins from non-specific effects. The most direct evidence is in non-neural cells. Experimental strategies to inhibit hsp70 synthesis, interfere with hsp70 and hsc70 function, or to increase hsp70 abundance generally support the hypothesis that members of the 70-kDa heat shock protein family can protect cells under some experimental conditions. Thus, micro-injection of rat fibroblasts with antibodies which interfered with the normal heat-induced nuclear translocation of the 70-kDa proteins and presumably their function rendered these cells thermosensitive (Riabowol *et al.*, 1988). A more specific protective effect for hsp70, was shown by Johnston and Kucey (1988). Cells which over-expressed the regulatory region of the hsp70 gene failed to synthesize hsp70 after heat shock and were rendered thermosensitive. Synthesis of hsc70 and a low molecular weight heat shock protein (approximately 27-kDa) were not significantly affected by the cell selection process. Direct micro-injection of hsp70 into CHO cells (Li, 1989) or hsp70 mRNA into mouse oocytes (Hendry and Kola, 1991) has also been reported to increase their thermal resistance. Increased accumulation of hsp70 in non-stressed cells was achieved by Angelides *et al.* (1991) who placed the hsp70 gene behind a human B-actin promoter and obtained a stably transfected CV1 cell-line which expressed hsp70 to a level about equivalent to that expressed in non-transfected CV1 cells after heat shock. The clone which overexpressed hsp70 demonstrated increased resistance to heat shock. This approach was also taken by Li *et al.* (1991) using transfected rat fibroblasts and Solomon *et al.* (1991) using *Drosophila* cells with similar results. As further evidence of specificity, *Drosophila* cells expressing antisense RNA achieved thermal protection at a slower rate (Solomon *et al.*, 1991).

Interestingly C-terminal mutations in the inducible heat shock protein localized to the region believed responsible for nucleolar localization interfered with the development of thermal resistance whereas N-terminal mutations localized in the ATP-binding domain did not (Solomon *et al.*, 1991; Li *et al.*, 1992c). There are exceptions and some cells seem to be able to acquire thermotolerance in the absence of protein synthesis (Lee and Dewey, 1986), in the absence of general heat shock protein induction (Smith and Yaffe, 1991), and in the specific absence of hsp70 synthesis (Aujame and Firko, 1988). The reasons for this are not understood in part because the mechanism/s involved in acquired thermotolerance are not understood. Perhaps hsc70 is available in sufficient amounts during the development of thermotolerance to replace uninduced hsp70.

In studies of this type the effects of other heat shock proteins must be considered. Hsp104 in yeast was found to have a dominating role in thermotolerance (Sanchez and Lindquist, 1990). (There does not seem to be an equivalent to hsp104 in *Drosophila*; in human cells the equivalent appears to be hsp110.) A role in thermoresistance in L cells has also been suggested for hsp90 (Bansal *et al.*, 1991), and, constitutively expressed human (Landry *et al.*, 1989) and *Drosophila* (Rollet *et al.*, 1992) hsp27 has been reported to confer thermal resistance in rodent cells. Many factors including cell type and heat shock protein profile likely contribute to the development of thermal resistance, and it is possible that the importance of any contributing factor will vary from cell to cell and on the experimental conditions employed. These considerations should be kept in mind during the discussion of neuroprotection which follows.

5.6 Heat shock proteins and neuroprotection

There are few studies dealing directly with the role of heat shock proteins in neuroprotection and these largely do not attack the specificity question. Khan and Sotelo (1989) reported that micro-injection of antibodies which presumably recognize hsp70 and hsc70 reduced survival of cultured mouse neurons after a 50°C, 15 min heat shock. An implicit assumption in the experimental design is that anti-hsp antibodies interfere with the functions of hsp70 and hsc70 during heat shock and recovery. Failure to establish this mechanism (or even that hsp70 was synthesized after heat shock) diminishes confidence that the striking increase in mortality reported in neurons microinjected with anti-hsp antibody (50% vs 90% in IgG controls) was a specific effect.

In a study from our laboratory the relative inability of cultured cortical neurons to synthesize hsp70 compared to astrocytes *in vitro* was correlated with increased thermosensitivity in neurons, assessed by trypan blue staining, and severe morphological changes assessed by electron microscopy (Nishimura *et al.*, 1991b). We concluded that synthesis of hsp70 may have a specific protective role in CNS cells since thermosensitive neurons as well as thermoresistant astrocytes have abundant amounts of constitutively synthesized hsc70 (Figure 2). One additional comment: Since our initial reports appeared (Nishimura *et al.*, 1986, 1987) we have consistently found under comparable conditions cultured cortical neurons do not mount as vigorous a heat shock response as astrocytes. This should not necessarily be interpreted to suggest that the neuronal heat shock response is inadequate. Assuming that hsp70 can provide stress protection it is necessary to establish what levels of hsp70 are protective in the cells being considered. This most likely will vary in different cells and depend on other factors including the presence of other potentially protective heat shock proteins and the basic biological differences between cells which contribute to sensitivity to injury.

Two studies have been reported in which a 'priming' heat shock was used and neuronal resistance to injury determined at a later time (Lowenstein *et al.*, 1991; Rordorf *et al.*, 1991). Lowenstein *et al.* (1991) showed that cerebellar granule cells 'primed' with a heat shock were more resistant to glutamate toxicity 14 h later. Their results suggest that tissue culture can be used to study the protective mechanisms of the *in vivo* 'priming' phenomenon, however, they do not establish a specific role for heat shock proteins in neuro-protection although heat shock protein accumulation was noted. Potentially protective non-specific effects of hyperthermia including altered membrane function could contribute to glutamate resistance. The concentration of glutamate in the medium (which was not measured) and the total volume of medium employed (which was not reported) are important in this experimental design. If glutamate is neurotoxic, then factors which remove glutamate from the medium may be neuroprotective. Astrocyte uptake of glutamate was shown over 15 years ago (Schousboe *et al.*, 1977) and is illustrated in Figure 4. The observation that glutamate uptake was more rapid in heat shocked astrocytes requires confirmation. More importantly, Figure 4 shows that a monolayer of astrocytes effectively removed almost 5 μmoles of glutamate (about 845 μg) from 5 ml

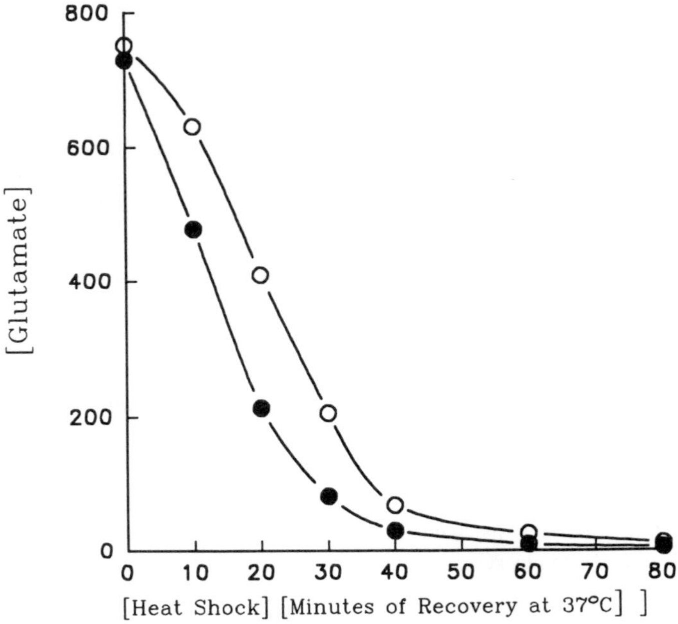

Figure 4 Glutamate uptake by control (∘) and heat shocked (•) rat astrocytes. Astrocytes were grown to confluence in 25cm² flasks. Cells were washed three times in methionine-deficient DMEM and placed in 5 ml for the experiment. One flask served as a control and one flask was heat shocked (45°C, 20 min) and returned to 37°C for 1 h. Serial samples (50 μl) were removed from each flask at the times indicated in the figure. Each point is a single measurement made from either the control or heat shocked culture. *L*-glutamate was added and a zero-time sample was removed before heat shock. Amino acids were derivatized with o-pthalaldehyde and separated by HPLC. Glutamate and other amino acids were identified and quantitated by comparison with known standards. *L*-glutamate was rapidly removed from medium, possibly more rapidly after heat shock. In addition, L-aspartate was rapidly removed from medium whereas the concentration of L-alanine in medium increased (results not shown). One implication is that intermediary metabolism is disturbed in heat shocked astrocytes.

of medium within 40 min. The report by Lowenstein *et al.* (1991) that hyperthermia-induced neuroprotection from glutamate toxicity was more pronounced at lower glutamate concentrations and could be overwhelmed at high glutamate concentrations, the length of time that cells were exposed to glutamate (6 h), and the small (about 3%) but possibly significant number of glial cells present in their neuronal cultures are consistent with the speculation that increased glutamate uptake in 'primed' cultures effectively resulted in neurons being exposed to lower and less lethal concentrations of glutamate over the course of the experiment. Thus, it may be prudent to measure glutamate concentrations in this type of experiment. However, the mechanism by which astrocytes afford neurons protection from injury

(Vibulsreth *et al.*, 1987; Rosenberg and Aizeman, 1989; Mattson and Rychlik, 1990) is not clear and other glial-effects may also be important.

This study also raises another point. A specific role for heat shock proteins in protection of cerebellar granule cells from glutamate toxicity would be strengthened if a mechanism of action was established. The 70-kDa heat shock proteins are synthesized and translocated to the nucleus in HeLa cells where they are believed to provide cell protection (Riabowol *et al.*, 1988). They relocate to the cytoplasm on recovery. It is not known if the 70-kDa heat shock proteins are translocated to the nucleus after heat shock in cerebellar granule cells or if they remained there 14 h later when these cells were challenged with glutamate. This is an important point because Lowenstein *et al.* (1991) report that glutamate itself does not induce heat shock protein synthesis and it is therefore questionable whether glutamate provides the stimulus for nuclear translocation of existing hsp70 and hsc70. Alternatively, nuclear translocation of heat shock proteins may not be needed to protect cerebellar neurons from glutamate toxicity. Many of the same comments hold true for another thermoprotection study in which glutamate toxicity is studied (Rordorf *et al.*, 1991), except that in their mixed neuronal-glial culture system confounding effects of astrocytes would be expected to be more severe.

5.7 The future

Our observed discrepancy in the heat shock response between cultured cortical neurons and astrocytes formed the basis of our hypotheses that failure to synthesize adequate amounts of hsp70 in a timely fashion after injury could provide a biochemical basis for selective neuronal vulnerability. Our interest was to establish whether stress-inducible hsp70 protects neurons and glia by selectively inhibiting hsp70 synthesis in resistant astrocytes in an effort to increase their vulnerability to injury and by selectively over-expressing hsp70 in neurons in an effort to make them more resistant to injury. Our first approach was to use antisense oligonucleotides to specifically inhibit induction of hsp70 and transformed rat astroglial cells were selected as test cells (Figure 5).

Clearly, antisense oligonucleotides can inhibit hsp70 synthesis. However, it is also clear that under the experimental conditions in Figure 5 that the antisense approached failed the criteria of selectivity: Inhibition of hsp70 synthesis was associated with increased synthesis of hsc70 and of a third band migrating around 60–65-kDa. This band may correspond to the recently reported 66-kDa protein in the mouse hsp70 family (Wadhwa *et al.*, 1993) but is as yet unidentified. If compensatory in nature, these changes may frustrate attempts to establish a specific role for hsp70 in cell protection. We would also note here that response of hsp70 synthesis to antisense oligonucleotides has been variable and to date these data have only been presented in abstract form (Dwyer and Nishimura, 1992b). In preliminary studies, an expected decrease in viability of transformed astroglial cells due to heat

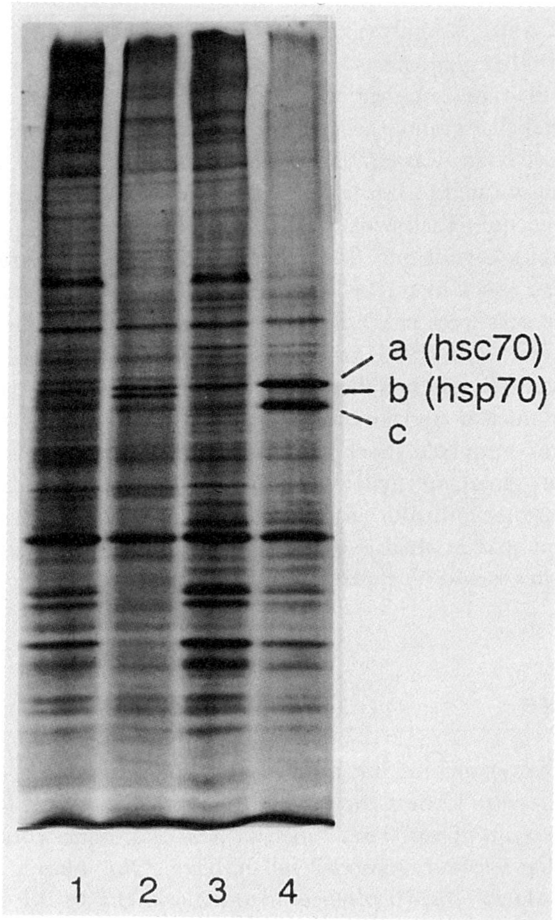

Figure 5 Autoradiograph of ^{35}S-methionine-labelled proteins in control and heat shocked transformed rat astroglial cells (ATs); antisense inhibition of hsp70 synthesis. An antisense oligonucleotide (3′-TAC CGG TTT CGG CGC-5′) containing the standard phosphodiester backbone targeting bases 1–15 of the hsp70 coding region (Hunt and Morimoto, 1985) was synthesized on an Applied Biosystems Model 391 PCR-Mate DNA synthesizer. ATs were grown in six-well dishes as previously described (Nishimura *et al.*, 1992a). ATs were washed three times in serum-free medium, antisense oligonucleotides were added to a final concentration of 50 μm as indicated below, and after 6 h serum was added back to all wells. Fifteen hours later ATs were heat shocked (45°C, 20 min). Control ATs (lanes 1,3) and heat shocked ATs (lanes 2,4) were radiolabelled with ^{35}S-methionine for 3 h and harvested for electrophoresis as described; equal cpm were applied to each lane. Lanes 1,2: ATs did not receive oligonucleotides and are typical of control and heat shocked ATs. Lanes 3,4: ATs were incubated with antisense oligonucleotide. Hsc70 (a), hsp70 (b), and an as yet unidentified third band (c) are indicated in the right margin.

shock in the presence of antisense oligonucleotides was not observed, but due to the variability of this antisense approach a direct correlation between the presence of antisense oligonucleotides and reduced hsp70 synthesis was not established either. We are currently evaluating different ways to introduce antisense oligonucleotides including liposome-based systems which are efficient and relatively gentle.

An alternative to direct addition of antisense oligonucleotides is the establishment of cell lines capable of expressing antisense RNA complementary to hsp70. Preliminary data from very few clones suggest that partial inhibition of hsp70 synthesis was obtained in transfected, neuroblastoma-derived B104 cell-lines (unpublished observation). However, the heat shock promoter is very strong and efforts may be frustrated by inability to generate adequate amounts of antisense RNA. While this approach may ultimately prove feasible in some cells, it is still not an option for neurons and primary cultures of astrocytes with limited ability to divide unless transient expression is desired. For these cells, viral vectors may be a useful alternative (Geller and Breakfield, 1988).

Preliminary experiments designed to overexpress hsp70 mRNA in neurons were successful. Figure 3 shows that cultured cortical neurons transiently transfected with a plasmid containing human hsp70 behind its promoter increased expression of hsp70 mRNA. No effort was made to detect synthesized hsp70 in this instance so that it is not known whether the increased hsp70 mRNA was translated into protein; this is the next step. This approach is inherently complex and many factors might frustrate it. For example, synthesis of stress protein mRNA and protein may require cell specific regulatory sequences for efficient transcription and translation. Furthermore, neuron transfection in this example was accomplished by liposome-mediated DNA transfer which may result in uneven loading of neurons. If increased hsp 70 synthesis is confirmed, the relationship between stressed neurons which die or survive and their hsp70 content would need to be established.

Direct micro-injection of purified hsp70 into a clearly defined cell may prove to be an effective way to establish specific protective effects for hsp70. However, this approach is not necessarily straightforward. It is necessary to establish that purified hsp70 is functional when it is injected into cells. This is also a consideration when purified heat shock proteins are applied topically as discussed elsewhere in this volume. To this end we are expressing recombinant stress proteins in bacterial cultures. Our initial efforts are focused on heme oxygenase (hsp32) as a test protein (Dwyer *et al.*, 1993). The advantage to using heme oxygenase is that it is an enzyme and has an activity that can be measured. The function of hsp70 is less clearly defined and makes more difficult the problem of establishing that recombinant hsp70 is in fact a functional protein.

Clearly, these are but a few of the increasing number of options becoming available to investigate specific functional effects of individual proteins. Most likely, the use of several approaches and possibly the 'knockout' of more than one heat shock protein may be necessary to establish specific protective effects. On the other hand it may be simplistic to assume that cell survival after a variety of injuries is solely the province of one or a small number of proteins.

Acknowledgements

The authors would like to acknowledge the Research Service of the Department of Veterans Affairs, The United Cerebral Palsy Research and Educational Foundation, Inc., and the American Heart Association for support. We would like to acknowledge the excellent technical help of Roger Baldwin, Raissa Marsigian, William Gardiner and Linda Esmaili. We would like to thank Drs William J. Welch and Richard Morimoto for their generosity in providing plasmids and antibodies used for much of our work. Lastly, we would like to thank Dr Ruben Mestril (Department of Medicine, UC San Diego) and Dr Frank Sharp (Department of Neurology, San Francisco VA Hospital) for making available plasmids and sequence information of rat HSP68 and for their comments on this work.

References

Anderson, R.L., Van Kersen, I., Kraft, P.E. & Hahn, G.M. (1989) Biochemical analysis of heat-resistant mouse tumour cell strains: A new member of the HSP70 family. *Mol. Cell. Biol.* **9**, 3509–3516.

Angelides, C.E., Lazarides, I. & Pagoulatos, G.N. (1991) Constitutive expression of heat shock protein 70 in mammalian cells confers thermoresistance. *Eur. J. Biochem.* **199**, 35–39.

Aujame, L. & Firko, H. (1988) The major inducible heat shock protein hsp68 is not required for acquisition of thermal resistance in mouse plasmacytoma cell lines. *Mol. Cell. Biol.* **8**, 5486–5494.

Banik, N.L., Chakrabarti, A.K. and Hogan, E.L. (1990) Purification of an endogenous 68-kDa inhibitor of calcium-activated neutral proteinase (CANP) from bovine brain: Immunoblot identification and characterization. *J. Neurosci. Res.* **25**, 119–124.

Barbe, M.F., Tytell, M., Gower, D.J. & Welch, W.J. (1988) Hyperthermia protects against light damage in the retina. *Science* **241**, 1817–1820.

Blake, M.J., Nowak, T.S. & Holbrook, N.J. (1990) *In vivo* hyperthermia induces expression of HSP70 mRNA in brain regions controlling the neuroendocrine response to stress. *Mol. Brain Res.* **8**, 89–92.

Blumenfeld, K.S., Welsh, F.A., Harris, V.A. & Pesenson, M.A. (1992) Regional expression of c-*fos* and heat shock protein-70 mRNA following hypoxia-ischaemia in immature rat brain. *J. Cereb. Blood Flow Metab.* **12**, 987–995.

Bressler, J.P. & de Vellis, J. (1985) Neoplastic transformation of newborn rat astrocytes in culture. *Brain Res.* **348**, 21–27.

Brown, C.R., Martin, R.L., Hansen, W.J., Beckmann, R.P. & Welch, W.J. (1993) The constitutive and stress-inducible forms of hsp70 exhibit functional similarities and interact with one another in an ATP-dependent fashion. *J. Cell Biol.* **120**, 1101-1112.

Brown, I.R. & Rush, S.J. (1990) Expression of heat shock genes (hsp70) in the mammalian brain: Distinguishing constitutively expressed and hyperthermia-inducible mRNA species. *J. Neurosci. Res.* **25**, 14–19.

Brown, I.R., Rush, S. & Ivy, G.O. (1989) Induction of heat shock gene at site of tissue injury in rat brain. *Neuron* **2**, 1559–1564.

Chopp, M., Chen, H., Ho. K.L., Dereski, M.O., Brown, E., Hetzel, F.W. & Welch, K.M.A. (1989) Transient hyperthermia protects against subsequent forebrain ischaemic cell damage in the rat. *Neurology* **39**, 1396–1398.

Chopp, M., Li, Y., Dereski, M.O., Levine, S.R., Yoshida, Y. & Garcia, J.H. (1991) Neuronal injury and expression of 72-kDa heat shock protein after forebrain ischaemia in the rat. *Acta Neuropathol.* **83**, 66–71.

Chopp, M., Tidwell, C.D., Lee, Y.J., Knight, R., Helpern, J.A. & Welch, K.M.A. (1989) Reduction of hyperthermic ischaemic acidosis by a conditioning event in cats. *Stroke* **20**, 1357–1360.

Cohen, M.V. & Downey, J.M. (1983) Ischaemic preconditioning: Can the protection be bottled? *The Lancet* **342**, 6.

Cosgrove, J.W. & Brown, I.R. (1983) Heat shock protein in mammalian brain and other organs after a physiologically relevant increase in body temperature induced by D-lysergic acid diethylamide. *Proc. Natl Acad. Sci. USA* **80**, 569–573.

Craig, E.A. (1990) Regulation and function of the HSP70 multigene family of *Saccharomyces cerevisiae.* In *Stress Proteins in Biology and Medicine* (eds Morimoto, R.I., Tissieres, A. & Georgopoulos, C), pp 301–321, New York, Cold Spring Harbor Laboratory Press.

Dwyer, B.E. & Nishimura, R.N. (1992a) Heat shock proteins in hypoxic-ischemic brain injury: A perspective. *Brain Pathol.* **2**, 245–251.

Dwyer, B.E. & Nishimura, R.N. (1992b) Antisense inhibition of HSP-68 synthesis. *Trans. Amer. Soc. Neurochem.* **23**, 263.

Dwyer, B.E., Nishimura, R.N., Clegg, K.B. & de Vellis, J. (1991) Regulation of heat shock protein synthesis in rat astrocytes. *J. Neurosci. Res.* **28**, 352–358.

Dwyer, B.E., Nishimura, R.N., de Vellis, J. & Yoshida, T. (1992) Heme oxygenase is a heat shock protein and PEST protein in rat astroglial cells. *Glia* **5**, 300–305.

Dwyer, B.E., Nishimura, R.N. & Yoshida, T. (1993) Expression of heme oxygenase, a CNS stress protein, in *E. coli. Soc. for Neurosci. Abstr.* **19**, (2):1650.

Ewing, J.F., Haber, S.N. & Maines, M.D. (1992) Normal and heat-induced patterns of heme oxygenase-1 (hsp32) in rat brain: Hyperthermia causes rapid induction of mRNA and protein. *J. Neurochem.* **58**, 1140–1149.

Ferriero, D.M., Soberano, H.Q., Simon, R.P. & Sharp, F.R. (1990) Hypoxia-ischaemia induces heat shock protein-like (HSP-72) immunoreactivity in neonatal rat brain. *Dev. Brain Res.* **53**, 145–150.

Geller, A.I. & Breakfield, X.O. (1988) A defective HSV-1 vector expresses *Escherichia coli* β-galactosidase in cultured peripheral neurons. *Science* **241**, 1667–1669.

Georgopoulos, C., Ang, D., Liberek, K. & Zylicz, M. (1990) Properties of the *Escherichia coli* heat shock proteins and their role in bateriophage lambda growth. In *Stress Proteins in Biology and Medicine* (eds Morimoto, R.I., Tissieres, A. & Georgopoulos, C.), pp 191–221. New York, Cold Spring Harbor Laboratory Press.

Gonzalez, M.F., Shiraishi, K., Hisanaga, K., Sagar, S.M., Mandabach, M. & Sharp, F.R. (1989) Heat shock proteins as markers of neural injury. *Mol. Brain Res.* **6**, 93–100.

Gower, D.J., Hollman, C., Lee, K.S. & Tytell, M. (1989) Spinal cord injury and the stress protein response. *J. Neurosurg.* **70**, 605–611.

Hendry, J. & Kola, I. (1991) Thermolability of mouse oocytes is due to the lack of expression and/or inducibility of hsp70. *Mol. Reprod. Dev.* **28**, 1–8.

Hunt, C. & Morimoto, R.I. (1985) Conserved features of eukaryotic hsp70 genes revealed by comparison with the nucleotide sequence of human hsp70. *Proc. Natl Acad. Sci. USA* **82**, 6455–6459.

Iggo, R.D. & Lane, D.P. (1989) Nuclear protein p68 is an RNA-dependent ATPase. *EMBO J.* **8**, 1827–1831.

Johnston, R.N. & Kucey, B.L. (1988) Competitive inhibition of HSP-70 gene expression causes thermosensitivity. *Science* **242**, 1551–1554.

Kawagoe, J., Abe, K., Sato, S., Nagano, I., Nakamura, S. & Kogure, K. (1992) Distributions of heat shock protein-70 mRNAs and heat shock cognate protein 70 mRNAs after transient global ischaemia in gerbil brain. *J. Cereb. Blood Flow Met.* **12**, 794–801.

Khan, N.A. & Sotelo, J. (1989) Heat shock stress is deleterious to CNS cultured neurons microinjected with anti-HSP70 antibodies. *Biol. Cell.* **65**, 199–202.

Kinouchi, H., Sharp, F.R., Hill, M.P., Koistinaho, J., Sagar, S.M. & Chan, P.H. (1993) Induction of 70-kDa heat shock protein and hsp70 mRNA following transient focal cerebral ischaemia in the rat. *J. Cereb. Blood Flow Met.* **13**, 105–115.

Kirino, T., Tsujita, Y. & Tamura, A. (1991) Induced tolerance to ischaemia in gerbil hippocampal neurons. *J. Cereb. Blood Flow Met.* **6**, 642–649.

Kitagawa, K., Matsumoto, M., Tagaya, M., Hata, R., Ueda, H., Niinobe, M., Handa, N., Fukunaga, R., Kimura, K., Mikoshiba, K. & Kamada, T. (1990) 'Ischemic tolerance' phenomenon found in the brain. *Brain Res.* **528**, 21–24.

Kitagawa, K., Matsumoto, M., Tagaya, M., Kuwabara, K., Hata, R., Handa, N., Fukunaga, R., Kimura, K. & Kamada, T. (1991) Hyperthermia-induced neuronal protection against ischaemic injury in gerbils. *J. Cereb. Blood Flow Met.* **11**, 449–452.

Landry, J., Chretien, P., Lambert, H., Hickey, E. & Weber, L.A. (1989) Heat shock resistance conferred by expression of the human HSP27 gene in rodent cells. *J. Cell Biol.* **109**, 7–15.

Lee, Y.J. & Dewey, W.C. (1986) Protection of Chinese hamster ovary cells from hyperthermic killing by cycloheximide or puromycin. *Radiat. Res.* **106**, 98–110.

Leung, T.K.C., Rajendran, M.Y., Monfries, C., Hall, C. & Lim, L. (1990) The human heat shock protein family. Expression of a novel heat-inducible HSP70 (HSP70B) and isolation of its cDNA and genomic DNA. *Biochem. J.* **267**, 125–132.

Li, G.C. (1989) HSP70 as an indicator of thermotolerance. In *Hyperthermic Oncology* (ed Sugahara, T. & Saito, M.), pp 256–259, London, Taylor and Francis.

Li, Y., Chopp, M., Garcia, J.H., Yoshida, Y., Zhang, Z.G. & Levine, S.R. (1992a) Distribution of the 72-kDa heat shock protein as a function of transient focal cerebral ischaemia in rats. *Stroke* **23**, 1292–1298.

Li, Y., Chopp, M. & Levine, S.R. (1992b) Distribution of 72-kDa heat shock protein in rat brain after hyperthermia. *Acta Neuropathol.* **84**, 94–99.

Li, G.C., Li, L., Liu, Y., Mak, J.Y., Chen, L. & Lee, W.M.F. (1991) Thermal response of rat fibroblasts stably transfected with the human 70-kDa heat shock protein-encoding gene. *Proc. Natl Acad. Sci. USA* **88**, 1681–1685.

Li, G.C., Li, L., Liu, R.Y., Rehman, M. & Lee, W.M.F. (1992c) Heat shock protein hsp70 protects from thermal stress even after deletion of its ATP-binding domain. *Proc. Natl Acad. Sci. USA* **89**, 2036–2040.

Longo, F.M., Wang, S., Narasimhan, P., Zhang, J.S., Chen, J., Massa, S.M. & Sharp, F.R. (1993) cDNA cloning and expression of stress-inducible rat hsp70 in normal and injured rat brain. *J. Neurosci. Res.* **36**, 325–335.

Lowe, D.G. and Moran, L.A. (1986) Molecular cloning and analysis of DNA complementary to three mouse M_r=68,000 heat shock protein mRNAs. *J. Biol. Chem.* **261**, 2102–2112.

Lowenstein, D.H., Chan, P.H & Miles, M.F. (1991) The stress protein-response in cultured neurons: Characterization and evidence for a protective role in excitotoxicity. *Neuron* **7**, 1053–1060.

Lowenstein, D.H., Simon, R.P. & Sharp, F.R. (1990) The pattern of 72-kDa heat shock protein-like immunoreactivity in the rat brain following flurothyl-induced status epilepticus. *Brain Res.* **531**, 173–182.

Manzerra, P. & Brown, I.R. (1992a) Expression of heat shock genes (hsp70) in the rabbit spinal cord: Localization of constitutive and hyperthermia-inducible mRNA species. *J. Neurosci. Res.* **31**, 606–615.

Manzerra, P. & Brown, I.R. (1992b) Distribution of constitutive- and hyperthermia-inducible heat shock-mRNA species (hsp70) in the Purkinje layer of the rabbit cerebellum. *Neurochem. Res.* **17**, 559–564.

Marcuccilli, C.J., Fox, L.E. & Miller, R.J. (1992) Deficient heat shock response in cultured CNS and PNS neurons. *Neurosci. Abstr.* **18** (part 2), 1104.

Marcuccilli, C.J., Mathur, S.K., Morimoto, R.I. & Miller, R.J. (1993) Heat shock factor 1 (HSF1) and HSF2 are expressed differently in cultured hippocampal pyramidal neurons and astrocytes. *Neurosci. Abstr.* (in press).

Marini, A.M.,, Kozuka, M., Lipsky, R.H. & Nowak, T.S. Jr. (1990) 70-kDa heat shock protein induction in cerebellar astrocytes and cerebellar granule cells *in vitro*: Comparison with immunocytochemical localization after hyperthermia *in vivo*. *J. Neurochem.* **54**, 1509–1516.

Matsumoto, M. & Fujimoto, H. (1990) Cloning of a hsp70-related gene expressed in mouse spermatids. *Biochem. Biophys. Res. Commun.* **166**, 43–49.

Mattson, M.P. & Rychlik, B. (1990) Glia protect hippocampal neurons against excitatory amino acid-induced degeneration: Involvement of fibroblast growth factor. *Int. J. Devl. Neurosci.* **8**, 399–415.

Mestril, R., Chi, S., Sayen, M.R. & Dillmann, W.H. (submitted) Isolation of a rat inducible heat shock protein (hsp70) gene and its regulation during ischaemia/hypoxia and heat shock.

Milner, C.M. and Campbell, R.D. (1990) Structure and expression of the three MHC-linked hsp70 genes. *Immunogenetics* **32**, 242–251.

Moss, S.E., Crompton, M.R. & Crumpton, M.J. (1988) Molecular cloning of murine p68, a Ca^{2+}-binding protein of the lipocortin family. *Eur. J. Biochem.* **177**, 21–27.

Nishimura, R.N. & Dwyer, B.E., (1993) Induction of HSP-68 in rat astrocytes by acidic pH and 1,10-phenanthroline. *J. Neurochem.* **61** (suppl.):540B.

Nishimura, R.N., Dwyer, B.E., Cole, R., & de Vellis, J. (1988a) Induction of the 68/72-kDa heat shock protein during hydrogen peroxide toxicity. In *NATO ASI Series Volume H22 Cellular and Molecular Aspects of Neural Development and Regeneration* (eds Gorio, A., de Vellis, J., Haber, B. & Perez-Polo, J.R.), pp 227–231. Berlin, Springer-Verlag.

Nishimura, R.N., Dwyer, B.E., Cole, R., & de Vellis, J. (1991a) Comparison of heat shock in cultured cortical neurons and astrocytes. *Mol. Brain Res.* **9**, 39–45.

Nishimura, R.N., Dwyer, B.E., Cole, R., de Vellis, J. & Picard, K. (1989) Induction of the major inducible 68-kDa heat shock protein after rapid changes of extracellular pH in cultured rat astrocytes. *Exp. Cell Res.* **180**, 276–280.

Nishimura, R.N., Dwyer, B.E., de Vellis, J. & Clegg, K.B. (1992a) Characterization of the major 68-kDa heat shock protein in a rat transformed astroglial cell line. *Mol. Brain Res.* **12**, 203–208.

Nishimura, R.N., Dwyer, B.E., Liotta, K., Cole, R. & de Vellis, J. (1986) Induction of stress proteins in astrocytes. *Tras. Amer. Soc. Neurochem.* **17**, 325.

Nishimura, R.N., Dwyer, B.E., Liotta, K., Cole, R. & de Vellis, J. (1987) Heat stress response in primary rat neuronal and glial cultures. *Neurology* **35 (suppl 1)**, 370.

Nishimura, R.N., Dwyer, B.E., Vinters, H.V., de Vellis, J. & Cole, R. (1991b) Heat shock in cultured neurons and astrocytes: Correlation of ultrastructure and heat shock protein synthesis. *Neuropath. Appl. Neurobiol.* **17**, 139–147.

Nishimura, R.N., Dwyer, B.E., Welch, W.J., Cole, R., de Vallis, J. & Liotta, K. (1988b) The induction of the major heat-stress protein in purified rat glial cells. *J. Neurosci. Res.* **20**, 12–18.

Nishimura, R.N., Dwyer, B.E. & Yoshida, T. (1992b) Induction of heat shock proteins in astrocytes and neurons after exposure to prostaglandin *A1*. *Soc. Neurosci. Abstr.* **18** (part 2), 1258.

Nowak, T.S. (1991) Localization of 70-kDa stress protein mRNA induction in gerbil brain after ischaemia. *J. Cer. Blood Flow Met.* **11**, 432–439.

Olney, J.W., Labruyere, J., Wang, G., Wozniak, D.F., Price, M.T. & Sesma, M.A. (1991) NMDA antagonist neurotoxicity: Mechanism and prevention. *Science* **254**, 1515–1518.

Pauli, D. & Tissieres, A. (1990) Developmental expression of the heat shock genes in *Drosophila melanogaster*. In *Stress Proteins in Biology and Medicine* (eds Morimoti, R.I., Tissieres, A. & Georgopoulos, C.), pp 361–378, New York, Cold Spring Harbor Laboratory Press.

Riabowol, K.T., Mizzen, L.A. & Welch, W.J. (1988) Heat shock is lethal to fibroblasts microinjected with antibodies against HSP-70. *Science* **242**, 433–436.

Rollet, E., Lavoie, J.N., Landry, J. & Tanguay, R.M. (1992) Expression of *Drosophila*'s 27-kDa heat shock protein into rodent cells confers thermal resistance. *Biochem. Biophys. Res. Comm.* **185**, 116–120.

Rordorf, G., Koroshetz, W.J. & Bonventre, J.V. (1991) Heat shock protects cultured neurons from glutamate toxicity. *Neuron* **7**, 1043–1051.

Rosenberg, P.A. & Aizenman, E. (1989) Hundred-fold increase in neuronal vulnerability to glutamate toxicity in astrocyte-poor cultures of rat cerebral cortex. *Neurosci. Lett.* **103**, 163–168.

Sanchez, Y. & Lindquist, S.L. (1990) HSP104 required for induced thermotolerance. *Science* **248**, 1112–1114.

Satoh, J., Yamamura, T., Kunishita, T. & Tabira, T. (1992) Heterogeneous induction of 72-kDa heat shock protein (HSP72) in cultured mouse oligodendrocytes and astrocytes. *Brain Res.* **573**, 37–43.

Schousboe, A., Svenneby, G. & Hertz, L. (1977) Uptake and metabolism of glutamate in astrocytes cultured from dissociated mouse brain hemispheres. *J. Neurochem.* **29**, 999–1005.

Schurr, A. & Rigor, B.M. (1992) The mechanism of cerebral hypoxic-ischaemic damage. *Hippocampus* **2**, 221–228.

Sharp, F.R., Jasper, P., Hall, J., Noble, L. & Sagar, S.M. (1991a) MK-801 and ketamine induce heat shock protein HSP72 in injured neurons in posterior cingulate and retrosplenial cortex. *Ann. Neurol.* **30**, 801–809.

Sharp, F.R., Lowenstein, D., Simon, R. & Hisanaga, K. (1991b) Heat shock protein HSP-72 induction in cortical and striatal astrocytes and neurons following infarction. *J. Cereb. Blood Flow Met.* **11**, 621–627.

Simon, R.P., Cho, H., Gwinn, R. & Lowenstein, D.H. (1991) The temporal profile of 72-kDa heat shock protein expression following global ischaemia. *J. Neurosci.* **11**, 881–889.

Sloviter, R.S. & Lowenstein, D.H. (1992) Heat shock protein expression in vulnerable cells of the rat hippocampus as an indicator of excitation-induced neuronal stress. *J. Neurosci.* **12**, 3004–3009.

Smith, B.J. & Yaffe, M.P. (1991) Uncoupling thermotolerance from the induction of heat shock protein. *Proc. Natl Acad. Sci. USA* **88**, 11 091–11 094.

Solomon, J.M., Rossi, J.M., Golic, K., McGarry, T. & Lindquist, S. (1991) Changes in hsp70 alter thermotolerance and heat shock regulation in *Drosophila. The New Biologist* **3**, 1106–1120.

Sprang, G.K. & Brown, I.R. (1987) Selective induction of a heat shock gene in fibre tracts and cerebellar neurons of the rabbit brain detected by *in situ* hybridization. *Mol. Brain Res.* **3**, 89–93.

Swanson, R.A. & Sharp, F.R. (1992) Zinc toxicity and induction of the 72-kD heat shock protein in primary astrocyte culture. *Glia* **6**, 198–205.

Tanno, H., Nockels, R.P., Pitts, Lawrence H. & Noble, L.J. (1993) Immunolocalization of heat shock protein after fluid percussive brain injury and relationship to breakdown of the blood–brain barrier. *J. Cereb. Blood Flow Met.* **13**, 116–124.

Vass, K., Berger, M.L., Nowak, T.S., Welch, W.J. & Lassman, H. (1989) Induction of stress protein HSP-70 in nerve cells after status epilepticus in the rat. *Neurosci. Lett* **100**, 259–264.

Vass, K., Welch, W.J. & Nowak, T.S. (1988) Localization of 70-kDa stress protein induction in gerbil brain after ischaemia. *Acta Neuropathol. (Berl)* **77**, 128–135.

Vibulsreth, S., Hefti, F., Ginsberg, M.D., Dietrich, W.D. & Busto, R. (1987) Astrocytes protect cultured neurons from degeneration induced by anoxia. *Brain Res.* **422**, 303–311.

Voellmy, R., Ahmed, A., Schiller, P., Bromley, P. & Rungger, D. (1985) Isolation and functional analysis of a human 70 000-dalton heat shock protein gene segment. *Proc. Natl Acad. Sci. USA* **82**, 4949–4953.

Wadhwa, R., Kaul, S.C., Ikawa, Y. & Sugimoto, Y. (1993) Identification of a novel member of mouse hsp70 family. Its association with cellular mortal phenotype. *J. Biol. Chem.* **268**, 6615–6621.

Wisniewski, J., Kordula, T. & Krawczyk, Z. (1990) Isolation and nucleotide sequence analysis of the rat testis-specific major heat shock protein (HSP70)-related gene. *Biochem. Biophys. Acta* **1048**, 93–99.

Wu, B., Hunt, C. & Morimoto, R. (1985) Structure and expression of the human gene encoding major heat shock protein HSP70. *Mol. Cell. Biol.* **5**, 330–341.

Zakeri, Z.F., Wolgemuth, D.J. & Hunt, C.R. (1988) Identification and sequence analysis of a new member of the mouse HSP70 gene family and characterization of its unique cellular and developmental pattern of expression in the male germ line. *Mol. Cell. Biol.* **8**, 2925–2932.

CHAPTER 6

HEAT SHOCK GENES AND CELL CYCLE REGULATION IN EARLY MAMMALIAN DEVELOPMENT

David Walsh, Karen Li, Frank Zeng, Li Zhe and Marshall Edwards

Mammalian Department B19, Department of Veterinary Clinical Sciences, University of Sydney, NSW 2006, Australia

Table of Contents

Summary

Elevated temperature at critical stages of neural plate development interrupts early brain and head formation causing major craniofacial defects. Exposure of the

Heat Shock Proteins in the Nervous System
ISBN 0-12-480960-X

presomite neural plate to heat results in death of precursor cells of the developing forebrain and eye situated around the anterior neuropore. In response to heat shock a highly conserved set of heat shock genes are activated and heat shock proteins (hsp) are induced. Hsp gene expression is tightly regulated during the cell cycle. The heat shock response affects cell survival; hsp induction can result in acquired thermotolerance with protection of the cells from a subsequent lethal heat shock. A delay at both G1 and G2 stages of the cell cycle accompanies thermotolerance. A lethal heat shock blocks cell progression at the G1 boundary and cells entering mitosis die within 3–5 h. The role of the hsp in neural plate formation and the mechanisms of thermotolerance may be critical for normal cellular differentiation in the developing nervous system.

6.1 Introduction

6.1.1 Formation of the CNS and a role for hsp genes

Evolution has provided mammals with the ability to maintain a relatively constant body temperature. Thermal stability during embryogenesis is important especially when developing organs such as the central nervous system (CNS) are undergoing rapid proliferation and differentiation (Edwards, 1989; Brown and Rush, 1990). However, thermal stability cannot always be assured during pregnancy. The incidence of febrile infections in women during the first 12 weeks of pregnancy (organogenesis) has been estimated to be between 2.5–8% (Layde *et al.*, 1980; Lipson, 1988). Hyperthermia during human pregnancy can result in a spectrum of adverse reactions, including embryonic or foetal death, prenatal growth retardation and developmental abnormalities (MacDonald, 1958; Edwards, 1986).

In all species, the initial formation of the CNS at neural tube closure appears to be very susceptible to damage by heat (Edwards, 1989). In human development this event occurs between three and five weeks of pregnancy. The development of the vertebrate nervous system begins at gastrulation with the induction of the neural plate on the dorsal surface of the embryo (Figure 1). Also linked to neural induction is the regionalization of the neuroectoderm along the anterior–posterior axis into prospective forebrain, midbrain, hindbrain and spinal cord.

As with all cells studied to date, the cells of the neural plate in early mammalian embryos react to heat shock with a rapid and characteristic heat shock response. Heat shock inhibits transcription and translation of normal protein synthesis but selectively enhances the synthesis of the multigene family of heat shock proteins (hsp) (Lindquist and Craig, 1988). During mammalian embryogenesis the capacity for induction of hsp genes and heat shock response varies with the developmental stage. Hsp activation occurs during early organogenesis in both

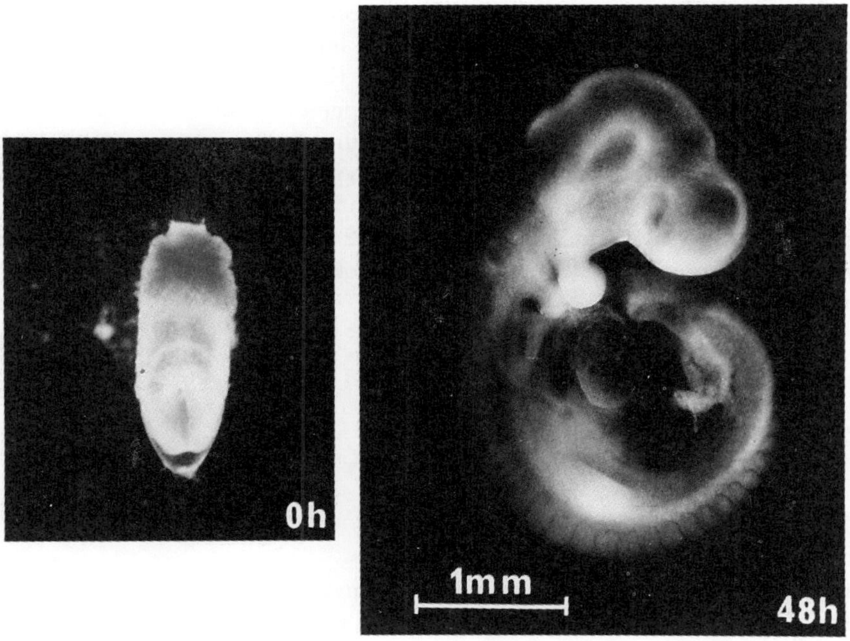

Figure 1 Developmental stages of *in vitro* 9.5 (0) to 11.5 (48) gestational day whole rat embryo in culture. The period of embryo culture embrace most of the major events of organogenesis, with the embryo developing from a relatively undifferentiated headfold stage (0 h) to have 24 somites, a beating heart, prominent optic vesicles, and a closed neural tube at the end of a 48 h culture.

vertebrates and invertebrates (Bienz, 1984; Walsh *et al.*, 1985; Heikkila *et al.*, 1989; Petersen, 1990). The transient expression and induction of hsp suggests an important cellular role in regulation and recovery of cells from the many stresses, including heat, which are known to activate the response (Hightower and Nover, 1991; Welch 1990a).

When the early neural plate of mammalian embryos is exposed to heat shock during gastrulation (Figure 2), several developmental defects of the forebrain and eye follow (Walsh *et al.*, 1985). Interruption or inhibition of gene expression and protein synthesis in the developing neural plate could result in neuroectoderm cell death (Figure 3). Heat shock induction and the heat shock response, however, need not cause developmental defects (Walsh *et al.*, 1991). A mild heat shock, while inducing hsp can result in a phenomenon, termed thermotolerance whereby cells appear protected against a further lethal heat stress.

During early formation of the CNS, control of the cell cycle length is critical for cell proliferation and closure of the neural tube. Throughout organogenesis a general lengthening of the cell cycle is observed particularly in the G1 and G2 phases. As heat shock causes rapid changes to the cell cycle phases at both G1 and G2 (Walsh *et al.*, 1993). Our interest has been to study the role of hsp in cell cycle regulation. Acquired thermotolerance and hsp induction is associated with a cell cycle delay with cells accumulating at the G1/S and the S/G2 boundaries (Walsh and Morris, 1989). The heat shock response appears to lengthen the cell cycle time and thermotolerance is associated with this. During neural tube closure many developmental genes, such as the homeobox-containing genes are expressed to establish regional patterning of the peripheral nervous system and craniofacial structures (Sham *et al.*, 1993; Simeone *et al.*, 1992). The ramifications of heat shock on cell cycle changes and the expression, control and regulation of these genes is not known.

6.2 Heat shock genes

6.2.1 Heat shock response

The heat shock response in the mammalian neural plate (Figure 4) is a well-studied biological phenomenon that is ubiquitous to all living cells and organisms (Hightower and Nover, 1991). The heat shock genes and their encoded proteins have been highly conserved throughout evolution. These features underscore the biological importance of hsps. Studies on the expression of hsp genes have provided evidence of a complex pattern of regulation. The expression of the human hsp70 gene is regulated in a cell cycle dependent manner (Milarski and Morimoto, 1986; Milarski *et al.*, 1989) and is responsive to many stresses, such as heat shock, serum and heavy-metal stimulation (Morimoto and Milarski, 1990; Wu and Morimoto, 1985; Wu *et al.*, 1986). The hsp genes are induced by a transacting heat shock transcription factor (HSF) following phosphorylation (Hueng-Sik *et al.*, 1991).

Some major hsp families are activated during early mammalian development (Table 1). The hsp90 family is associated with membrane stability and glucocorticoid receptor functions (Lindquist and Craig, 1988; Hickey *et al.*, 1989). The hsp70 family comprises the heat shock cognate, hsc73, that is constitutive during neural tube closure and an inducible hsp71 (Walsh *et al.*, 1987; Mirkes and Doggett, 1992) and hsp72 that are developmentally expressed during differentiation of the mammalian germ line (Zakeri *et al.*, 1990). Recently, hsp47 was identified in developing embryos (Li *et al.*, 1992) and is reported to be a specific chaperone for collagen IV (Nagata *et al.*, 1988), finally, the small hsp27 has functions which appear to involve the organisation of several cytoskeletal proteins (Gernold *et al.*, 1993). All the hsps appear to have important 'chaperone' and protective roles that would be required for many embryological events, including neurogenesis. We propose that the function for hsp chaperones during mammalian development may be to dramatically affect the control of the cell-cycle checkpoints and thus ultimately the cell fate of the neuroectoderm.

Table 1 Heat shock proteins in early mammalian embryos.

Hsp	Location 37°C	Location 42°C	Function
Hsp110	Nucleosomes	Nucleosomes	
Hsp100	Endoplasmic reticulum	Endoplasmic reticulum	
Hsp90	Cytoplasm	Cytoplasm	Bind to actin filaments Bind to steroid receptors
Grp78	Endoplasmic reticulum	Endoplasmic reticulum	Attach to nascent proteins Glucose deprivation response
Grp75	Mitochondria	Mitochondria	
Hsc73	Cytoplasm ATP/activation	Cytoplasm Nuclear	Uncoating of clathrin cages Protein degradation/transport
Hsp71	cell cycle (G2+M)	Lysosome	Bind to new protein from ribosomes
		Nuclear	Bind to HSF/DNA binding
Hsp60	Mitochondria	Mitochondria	Intermembrane protein transport
Hsp47	Endoplasmic reticulum	Endoplasmic reticulum	mRNA splicing, cell division Collagen transformation,
Hsp27	Cytoplasm Golgi bodies	Nuclear Cytoplasm	DNA regulation
Ubiquitin	Cytoplasm	Cytoplasm	Protein degradation

6.2.2 The hsp gene families

6.2.2.1 Hsp90

The hsp90 family is known to form complexes with steroid hormone receptors, preventing proteolytic degradation and loss of hormone binding activity (Picard *et al.*, 1990). The major cytosolic hsp90 is also associated with several other classes of proteins, including oncogenic proteins and several tyrosine, serine and threonine kinases. The specific binding to casine kinase II to maintain its optimum folding and conformation indicates its chaperone activity to this target protein. Hsp90 has an important role in binding to and changing the endoplasmic reticulum responsive element to an active form. The mouse hsp90 consists of an α and β isoform: both isoforms are capable of binding to actin filaments and this interaction is inhibited by Ca^{2+} calmodulin complexes.

Analysis of hsp90 in early neural plate development revealed that hsp90 expression may be required to maintain cells in G0 thus affecting cell progression at the start of the cell cycle (Walsh *et al.*, 1993).

6.2.2.2 Hsp70

The hsp70 family comprises about 1% of the total protein in mammalian cells. They are required for a multitude of cellular functions, including the translocation of proteins into the endoplasmic reticulum and mitochondria, the uncoating of clathrin-coated pits during receptor mediated endocytosis, and the processing of antigens in T-cells and the targeting of proteins to lysosomes for degradation (Pelham, 1990). Mammalian embryos constitutively express hsc73 (Walsh *et al.*, 1989). In response to heat shock and many other stresses, increased expression of an inducible hsp71 is observed. The constitutively expressed hsc73 and the hsp71 are related proteins and, following heat shock, hsc73 rapidly forms a stable complex with newly synthesized stress inducible hsp70 (Welch and Suhan, 1986). The function of this novel stress-inducible hsp71 in eukaryotic cells is unclear. Either it appears to bind specifically to proteins damaged directly or indirectly by stressors (Pelham, 1989). After induction, both hsp 71 and hsc 73 bind and co-precipitate with newly synthesized proteins, the hsp71 binding to proteins and increasing their solubility by ATP-dependent mechanisms (Pelham, 1984). The hsps have also been shown to transiently interact to facilitate protein folding and assembly through membranes (Beckmann *et al.*, 1990). Furthermore, the function of hsc73 may be to bind to new proteins and aid in transportation to their normal cellular compartments (Welch, 1990b).

The heat shock response also involves transcriptional activation mediated by the heat shock transcription factor (HSF). In unstressed and early embryonic cells HSF is present in both the cytoplasm and nucleus in a monomeric form that has no DNA-binding activity. At critical stages of development and in response to heat shock and other stresses, the HSF forms a trimer and accumulates in the nucleus. When the HSF trimer becomes activated it binds to a specific 5′ consensus sequence of DNA [GAAnnTTCnnGAA] known as the heat shock element (HSE) (Morimoto, 1993).

The hsp70 members may also be important autoregulators of their own negatively regulated hsp70 gene. The hsp71 may respond under stress to the increased damaged, misfolded and abnormal proteins in the cell that affect HSF binding and activation by phosphorylation of the hsp70 complex. The recent cloning and identification of a family of three HSF genes also offers an explanation for heat shock gene expression and induction during specific stages of development and differentiation. The activation of the HSF1 gene by different chemical and physical stresses and the developmental expression of the HSF2 gene responding to specific stresses could explain the modulation of the heat shock response in embryos (Morimoto, 1993).

6.2.2.3 Hsp47

A novel hsp47 identified in the endoplasmic reticulum (ER) of collagen-secreting cells is associated with collagen metabolism either by binding to the native collagens I, III and IV or by procollagen processing (Nakai *et al.*, 1992). Collagens are major components of the extracellular matrix and basal laminae. The hsp47 is responsive

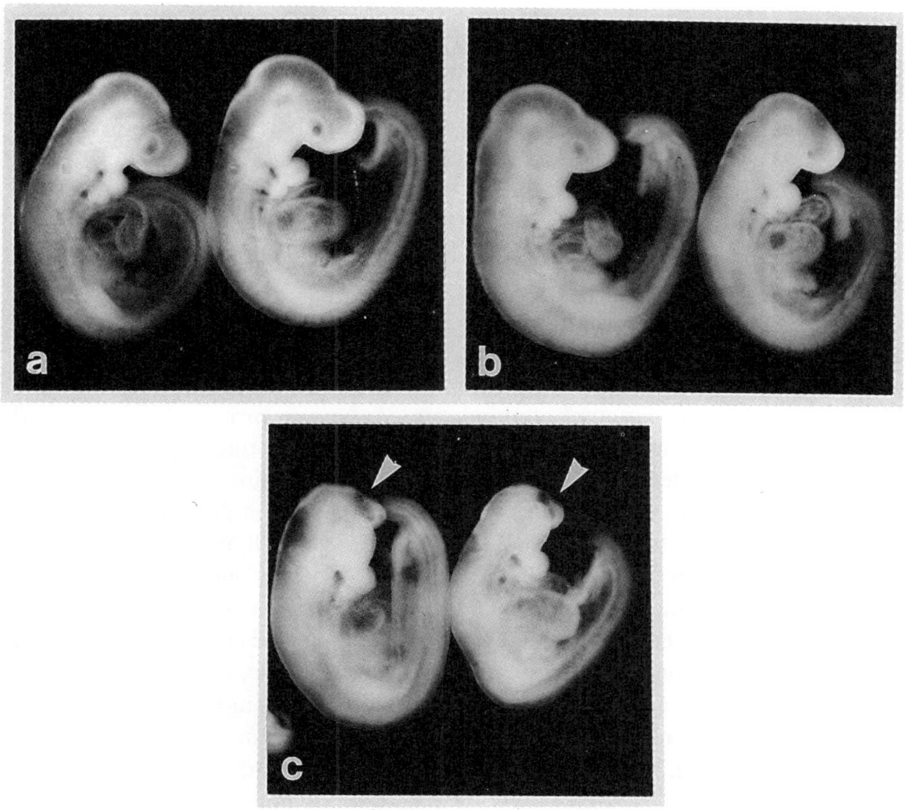

Figure 2 11.5-day rat embryos were cultured for 48 h after various heat shock regimes at the neural plate stage. (a) Embryos exposed to a mild heat shock of 42°C for 10 min with 22 somites representing a general growth delay of 4 h. No obvious developmental defects were observed. (b) Embryos subjected to 42°C for 10 min followed by a recovery period of 60 min at 38.5°C, then a further heat shock of 43°C for 7.5 min. Although growth delay was observed, there were no morphological defects. (c) Embryos exposed to 43°C for 7.5 min displaying developmental defects and absence of both the eye and forebrain (arrows). The midbrain, branchial arch and hindbrain were unaffected.

to heat shock and preliminary results show hsp47 may affect the splicing mechanism of collagen mRNA and protein synthesis in the neural plate (Nakai *et al.*, 1992; Li *et al.*, 1992). Hsp47 may also incorporate a cellular mechanism to protect the basement membrane following heat shock, affecting both induction and formation of the mesoderm and neuroectoderm in the developing brain.

6.2.2.4 Hsp27

Recent evidence shows that hsp27 is associated closely with the metabolism and organization of several structures such as muscle and eye. Hsp27 synthesis in normal development is differentially expressed with an accumulation of hsp27 after organogenesis in muscle, heart, Purkinje cells and neurons of the brain (Gernold *et al.*, 1993). In mammalian cells, heat shock induces the phosphorylation of the small hsp27 within minutes, and it accumulates over 2–3 h (Crete and Landry, 1990). The phosphorylation of hsp27 does not directly involve known protein kinases but selective induction of the hsp27 confers thermotolerance. Evidence suggests that hsp27 mediated thermotolerance at the whole cell level results from a direct interaction of hsp27 with actin components and that hsp27 inhibits polymerization of actin (Miron *et al.*, 1991).

6.2.2.5 Ubiquitin

The heat shock response causes accumulation of aberrant proteins in the cell and also the induction of ubiquitin genes. In mammalian cells this response is recognized as a ubiquitin-dependent degradation system. Ubiquitin, as its name implies, has multiple functions, has an HSE and plays a major role in the cellular response to stress (Bond and Schlesinger, 1985). Found free in the cell, it conjugates to proteins, including histones, as a cellular tag in the ATP-dependent proteolytic degradation of damaged and denatured proteins. The expression of the poly-ubiquitin gene has been observed in the accumulation of ubiquinated filamentous inclusion bodies found in neuronal disorders of the brain, liver disease, myopathy and some viral diseases (Mayer *et al.*, 1989). Most eukaryotic cells contain three major ubiquitin transcripts. In human cells the Ub C transcript codes for a tandem repeat of the 76-aa coding unit, while the Ub B codes for a 52-aa unit, and a third Ub A, a lower weight transcript (0.6 kb) codes for a single copy ubiquitin fusion protein(s) (Fornace *et al.*, 1989).

6.3 Neural plate development

6.3.1 Heat shock gene expression during neural tube closure

The formation of the mammalian forebrain is one of the least understood processes in embryology and is extremely sensitive to heat shock (Walsh *et al.*, 1991). The structure of the early mammalian head and forebrain is predetermined in the neural plate during gastrulation. The development of the vertebrate nervous system begins with the induction of the neural plate on the dorsal surface of the embryo at about the completion of gastrulation. A process closely linked to neural induction is the regionalization of the neural ectoderm along the anterior–posterior (AP) axis into prospective forebrain, midbrain, hindbrain and spinal cord. The neural plate can also be divided along the AP axis according to the character of the underlying mesoderm (Saxen, 1989). The part of the nervous system posterior to, and

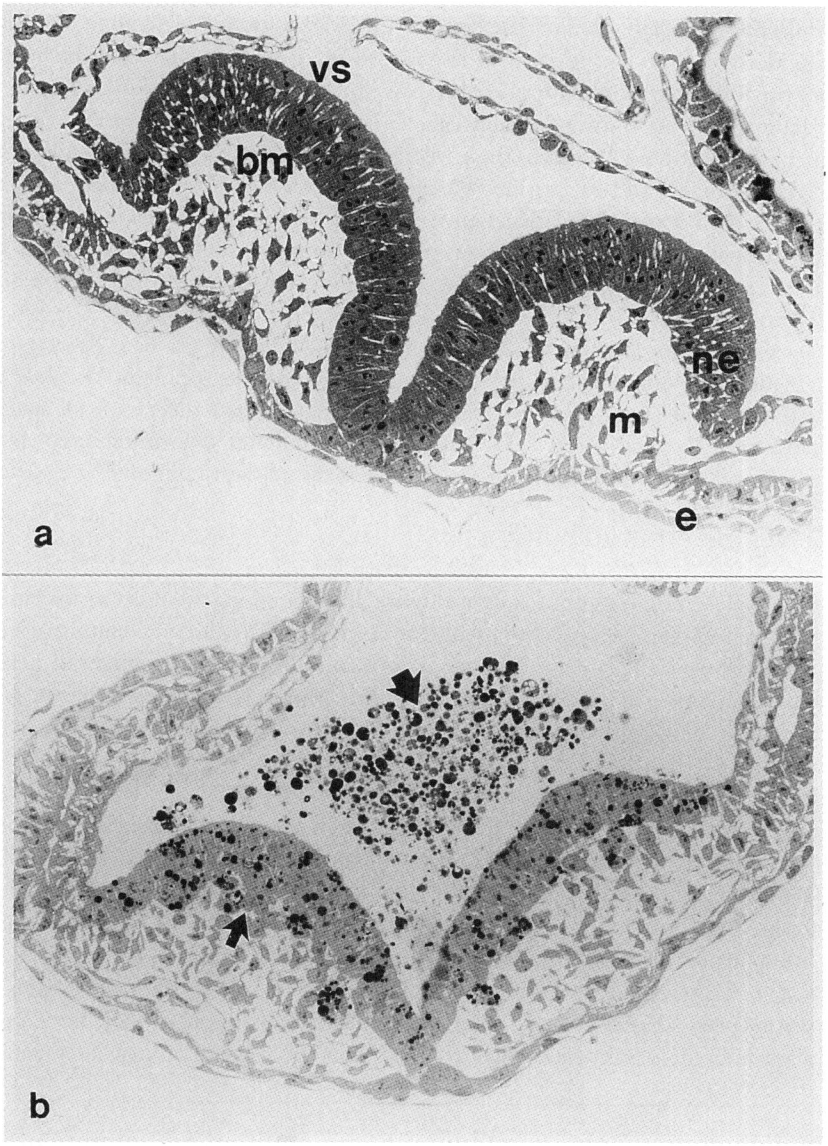

Figure 3(a) (×600) Transverse plastic section through the anterior region of 9.5 day-neural plate 5 h after a heat shock at 42°C followed 60 min later by a heat shock at 43°C. Acquired thermotolerance is evident. The mesoderm (m), neuroectoderm (ne), endoderm (e), basement membrane (bm) and ventricular surface (vs) are normal. (b) (×600) Section from a neural plate 5 h after heat shock at 43°C. Large regions of cell death are seen throughout the neuroectoderm and these dead cells expelled into the amnion (arrow). Major disruption to the basement membrane (arrow) is shown.

including the midbrain, overlies the chordal mesoderm (notochord), while the forebrain is derived from the portion of the neural plate that overlies the prechordal mesoderm. The migratory patterns of the prechordal mesoderm and overlying forebrain neuroectoderm during gastrulation are very different as the formation of the basement membrane is absent in the early anterior forebrain region (Yamada *et al.*, 1991). Hsp kinetics and expression (Figure 4) in the neural plate is tissue specific during this period of normal development. High expression of the inducible hsp gene, specifically the hsp71, was observed in the anterior regions of the forebrain processes and mesoderm, while the endoderm showed little expression and no heat shock response (Walsh *et al.*, 1989).

One of our major interests has been to study how and why the heat shock genes are activated and how they relate to normal embryonic development (Walsh *et al.*, 1987, 1989). Heat shock can result in severe developmental defects in all species (Peterson and Mitchell, 1991). The major developmental deformities appear to result from cell death but details of why the cells die remain unknown.

6.3.2 Whole embryo culture

Using a whole embryo culture system (Figure 1), normal gastrulation in mammalian development can be observed (Walsh *et al.*, 1987). The embryo culture system allows a controlled and precise heat shock to be applied at specific stages of neural development (New *et al.*, 1973; Klein *et al.*, 1980). Maternal influences are removed and the embryonic response to heat shock and neural plate development can be studied directly. Induction of the neuroectoderm and closure of the neural tube occurs between days 9.5 to 10.5 in rat embryos.

In embryogenesis the chaperone functions of hsps for stability of critical developmental proteins during cellular proliferation, differentiation and migration would be critical. We have examined the heat shock response using known lethal and nonlethal heat shock regimes in whole cultured rat embryos at the presomite neural plate stage. The influence of hsp expression on cell fate at specific phases of the cell cycle in relation to either cell death or acquired thermotolerance was investigated (Figure 3). Our studies with rat embryos in culture (Figures 2 and 3) have shown that a temperature of 43°C for 7.5 min causes severe brain abnormalities (Figure 2c), resulting from death of specific progenitor

---

Figure 4 The kinetics and Northern blot analysis of hsp27 (1.6 kb), hsp71 (2.8/2.6 kb) and hsp88 (2.95 kb) mRNA 60 min after heat shock of 9.5-day embryos. Lines represent the molecular weights as measured by mobility of Hind 111–123 digested Lambda DNA. Total mRNA was subjected to electrophoresis, transfered and the mRNA bands identified by the appropriate probes. Northern analysis: hsp27, lanes (a) control; (b) 42°C; (c) 43°C; (d) 42°C/43°C. Hsp71, lanes (A) control; (B) 42°C (15 min); (C) 42°C/43°C; (D) 43°C. Hsp 88, lanes (A) control; (B) 42°C; (C) 42°C/43°C; (D) 43°C. The induction and kinetics of hsp27, 71 and 88 were examined following exposure of the neural plate to the three heat shock regimes used and levels of mRNA determined over a 6-h recovery period to observe the heat shock response: symbols used were (■) 42°C for 10 min; (♦) 42°C for 10 min followed 60

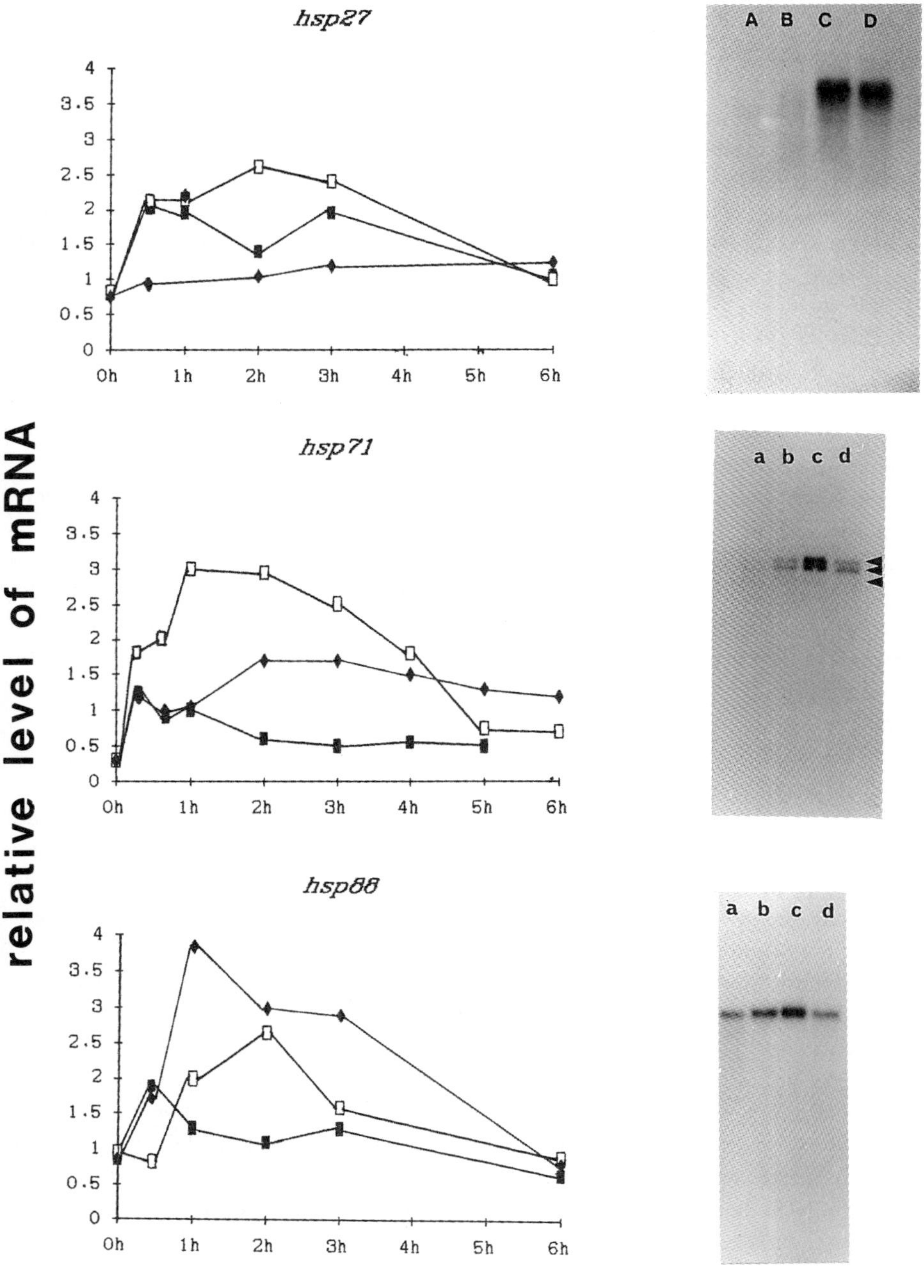

min by a further 43°C heat shock; (□) 43°C for 7.5 min. The response to a heat shock of 42°C followed by a further 43°C thermotolerance in both the hsp27 and 71 showed a down-regulation of the genes when compared with a 43°C heat shock alone. However the response to hsp88 showed a quantitative effect in the response to heat shock.

cells in the anterior neuropore of the neural plate (Walsh *et al.*, 1991). If pre-somite neural plates are exposed to a mild heat shock of 42°C for 10 min alone (Figure 2a) and compared to controls a growth delay of 4 h is observed after 48 h of culture but no developmental defects were seen. Neural plates exposed to a 42°C heat shock (thermotolerant) following after a recovery period of 15 min at 38.5°C, embryos were then further heat shocked at 43°C for 7.5 min (Figure 2b) and showed a mild growth delay, but no craniofacial defects were observed. The heat shock response and hsp expression response lasted for 6–7 h, corresponding to the period of thermotolerance and neuroectoderm cell cycle time.

6.3.3 Hsps and developmental genes in early CNS formation

In embryonic development and early CNS formation, the hsps act as both chaperones and regulators of biochemical molecules that function as primary determinants in development. Induction of the neural plate depends on the stability of proteins that interact and cause the known activation and cascade of developmental genes. The role of hsp and its effect on other developmental genes related to the cytoskeletal structures, extracellular matrix components and possible morphogenic fields are unknown. However, the regulation of the hsp71 promoter by the proto-oncogene and tumour suppressor p53 (a nuclear phosphoprotein) has been reported. The p53 also binds hsp71 and is involved in the cell cycle control of the proliferating brain during early mouse embryogenesis (Agoff *et al.*, 1993; Schmid *et al.*, 1991).

The onset of gastrulation and the mechanical events of neural tube closure signal the beginning of CNS formation and activation of many cytoskeletal proteins. The association of the hsps as chaperones to cytoskeletal proteins, and specifically the hsp47 with collagen, would be required for these morphogenic events and for protein stability and degradation. The rate of denaturation of protein increases with increasing temperature and this is obviously of importance in the effects of abnormally high temperatures on embryos.

Recently, several new homeobox-containing genes (Krumlauf, 1993; Simeone *et al.*, 1992) have been identified in the developing fore-, mid- and hindbrain and are known to be involved in the development of the antero–posterior and dorso–ventral axes in embryos. They reveal graded distribution of activity. Gene transcripts affect and influence the spatial patterning. Our preliminary experiments show that heat shock interrupts the normal spatial patterning and regionality of some of the homeobox containing transcripts found in the forebrain. It would be of interest to examine whether heat shock and thermotolerance regulate important developmental genes such as the homeobox genes.

6.3.4 Spatial identification of the hsp in the neural plate

In situ hybridization at various stages of development using hsp cDNA probes has allowed us to examine gene expression in specific regions near the anterior neuropore. The mRNA expression during all the cell cycle stages in the anterior neuropore has been studied (Figure 5). We have identified expression of hsp71 during neural tube closure; the gene is actively induced in both the neuroectoderm and anterior mesoderm of the neural plate but not in the endoderm (Walsh *et al.*, 1991). The developmental activation and tissue specificity of heat shock genes in the neural plate may affect the ability of the anterior neuropore to recover and survive heat shock.

6.4 The cell cycle

6.4.1 Cell cycle in rat embryos

The neuroectoderm nuclei in the neural plate undergo intracellular migration from a position near the basement membrane (G0/G1, early S-phase) to the ventricular surface (Figure 3). At this surface, G2+M (mitosis) occurs and the two cells separate (Walsh *et al.*, 1989). The function of hsps during cell cycle progression is not clear. The major hsp genes appear to be expressed at different phases and highly regulated throughout the cell cycle in control neuroectoderm cells (Figure 5). The inducible hsp71 may have a role in non-stressed mammalian cells during the G2+M phase. This is supported by the finding that hsp71 is expressed in non-stressed HeLa cells (Milarski *et al.*, 1989). The molecular controls and regulating mechanisms in acquired thermotolerance are also unknown. We have examined the association of the hsp expression and the ability of the embryo to recover and return to the normal body plan. Following all heat shock regimes, neuroectodermal cells accumulate at either the G0/G1 and S/G2 boundaries or both.

The hsps through their chaperone function, could maintain and regulate a critical rate-limiting step or checkpoint in the cell cycle and affect the ability of the cell to acquire thermotolerance. The stability of many of the cyclins (ABD and E) and their degradation at specific phases such as G0, G1, G2 and M is critical (Hunt, 1989). Cyclin B synthesis, activation and breakdown by phosphorylation are required to allow cells to progress through late S-phase and also to complete G2+M. Cyclin degradation is a key step in controlling the exit of cells after completing mitosis and the re-start into the next cell cycle. During the course of cyclin B degradation it forms conjugates with the hsp ubiquitin in anaphase and becomes susceptible to proteolysis (Glotzer *et al.*, 1991).

6.4.2 Hsp expression in the cell cycle phases

The kinetics and expression of the hsp genes were examined in neuroectodermal cells by flow cytometry with the neuroectoderm cells being sorted into G0/G1, S,

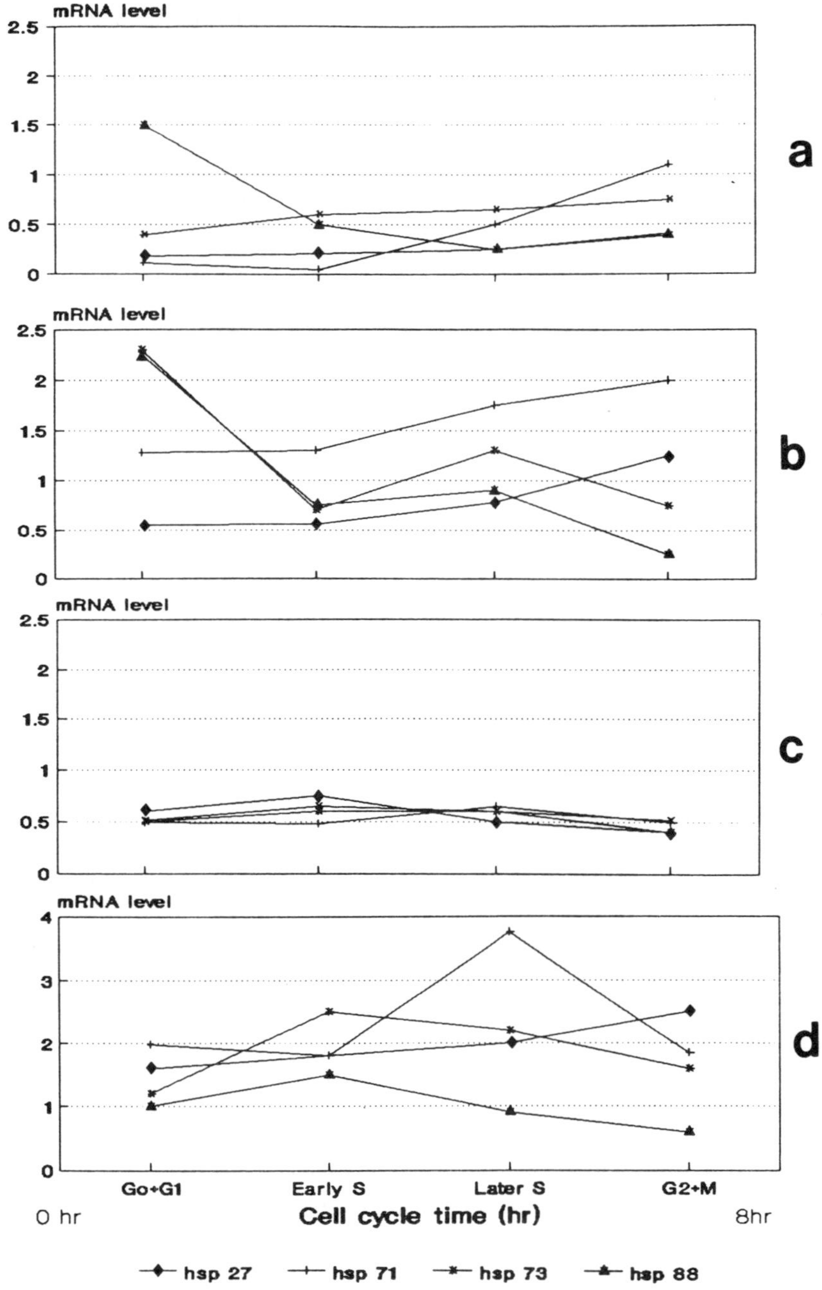

Figure 5 The regulation of hsp mRNAs in the cell cycle. Hsp27, 71, 73 and 88 mRNA expression was determined at G0+G1, early S, late S and G2+M phases of the cell cycle following exposure to heat shock and after 1 h cells were sorted into the four phases

G2+M and the levels of hsp mRNA 27, 71, 73 and 88, identified by dot-blot and Northern analysis (Figure 4). Using flow cytometry and a cell cycle inhibitor ICRF159 to block the cycle at mitosis, progression and changes to the cell cycle phases were determined (Figure 6), after exposure of the neural plate to 42°C (induced heat shock response), 43°C (lethal) and 42°C–43°C (thermotolerant) heat shock regimes. In control cells high hsp 88 expression was observed in G0 and hsp71 was expressed at G2+M of the cell cycle (Figure 5). Cells exposed to a heat shock of 42°C had raised hsp71 mRNA levels in all phases of the cell cycle, with hsc 73 specifically increased in G0. The general mRNA levels of hsp27 and 88 were also increased two-fold, remaining relatively constant through the cell cycle and returning to control levels at the G2+M phase. Following a 43°C heat shock, dramatic increases were seen, especially the enhanced hsp71 induction in late S-phase. No expression of the hsp88 or the hsc73 was observed in G0. A thermotolerant heat shock of 42°C followed by 43°C caused a down-regulation of all the hsp through all phases of the cell cycle. The regulated expression of hsps during the cell cycle at various phases could play an important role in the fate and recovery of neuroectoderm cells during early mammalian embryo development.

6.4.3 Heat shock and cell death

Why do neuroectoderm cells in the anterior neuropore region die following severe heat shock? The fate of cells after heat shock may depend on their hsp response and to the length and state of general translational and transcriptional inhibition. The normal cell cycle time in rapidly dividing neuroectoderm cells is 7–8 h (Walsh and Morris, 1989; MacAuley *et al.*, 1993). These cells show similar cell cycle patterns as CHO cells and other cultured cell lines following heat shock (Kung *et al.*, 1990; Roti Roti *et al.*, 1992).

The first neuroectoderm cell death in neural plate occurs in cells heat shocked during G2+M. Cells die within 3–4 h. The second type of death occurs after 12–15 h in cells heat shocked in G0/G1 or mid- to late S-phase. Cells appear to complete one cell cycle and then die during the next mitosis. Two primary types of cell death are implicated, apoptosis and mitotic damage.

Apoptosis is a complex set of events including activation of cell surface movement and membrane disruption, alterations in cyto-architecture, chromatin hypercondensation and DNA-degradation. Heat shock appears to prevent the cells from completing mitosis irrespective of where the cells are in the cell cycle (Walsh *et al.*,

---

(a) control; (b) 42°C; (c) 42/43°C; (d) 43°C for 1 h. Note: the high-expression levels of hsp88 in G0 and hsp71 expression in G2+M of controls. Following a heat shock of 42°C a 4-fold increase of all hsp was observed. A heat shock of 42°C/43°C resulted in a down-regulation of all mRNA hsps when compared to a 43°C heat shock. Exposure to a lethal 43°C caused a high over expression of all hsp especially hsp71 in late S-phase.

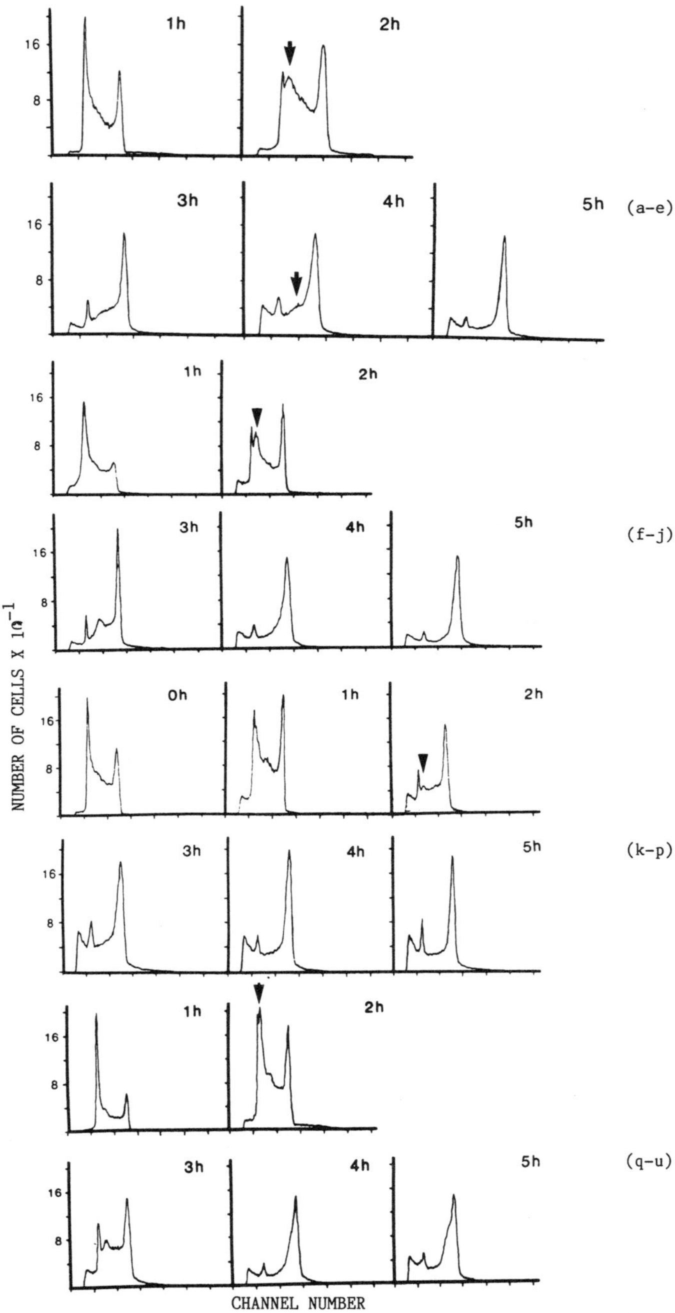

Figure 6 DNA histograms and profiles of cell populations from 0.5–5 h following heat shock and ICRF 159. Controls (a–e); mild heat shock (f–j); thermotolerant heat shock (k–p); and a teratogenic heat shock (q–u). Percentage of cells in the 2C peak in b (arrow) was estimated

1990). The cell death seen in embryos has been postulated to depend on the time and phase of cell cycle inhibition.

A rapid inset of apoptosis occurs after a prolonged state of cell arrest following a 43°C heat shock, resulting in a specific 2–3 h G1 block. A second type of apoptosis is seen after a mild heat shock regime of 42°C (thermotolerant) where, following a short cell cycle delay of 2–3 h in G1, cells complete a division. Cell death occurs 14 h after the heat shock exposure. It appears that once cells are committed to a cell cycle (start) an apoptotic programme is initiated, and cancelled at the time of successful exit from mitosis (Kung *et al.*, 1990). This suggests that one determinant of cell death may be the time of cell cycle arrest and the time taken to initiate the apoptotic programme. Another interesting question is the developmental ability of the cells to undergo apoptosis. Although suicide genes such as the ced and Bc12 genes have been reported, similar events at neurulation in mammals has not. Apoptosis may be a mechanism during development to ensure cell death if progression through a cell cycle is inhibited, and proliferation and synchrony in the neural plate are incorrect.

Neuroectoderm cells are most susceptible to a lethal heat shock exposure during late S-phase and G2+M, responding by cell death within 3–4 h (Walsh *et al.*, 1993). Exposure to a lethal heat shock during other stages of the cell cycle results in a dissociation of orderly cycle progression and aberrant mitotic events (mal motisis) occur. These result in aberrant segregation of chromosomes into daughter cells. These cells are unable to progress through a subsequent cell cycle because of incomplete chromosome complements and die. A sensing mechanism termed a mitotic checkpoint in yeast prevents and delays mitosis if chromosomes are broken, if DNA synthesis is not completed, or if spindles are incorrectly formed (Hartwell and Weinert, 1989). In mammalian neuroectodermal cells similar mechanisms in mitotic checkpoints may exist but have not been identified. Neuroectoderm cells delayed at the G1/S boundary appear not to die but may be sensitive to further stress. Differences in the susceptibility to cell death in the anterior forebrain region compared with the more posterior mid- and hindbrain region of the neuroectoderm could be due to differences in cell cycle time, and perhaps differences in mitotic checkpoint mechanisms. These mitotic checkpoints would have important survival benefits to the developing embryo and an effect on acquired thermotolerance.

6.4.4 Hsp and thermotolerance

In the thermoprotective 42°C regime, cells at the G1 boundary accumulate for 1–2 h and entry of G0 cells into S-phase was delayed. Upon recovery a synchron-

---

at 23.4% and illustrates the relative rate of exit from the 2C peak into S-phase at 2 h after treatments. The numbers in each column are hours after the first heat shock and ICRF 159. Arrows identify the synchronized peak of cells progresssing through S phase. Peak highs are plotted approximately relative to the proportions in the 2C peak.

ized wave of cells from G0/G1 entered S-phase and progressed slowly through the cell cycle (Figure 6). A teratogenic heat shock of 43°C blocked the cell cycle. Cells at the G1/S boundary were stopped for 2–3 h. However, cells from mid- to late S phase appeared to progress and enter mitosis and die. The cell population of the neuroectoderm in S-phase was therefore depleted (Walsh *et al.*, 1993). Thermotolerance resulted in high expression of all hsp in all phases of the cell cycle. This also corresponded to a general lengthening of the cell cycle by 2–3 h. Neuroectoderm cells displayed a longer thermoprotection when heat shocked in G1 and early S-phase compared with the other stages of the cell cycle. Cells in G2+M appear to be the most susceptible to a lethal heat shock. This is associated with high expression of inducible hsp71 in mid- to late S-phase. It appears that the transcript may not be translated. The hsp71 is normally expressed only during G2+M. Under stress, delay of the cell cycle and slowed progression of the cells may prevent initiation of DNA synthesis and differentiating at the sensitive stages of the cycle.

6.5 Future strategy

6.5.1 Antisense strategy

One of the classic methods of studying the function of a particular gene is to inhibit its function. The use of antisense cDNA is a particularly elegant way of studying a gene. Using antisense phosphorothiolated oligonucleotides to inhibit gene function, antisense RNA can be micro-injected into the amnion at early head-fold stages of embryo development. The RNA penetrates the cell and hybridizes to mRNA. This method has recently been used successfully to study the function of the wnt-1 gene in a whole mouse embryo culture system (Augustine *et al.*, 1992).

Antisense experiments can explore the combined functions of the hsp and its sequential regulation during neural tube closure. By inhibiting specific hsp genes, their role with other hsp genes can be investigated following heat shock. The role of the hsp genes during early forebrain development could also be examined. We have established a COS cell line transfected with an antisense hsc73 construct, activated by a metalothionine promoter. The cell cycle response of these cells to stress and the role of the hsc73 are now being examined.

6.5.2 Expression of hsp 27, 71, 73, 88, and cyclin in the cell cycle

Specific antibodies, now available for the hsp27, 47, 71, 73, and 88, will allow identification of the hsps at specific phases of the cell cycle. By using hsp fluorescent probes, translational regulation and synthesis in neuroectoderm cells in G0/G1, early and late S-phase and G2+M phases can be observed following heat shock. The role of hsp on cyclin and its degradation during development would also be of interest in understanding the mechanisms controlling cell cycle progression.

6.6 Discussion and conclusion

Mild heat shock during neural tube closure interrupts and delays the normal developmental growth pattern of the CNS. The significance of this delay in terms of neurological development is unclear. In terms of overall growth the mildly stressed embryo appears to recover fully by the time of birth (Walsh *et al.*, 1987) however, illustrations of the head seem to indicate that the brains of the heated embryos are smaller (Edwards, 1986). Severe heat shock also causes cell death resulting in lethal developmental defects of the forebrain and eye.

It has generally been acknowledged that hsps play a role in providing acquired thermotolerance. Pre-heating embryos activates the hsp genes, resulting in acquired thermotolerance against a lethal heat shock. Thermotolerance is acquired following an initial 42°C heat shock. The rapid induction of hsp71 and all other hsp occur, and protection persists for 7–8 h (Walsh *et al.*, 1987). We have shown that thermotolerance is associated with cell cycle delay in S-phase and specific hsp expression at the G1/S and S/G2 boundaries.

A heat shock of 43°C causes death of cells in late S-phase characterised by an over expression of hsp71. A second mode of cell death is observed 12–15 h following exposure to a 42°C heat shock. Cells from G0 appear to be pushed into G1 and the start of the cell cycle. This synchronized wave of prematurely committed cells may be unable to respond to developmental induction and DNA initiation. The outcome following heat shock of neuroectoderm cells appears to be dependent on the stage of the cell cycle during which the cells are exposed (Mackey and Dewey, 1988; Roti *et al.*, 1992). Cells exposed to heat shock in early S-phase remain thermotolerant for a longer period. Cells appear to lose acquired thermotolerance after completion of mitosis.

The suggestion that hsps, particularly the hsp71, play an important role in thermoprotection has been proposed for many systems (Lindquist and Craig, 1988; Riabowol *et al.*, 1988; Landry *et al.*, 1989). HeLa cells transfected with the hsp71 gene and over-expressing the hsp71 appear to have an increased thermotolerance (Welch *et al.*, 1990b). Induction of hsp71 thus serves a protective function if present before a damaging exposure. However, it does not repair cells already damaged by a severe heat shock. The proposed role for hsp71 in thermotolerance as a chaperone protein for damaged proteins is consistent with our findings (Ellis, 1987; Pelham, 1989).

Heat shock causes rapid and dramatic changes to the cell cycle. Thermotolerance may be due to a mechanism which inhibits DNA synthesis at the most sensitive stages of the cell cycle (Walsh and Morris, 1989). Cell cycle changes following stress and hsp induction may be associated with other known cell cycle regulators such as p53, a tumour suppressor gene, in proliferating cells (Lane, 1992). The significance of hsp expression at specific phases is unknown, the cell cycle delay seen may prolong the neuroectoderm cell division until recovery and normal development can continue.

References

Agoff, S.N., Hou, J., Linzer, D.I.H. & Wu, B. (1993) Regulation of the human hsp70 promoter by p53. *Science* **259**, 84–87.

Augustine, K., Lui, E. & Sadler, T. (1992) Examining the role of the wnt-1 proto-oncogene in murine embryogenesis. *Teratology* **45**, 5.503.

Beckmann, R., Mizzen, L.A. & Welch, W.J. (1990) Interaction of hsp70 with newly synthesized proteins: Implications for protein folding and assembly. *Science* **235**, 305–311.

Bienz, M. (1984) Developmental control of the heat shock response in *Xenopus*. *PNAS* **81**, 3138–3142.

Brown, I.R. & Rush, S.J. (1990) Expression of heat shock genes (hsp70) in the mammalian brain: distinguishing constitutively expressed and hyperthermia-inducible species. *J. Neurosci. Res.* **25**, 14–19.

Bond, U. & Schlesinger, M.J. (1985) Ubiquitin is a heat shock protein in chicken embryo fibroblasts. *Mol. Cell Biol.* **5**, 949–956.

Crete, P. & Landry, J. (1990) Induction of hsp27 phosphorylation and thermoresistance in Chinese hamster cells by arsenite, cycloheximide. *Radiat. Res.* **121**, 320–327.

Edwards, M.J. (1986) Hyperthermia as a teratogen: a review of experimental studies and their clinical significance. *Teratogenesis Carcinog. Mutagen.* **6**, 563–582.

Edwards, M.J. (1989) Hyperthermia and the developing central nervous system. In *Ann. of the Res. Inst. Environ. Med.*, Nagoya University **40**, pp 355–364.

Ellis, J. (1987) Proteins as molecular chaperones. *Nature* **328**, 378–379.

Fornace, A.J., Alamo, I., Hollander, M.C. & Lamoreaux, E. (1989) Ubiquitin mRNA is a major stress-induced transcript in mammalian cells. *Nucleic Acids Res.* **173**, 1215–1230.

Gernold, M., Knauf, U., Gaestel, M., Stahl, J. & Kloetzel, P.M. (1993) Tissue-specific distribution of mouse small heat shock protein hsp25. *Dev. Gen.* 2, **14**, 103–111.

Glotzer, M., Murry, A.W. & Kirschner, M.W. (1991) Cyclin is degraded by the ubiquitin pathway. *Nature* **349**, 132–138.

Hartwell, L.H. & Weinert, T. (1989) Checkpoints: controls that ensure the order of cell cycle events. *Science* **246**, 629–634.

Hightower, L.E. & Nover, L. eds. (1991) Heat Shock : Results and problems. In *Cell Differentiat.*, pp 000 Heidelberg, Springer-Verlag.

Heikkila, J.J., Krone, P.H. & Ovsenek, N. (1991) Regulation of heat shock gene expression during *Xenopus* development. *Development* **17**, 120–137.

Hickey, E., Brandon, S.E., Smale, G., Lloyd, D. & Weber, L.A. (1989) Sequence and regulation of a gene encoding a human 89-kDa hsp. *Mol. Cell Biol.* **9**, 2615–2626.

Heung-Sik, C., Li, B., Lin, Z., Huang, E. & Liu, A. (1991) cAMP dependent protein kinase regulates the human heat shock protein 70 gene promoter activity. *J. Biol. Chem.* **26618**, 11 858–11 865.

Hunt, T. (1989) Under arrest in the cell cycle. *Nature* **342**, 483–484.

Klein, N.W., Vouler, M.A., Chatot, C.L. & Pierro, L.J. (1980) The use of cultured rat embryos to evaluate the teratogenic activity of serum: Cadmium and cyclophamide. *Teratology* **21**, 199–208.

Krumlauf, R. (1993) Hox genes and pattern formation in the branchial region of the vertebrate head. *Trends Genet.* **9**, 106–112.

Kung, A.L., Sherwood, S.W. & Schimke, R.T. (1990) Cell line specific differences in the control of cell cycle progression in the absence of mitosis. *Proc. Natl Acad. Sci. USA* **87**, 9553–9557.

Lane, D.P. (1992) P53, guardian of the genome. *Nature* **358**, 15–16.

Landry, J., Chretien, P., Lambert, H., Hickey, E. & Weber, L.A. (1989) Heat shock resistance conferred by expression of the human hsp27 gene in rodent cells. *J. Cell Biol.* **109**, 7–15.

Layde, P.M., Edmonds, L.D. & Erickson, J.D. (1980) Maternal fever and neural tube defects. *Teratology* **21**, 105–108.

Li, Z., Zeng, F., Nagata, K. & Walsh, D. (1992) Hsp47 expression during early mammalian development: heat shock proteins *UOEH Japan* **12**, 93.

Lindquist, S. & Craig E.A. (1988) The heat shock proteins. *Annu. Rev. Genet.* **22**, 631–677. Landry *et al.*, 1992.

Lipson, A. (1988) Hirschsprung disease in the offspring of mothers exposed to hyperthermia during pregnancy. *Am. J. Med. Genet.* **29**, 117–124.

MacAuley, A., Werb, Z. & Mirkes, P. (1993) Characterization of the unusually rapid cell cycles during rat gastrulation. *Development.* **117**, 873–883.

Mackey, M.A. & Dewey, W.C. (1988) Time-temperature analyses of cell killing of synchronous G1 and S-phase CHO cells *in vitro. Radiat. Res.* **133**, 318–329.

Mayer, R.J., Landon, M., Doherty, F.J., Lowe, J.S., Reynolds, G.P., Byrne, E.J., Lennox, G.G., Jefferson, D. & Godwin-Austen, R.B. (1989) Ubiquitin and dementia. *Nature* **340**, 193.

McDonald, A.D. (1958) Maternal health and congenital defects. *N. Eng. J. Med.* **258**, 767–773.

Milarski, K.L. & Morimoto, R.I. (1986) Expression of human hsp70 during the synthetic phase of the cell cycle. *Proc. Natl Acad. Sci. USA* **83**, 9517–9521.

Milarski, K.L., Welch, W.J. & Morimoto, R.I. (1989) Cell cycle-dependent association of hsp70 with specific cellular proteins. *J. Cell Biol.* **108**, 413–423.

Mirkes, P.E. & Doggett, B. (1992) Accumulation of heat shock protein 72 (hsp72) in postimplantation rat embryos after exposure to various periods of hyperthermia *in vitro. Teratology* **46**, 301–309.

Morimoto, R.I. & Milarski, K.L. (1990) Expression and function of vertebrate hsp70 genes. In *Stress Proteins in Biology and Medicine* (eds Morimoto, R.I., Tissieres, A. & Georgopoulas, C.), pp 323–359. New York, Cold Spring Harbor Lab Press.

Morimoto, R. (1993) Cells in stress: transcriptional activation of heat shock genes. *Science* **259**, 1409–1410.

Miron, T., Vancompernolle, K., Vanderkerckhove, J., Wilchek, M. & Greiger, B. (1991) A 25-kDa inhibitor of actin polymerization is a low molecular mass heat shock protein. *J. Cell Biol.* **114**, 2.255–261.

Nagata, K., Hirayoshi, K., Obara, M., Saga, S. & Yamada, K. (1988) Biosynthesis of a novel transformation-sensitive heat shock protein that binds to collagen. *J. Biol Chem.* **263**, 17 8344–8349.

Nakai, A., Satoh, M., Hirayoshi, Z. & Nagata, K. (1992) Involvement of the stress protein hsp47 in procollagen processing in the endoplasmic reticulum. *J. Cell. Biol.* **17**, 4903–914.

New, D.A.T., Coppola, P.T. & Terry, S. (1973) Culture of explanted embryos in rotating tubes. *J. Reprod. Fertil* **35**, 135–138.

Petersen, N.S. (1990) Effects of heat and chemical stress on development. In *Advances in Genetics: Responses to Environmental Stress* (ed Scandalios), pp 28. San Diego CA, Academic Press.

Peterson, N.S. & Mitchell, H.K. (1991) Environmentally induced development defects in *Drosophila.* In *Heat Shock and Development* (eds Hightower, L. & Nover, L.), **17**, pp 29–43. Berlin, Springer-Verlag.

Pelham, H.R.B. (1984) Hsp70 accelerates the recovery of nucleolar morphology after heat shock. *EMBO J* **3**, 3095–3100.

Pelham, H.R.B. (1989) The selectivity of secretion: protein sorting in the endoplasmic reticulum. *Biochem. Soc. Transactions.* **17**, 795–802.

Pelham, H. (1990) Functions of the hsp70 protein family: An overview. In *Stress Proteins in Biology and Medicine* (eds Morimoto, R., Tissieres, A. & Georgopoulos, C.), pp 287–299. Cold Spring Harbor.

Picard, D., Khursheed, B., Garabedian, M.J., Fortin, M.G., Linquist, S., Yamamoto, K.R. (1990) Reduced levels of hsp90 compromise steroid receptor action *in vivo*. *Nature* **348**, 166–168.

Roti Roti, J.L., Mackey, M.A. & Higashikubo, R. (1992) The effects of heat shock on cell proliferation. *Cell Prolif.* **25**, 89–99.

Riabowol, K.T., Mizzen, L.A. & Welch, W.J. (1988) Heat shock is lethal to fibroblasts microinjected with antibodies against hsp70. *Science* **242**, 433–436.

Saxen, L. (1989) Neural Induction. *Int. J. Dev. Biol.* **33**, 21–48.

Schmid, P., Lorenz, H., Hameister, H. & Montenarh, M. (1991) Expression of p53 during mouse embryogenesis. *Development* **113**, 857–865.

Simeone, A., Acampora, D., Gulisano, M., Stornaiuolo & Bocinelli, E. (1992) Nested expression domains of four homeobox genes in developing rostral brain. *Nature* **358**, 687–690.

Sham, M.H., Vesque, C., Nonchev, S., Marshall, H., Frain, M., Gupta, R., Whiting, J., Wilkinson, D., Charnay, P. & Krumlauf, R. (1993) The Zinc finger gene Krox20 regulates HoxB2 during hindbrain segmentation. *Cell* **72**, 183–196.

Walsh, D.A., Hightower, L.E., Klein, N.W. & Edwards, M.J. (1985) The induction of the heat shock proteins during early mammalian development. Heat shock *Cold Spring Harbor Symposium* **2**, 92.

Walsh, D.A., Klein, N.W., Hightower, L.E. & Edwards, M.J. (1987) Heat shock and thermotolerance during early rat embryo development. *Teratology* **36**, 181–191.

Walsh, D.A., Li, K., Speirs, J., Crowther, C.E. & Edwards, M.J. (1989) Regulation of the inducible heat shock 71 genes in early neural development of cultured rat embryos. *Teratology* **40**, 321–334.

Walsh, D.A. & Morris, V.B. (1989) Heat shock affects cell cycling. *Teratology* **40**, 583–592.

Walsh, D.A., Li, K., Crowther, C., Marsh, D. & Edwards, M.J. (1991) Thermotolerance and heat shock response during early development of the mammalian embryo. In *Heat Shock and Development* (eds Hightower & Nover), **17**, pp 58–69. Springer-Verlag.

Walsh, D.A., Li, K., Wass, J., Dolnikov, A., Zeng Li, Z and Edwards, M. (1993) Heat shock gene expression and cell cycle changes during mammalian development. *Dev. Genetics* **14**, 127–136.

Welch, W.J. & Suhan, J.P. (1986) Cellular and biochemical events in mammalian cells during and after recovery from physiological stress. *J. Cell Biol.* **103**, 2035–2052.

Welch, W.J. (1990a) The mammalian stress response: cell physiology and biochemistry of stress proteins. In *Stress Proteins in Biology and Medicine* (eds Morimoto, R.I., Tissieres, A. & Georgopoulos, C.), pp 223–278. Cold Spring Harbor Laboratory Press.

Welch, W.J. (1990b) Interaction of hsp70 with newly synthesized proteins: implications for protein folding and assembly. *Science* **235**, 305–311.

Williams, G.T., McClanahan, T.L. & Morimoto, R.I. (1989) Association of the E1A promoter with the hsp genes. *Mol. Cell Biol.* **9**, 2574–2587.

Wu, B.J. & Morimoto, R.I. (1985) Transcription of the human hsp 70 gene is induced by serum stimulation. *Proc. Natl Acad. Sci. USA* **82**, 6070–6074.

Wu, B.J., Kingston, R.E. & Morimoto, R.I. (1986) Human hsp 70 promoter contains at least two distinct regulatory domains. *Proc. Natl Acad. Sci. USA* **83**, 629–633.

Yamada, T., Placzek, M., Tanaka, H., Dodd, J. & Jessell, T.M. (1991) Control of cell pattern in the developing nervous system: Polarizing activity of the floor plate and notochord. *Cell* **64**, 635–647.

Zakeri, Z.F., Welch, W.J. & Wolgemuth, D.J. (1990) Characterization and inducibility of the hsp70 proteins in the male mouse germ line. *J. Cell Biol.* **111**, 1785–1792.

CHAPTER 7

EXPRESSION, CELLULAR LOCATION AND FUNCTION OF LOW MOLECULAR WEIGHT HEAT SHOCK PROTEINS (Hsp20s) DURING DEVELOPMENT OF THE NERVOUS SYSTEM

Andre-Patrick Arrigo and Patrick Mehlen

Laboratoire du Stress Cellulaire, Centre de Génétique Moleculaire et Cellulaire, CNRS UMR-106, Claude Bernard University, Lyon-1, 69622 Villeurbanne, France

Table of Contents

Heat Shock Proteins in the Nervous System
ISBN 0-12-480960-X

7.1 Introduction

Heat shock (or stress) proteins (hsps) are defined as conserved proteins that are specifically induced when organisms are exposed to supra-optimal temperatures or to various other types of stress. These proteins can be subdivided in two classes: the high molecular weight hsps (100-, 90–83, 70-, and 60-kDa) and the small hsps (16–30-kDa). Most of the early studies concerning these proteins were made in *Drosophila melanogaster.* In tissue culture cells of this organism, Mirault *et al.* (1978) were the first to suggest that some of the hsps are constitutively expressed. This was confirmed by Chomyn and Mitchell (1982) who demonstrated the presence of high levels of the hsp83 gene product in non-stressed *Drosophila* cells. A similar observation was made for cognates of hsp70 (Ingolia and Craig, 1982a; Craig *et al.*, 1983) and for the small hsps during development and cell differentiation (Sirotkin and Davidson, 1982; Buzin and Bournias-Vardiabasis, 1982). Thus, it appears obvious from these instances that some of the hsps are expressed at normal temperatures at least in some tissues and during specific periods of the development (reviews: Bond and Schlesinger, 1987; Pauli and Tissières, 1990; Arrigo and Tanguay, 1991). Here, we review studies on the expression of the small hsps during development with a special emphasis on their accumulation in the nervous system of *Drosophila.* We then analyse the expression of these proteins in other species, particularly mammals. We conclude by discussing the role played by these proteins.

7.2 Gene structure of the *Drosophila* small hsps and control of their expression

Among species, the small hsps show greater variation than the high molecular weight hsps (i.e. hsp70), in sequence, in number (four in *Drosophila*, one in mammals and yeast, >20 in plants) and in molecular masses. In *Drosophila*, the four small heat shock proteins are characterized by apparent molecular masses of 22-, 23-, 26-, and 27-kDa (also indicated as hsp22, 23, 26 and 27) (McKenzie *et al.*, 1975; Mirault *et al.*, 1978). The genes encoding these polypeptides are clustered within 12 kb of DNA at the *locus* 67B of chromosome 3L (Petersen *et al.*, 1979; Corces *et al.*,

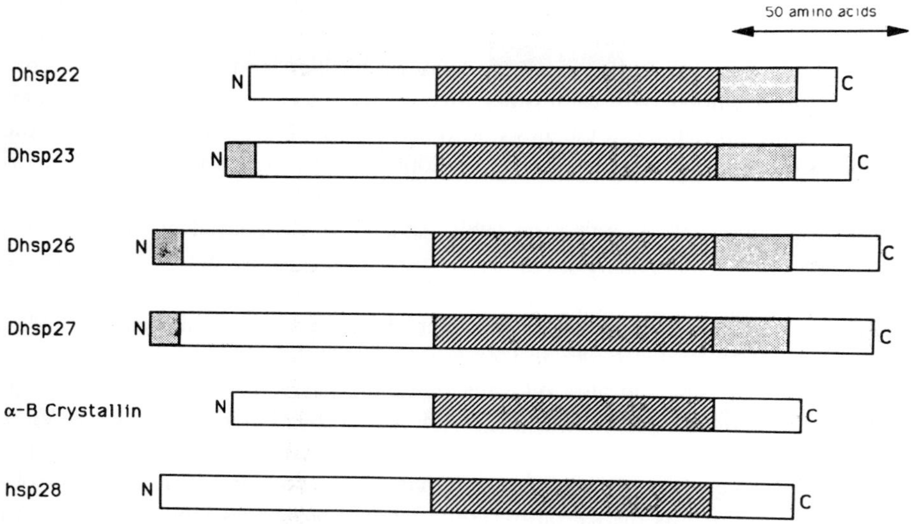

Figure 1 Schematic illustration of the various amino acid similarities between the four *Drosophila* small hsp (hsp22, 23, 26 and 27), α-crystallin and the unique mammalian small hsp (hsp28). The hatched box represents the conserved amino acids stretch between the small hsps and α-crystallin. The light-grey boxes represent the sequence similarities between the four small *Drosophila* hsps. The open boxes represent little or no sequence similarity. Based on the sequences published by Ingolia and Craig (1982b); Southgate *et al.* (1983) and Hickey *et al.* (1986).

1980; Craig and McCarthy, 1980; Wadsworth *et al.*, 1980; Voellmy *et al.*, 1981; Ayme and Tissières, 1985; and reviews: Southgate *et al.*, 1985; Pauli and Tissières, 1990; Arrigo and Tanguay, 1991; Pauli *et al.*, 1992). These genes share sequence homology in several domains of their coding region (Figure 1). The first conserved region comprises the sequence encoding the first 15 N-terminal amino acids. These amino acids are very hydrophobic and resemble those observed in signal peptides (Arrigo and Pauli, 1988). The second domain consists of 80 residues in the second half of the genes. This domain is similar to a carboxy-terminal region of αA,B-crystallins from bovine lens (Ingolia and Craig, 1982b). αA,B-crystallins (molecular masses of 19–20-kDa) are major polypeptides which accumulate in the cytoplasm of the lens cells. This similarity to α-crystallin exists in most of the small hsps so far investigated. In addition, after the crystallin-like domain, the *Drosophila* small hsps share a conserved region of 25 residues.

Synthesis of the small hsps is co-ordinately induced by heat shock and by other physical or chemical stresses. As observed for all heat shock genes, the stress-dependent induction of the small hsps is under the control of specific regulatory sequences (HSE: heat shock element) localized upstream of the genes. A specific transcription factor, HSF, specially binds to HSE elements and activates the transcription of heat shock genes (Morimoto *et al.*, 1990).

As mentioned before, in contrast to their expression following heat shock, the small *Drosophila* hsps genes are expressed also in the absence of stress during development. This observation, first made in 1982 by Sirotkin and Davidson, was confirmed and extended by several other investigators (review: Arrigo and Tanguay, 1991). At the protein level, Cheney and Shearn (1983) were the first to report the synthesis of these polypeptides during development. Further studies included quantitative analysis of the small hsps expression using specific antibodies (Arrigo, 1987; Arrigo and Pauli, 1988; Pauli *et al.*, 1989; Arrigo and Tanguay, 1991; Marin *et al.*, 1993; see Section 7.3). During late third instar-larvae young pupae development, the expression of several small hsps corresponds to a peak of accumulation of the moulting hormone β-ecdysterone (Handler, 1982; Thomas and Lengyel, 1986; Mason *et al.*, 1984; Dubrovsky and Zhimulev, 1988). In addition, the synthesis of these proteins is observed also following treatment of tissue culture cells or imaginal discs with this hormone (Ireland and Berger, 1982; Ireland *et al.*, 1982; Vitek and Berger, 1984). It has been found that the heat shock element (HSE) does not play a major role in this phenomenon and that other regulatory sequences are probably involved, such as the steroid-receptor binding sequences found far upstream of the beginning of these genes (Mestril *et al.*, 1986; Riddihough and Pelham, 1986a,b). Other studies using P-element transformation indicate the presence of multiple cis- and trans-regulatory elements controlling the transcription of these *Drosophila* hsp genes (Cohen and Meselson, 1985; Glaser *et al.*, 1986; Klemenz and Gehring, 1986; Hofman *et al.*, 1987; Glaser and Lis, 1990). Taken together, these studies show that the coarse regulatory sequences of several small hsps genes differ from one tissue, or one stage to another. For example, in the case of hsp26 gene, sequences located 350–500 bp upstream of the beginning of the gene appear essential for its developmental induction in the female germ line (Cohen and Meselson, 1985). Contrasting with this result, the expression of this gene in the male germ line depends on specific sequences residing in the first 278 bp upstream of the gene (Glaser *et al.*, 1986).

Using specific DNA probes and antibodies, quantitative analysis of hsp23 and hsp27 mRNAs and proteins were performed during specific stages of the developing fly (Pauli *et al.*, 1989). It is seen in Table 1 that these two proteins differ by the fact that only hsp27 is strongly expressed during the embryogenesis. Transcripts from both genes are maximally abundant in white prepupae, but their corresponding polypeptides are only weakly detectable. The maximal abundance of these proteins usually corresponds to the middle of the pupal stage, when their corresponding mRNAs have almost completely disappeared. A similar lag in the expression of hsp27 is observed during the embryogenesis (Arrigo and Pauli, 1988; Pauli *et al.*, 1989). These observations suggest that the small hsps are more stable than their corresponding messages. Hsp23 and 27 may then play a role during long periods of the development and are not restricted to early embryogenesis (Zimmerman *et al.*, 1983) or to the prepupal stage (Pauli *et al.*, 1990). Taken together, these observations suggest that each individual small hsp has its own characteristic pattern of expression during development.

7.3 Tissue-specific expression of the small hsps during *Drosophila* development: presence of these proteins in the central nervous system

Contrasting with the co-ordinated synthesis of the hsps in almost all tissues following heat shock (Arrigo and Tanguay, 1991), the expression of the small hsps during development is tissue-specific (Glaser *et al.*, 1986; Glaser and Lis, 1990; Pauli *et al.*, 1990; Hass *et al.*, 1990; Arrigo and Tanguay, 1991; Marin *et al.*, 1993). For example, the transcription of hsp26 and 27 genes is detected in the ovarian nurse cells, with transcripts accumulating in the developing oocyte (Zimmerman *et al.* 1983). Using P-element transformation of a fusion gene hsp26–lacZ, Glaser *et al.* (1986) were the first to detect the presence of this hsp in neurocytes. The immunoblots presented in Figure 2 illustrates the different patterns of accumulation of hsp27 and 23 in dissected head, thorax and abdomen of late pupae. Hsp23 appears more abundant in the abdomen, while hsp27 is equally detected in the different parts of the pupal body. By taking into account the protein content of each part of the pupal body, it can be concluded that hsp27 is particularly abundant in the head of the developing fly. Recent findings by Marin *et al.* (1993) indicate that the distribution of hsp23 in the pupal body is dependent on the sex of the fly: this protein is more abundant in the head of the female compared to its level in the male. Hsp26 is only faintly detected in the head of the developing fly and appears more concentrated in the abdomen (Marin *et al.*, 1993). No specific expression of hsp22 is observed during the development, in contrast to its strong synthesis when the fly is exposed to heat shock or other forms of stress (Arrigo and Tanguay, 1991). As described in detail below (Section 7.3.1) the developmental expression of hsp27, 26 and 23 is tissue-specific. This is particularly obvious at the level of the central nervous system, gonads and imaginal discs.

7.3.1 Tissue-specific expression of hsp27

In a recent study using immunological detection on thin sections of the developing fly, we have analysed the tissue-specific accumulation of hsp27 during *Drosophila* development (Pauli *et al.*, 1990). Hsp27 is uniformly distributed in early embryos (see Pauli and Arrigo, 1988). Since the level of hsp27 mRNA is no more detectable after the pre-blastoderm period (Zimmerman *et al.*, 1983; Table 1), hsp27 polypeptide is probably synthesized during oogenesis and early development. During gastrulation, this protein is particularly abundant in the developing nervous system. In late embryos, decreased levels of hsp27 are observed suggesting that this polypeptide is slowly degraded. At the larval stages, hsp27 is detected only in the central nervous system (CNS) and the gonads of the first, second and early third-instar larvae (Figure 3A and 4A). Note that following a heat shock treatment all the larval tissues synthesize hsp27 which accumulates in the cell nucleus (Figure 3B). Of interest is the fact that the CNS

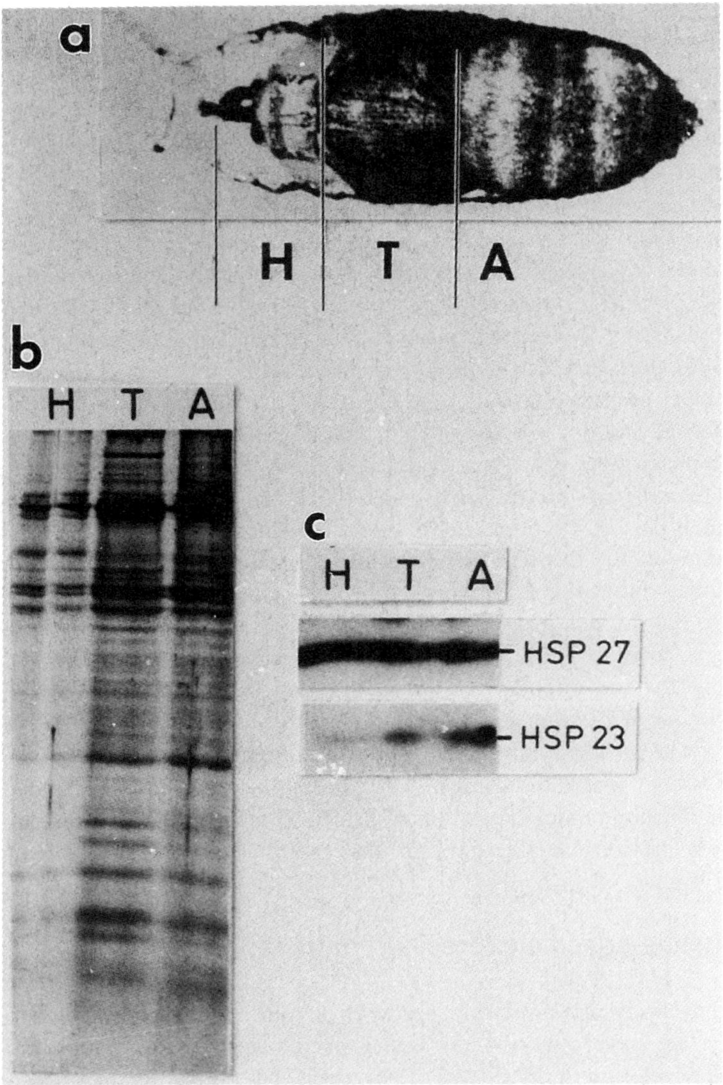

Figure 2 Analysis of hsp27 and hsp23 levels in the different parts of the body of late *Drosophila* pupae. Several late pupae were dissected and the protein content of the head, thorax and abdomen was analysed in SDS polyacrylamide gels. Hsp27 and hsp23 levels were determined in Western immunoblot using specific antisera. (a) late pupa with the different parts of the body analysed; (b) Coomassie stained gel of the protein content of the different parts of the pupal body; (c) Immunoblots. H: head, T: thorax, A: abdomen.

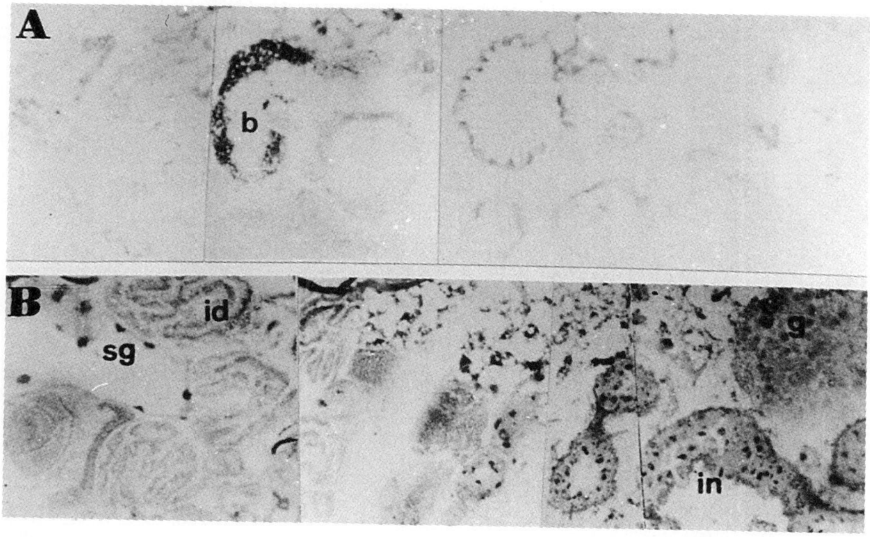

Figure 3 Presence of hsp27 in the CNS of unstressed larvae and in most of the larval tissues following heat shock. Early third-instar larva were either kept at normal temperature (A) or exposed to a heat-shock treatment of one hour at 37°C and allowed to recover for 2 h at 20°C (B). Thin sagittal sections of the larvae were processed for immunohistology using an hsp27 antibody. The figures are composed of different panel segments which represent adjacent photomicrographs which cover the entire larval body. Abbreviations: b, brain; id, imaginal discs; sg, salivary glands; in, intestine; g, gonad. The dark colouration reveals the presence hsp27. Note in (A) the specific accumulation of hsp27 in the brain and thoracic ganglion of the unstressed larvae. The gonads are not seen in this section. In (B) note the strong nuclear accumulation of hsp27 in the cells of most of the tissues of the heat treated larvae. Reproduced from Pauli *et al.*, *Journal of Cell Biology* (1990), 111, 817, 828, by copyright permision of the Rockefeller University Press.

and the gonads are characterized by high levels of mitotic cells. The imaginal discs are the only other larval tissue that contain cells active in division. Three tissues remain devoid of hsp27 until the late third-instar larval stage, when their cells stop to divide and begin to differentiate. At that time all the imaginal disc cells strongly accumulate hsp27. This phenomenon, particularly visible in the eye imaginal disc, is presented in Figure 4B. During the pupal development of the eye the level of hsp27 slowly decreases. In the eye of late pupae this protein is only present at the top of the omatidia (Figure 4C) and has disappeared from the eye of the new-born fly. Similarly, we observed that hsp27 is absent from all the disc-derived adult tissues. During the larval-pupal development, hsp27 is still detected in the cortex of the brain and thoracic ganglion where most of the neurocytes are localized (Figure 4 C,D). In late pupae and new-born flies, the neurocytes divide no longer and large portions of the cortex of the brain are already devoid of this protein. These observations suggest important

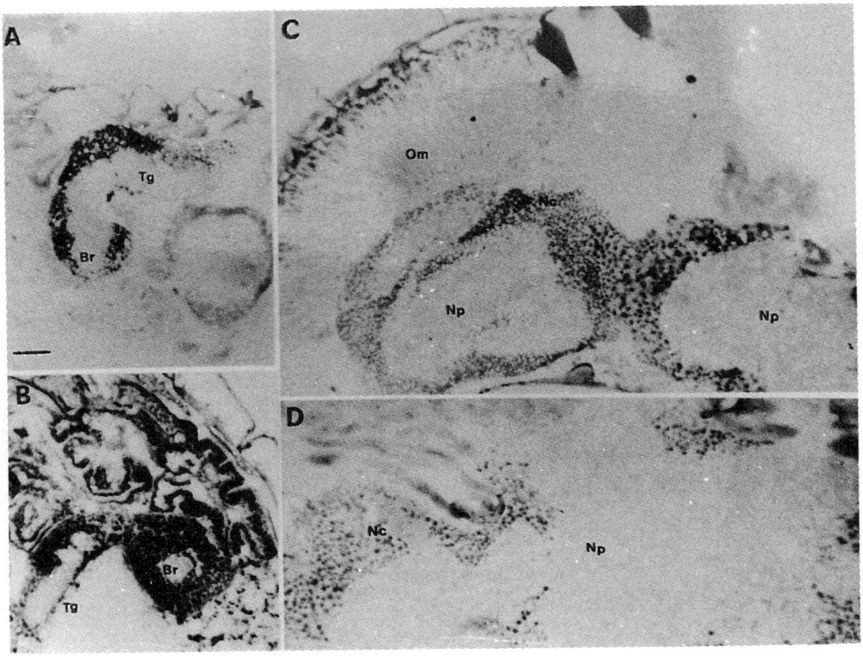

Figure 4 Immunolocalization of hsp27 in the central nervous system during *Drosophila* development. Immunohistological localization of hsp27 in the central nervous system detected in thin section of the developing insect. Enlargement of the regions showing the central nervous system is presented. The presence of hsp27 is visualized by a dark colouration. (A) sagittal section of second-instar larva, hsp27 is present in the CNS, Bar:50 μm; (B) sagittal section of late third-instar larva, note hsp27 staining in the CNS and imaginal discs. The eye disc is localized on the external side of the larval body on the right side of the picture; (C) partial view of a longitudinal section of the head of a late pupa, note the hsp27-staining at the top of the eye omatidies and at the level of the neurocytes in the cortex of several brain lobes; (D) longitudinal section of the thoracic ganglion of a late pupa, hsp27 is localized in the neurocytes of the cortex. Abbreviations: Br, brain; Tg, thoracic ganglion; R, retina; Nc, neurocyte; Np, neuropile; Om, ommatidia. Reproduced from Pauli *et al.*, *Journal of Cell Biology* (1990), 111, 817–828, by copyright permission of the Rockefeller University Press.

roles for hsp27 in cells that are actively dividing and in cells that have the capability to differentiate.

During the three larval stages, the CNS transcribes the hsp27 gene but transcripts are not detected in the adult fly. However, as mentioned above, the protein is still present in several parts of the brain and thoracic ganglion of the new-born fly. This indicates that the hsp27 present in the CNS of pupae and adult flies is synthesized during the larval stages. Thus, it can be concluded that the tissue-specific accumulation of hsp27 during development depends also on the stability of this protein.

Table 1 Quantitation of hsp23 and hsp27 transcripts and proteins during *Drosophila melanogaster* development.

	hsp23		**hsp27**	
	mRNA	Protein	mRNA	Protein
Embryos (h):				
6	+	+	+++++++++	+++++++++
12	+	+	–	+++++++
18	–	–	–	+++++
Instar Larvae:				
1st	+	–	+	++
2nd	–	–	–	++
3rd	+	+	+	++
Pupae (h):				
W	+++++++++	+	+++++++++	+++
25	++++	+++++	++++++	++++++
50	++	+++++++++	++++	+++++++
75	+	+++++++++	+	+++++++++
100	–/+	+++	+	+++++++++
125	–/+	+	–/+	++++
150	–/+	+	–/+	++
Adults:	ND	ND	–/+	++

Transcripts were analysed in Northern blots and the protein levels were determined in Western immunoblots using specific probes and antibodies. Note: – not detectable; –/+, barely detectable. Each additional + means a one-fold increase in the level of the mRNAs or their corresponding proteins. W, white prepupae. Adapted from our data (Pauli *et al.*, 1989).

7.3.2 Tissue-specific expression of hsp26

Analysis of the accumulation of hsp26–lacZ fusion protein during *Drosophila* development reveals that this protein has a pattern of tissue-specific expression which resembles that of hsp27 (Glaser *et al.*, 1986; Glaser and Lis, 1990). However, there are some differences, particularly at the quantitative level. In the CNS, low levels of this protein are detected in the neurocytes of the brain and thoracic ganglion (Marin *et al.*, 1993). The major tissue where this protein may play a role appears to be the gonads where it is very abundant throughout larval and pupal development.

7.3.3 Tissue-specific expression of hsp23

The developmental expression of small hsps was investigated by immunocytology using specific antibodies. In comparison to hsp27, only a weak accumulation of hsp23 is observed during the embryogenesis (see Table 1). This protein accumulates maximally during the pupal phase of the insect with an increase of about 30-fold (Arrigo, 1987). However, during the embryogenesis, interesting patterns of hsp23 expression were observed by different investigators (Tanguay, 1989; Hass *et al.*, 1990; Arrigo and Tanguay, 1991). At this stage of development, this protein accumulates in a restricted set of cells located at regular intervals along the CNS, which are identified as midline precursor cells (MPC). None of these cells stain with an anti-hsp27 antiserum. In new-born adults, hsp23 is found in the gonads and in neurocytes of the brain and thoracic ganglion (Marin *et al.*, 1993). In addition, hsp23 is also found in the central neuropile of these two organs (Marin *et al.*, 1993) as well as in glia cells and in the leg nerves (Hass *et al.*, 1990). As mentioned previously, hsp23 appears to be more abundant in the brain of new-born female fly (Marin *et al.*, 1993). Similarly to hsp27 and hsp26, this protein disappears rapidly from the brain during ageing. The pattern of expression of this protein in differentiating imaginal discs is unknown.

Taken together, these observations demonstrate that at least three of the small hsps display a specific spatial and temporal pattern of accumulation during *Drosophila* development. This suggests that these proteins may play similar or related function(s) in different tissues of the developing fly. At least two of these proteins, hsp27 and hsp23, appear to play a major role during the development of the central nervous system.

7.4 Cellular localization, biochemical properties and function(s) of the *Drosophila* small hsps

7.4.1 Cellular locale and biochemical properties

An interesting feature of the *Drosophila* small hsps concerns their different intracellular localization in unstressed cells. For example, hsp27 is nuclear following ecdysterone stimulation of embryonic tissue culture cells (Beaulieu *et al.*, 1989). During *Drosophila* development, hsp27 is also detected at the level of the nucleus, except in pre-blastoderm and blastoderm embryos, where it is cytoplasmic as well as nuclear (Pauli *et al.*, 1990). In new-born adults, this protein accumulates also in the nucleus of vasa-deferent, nurse and neuronal cells. Cellular fractionation analysis indicates that a large fraction of hsp27 is easily extracted from the nuclear pellet by non-ionic detergents, while after heat stress this is no longer the case (Arrigo and Pauli, 1988; Mehlen *et al.*, 1993). In contrast, hsp23 is cytoplasmic (Arrigo and Ahmad-Zadeh, 1981; Marin *et al.*, 1993) and appears concentrated in cytoplasmic granules (Tanguay *et al.*, 1985; Duband *et al.*, 1966); a situation which is reminiscent of the cellular locale of the small hsps of tomato cells (Nover *et al.*, 1989).

Biochemical analysis has demonstrated that, in unstressed tissue culture cells or during development, the *Drosophila* small hsps form large oligomers which sediment at 15–20S in a sucrose gradient (Arrigo and Ahmad-Zadeh, 1981; Arrigo *et al.*, 1985; Arrigo, 1987; Arrigo and Pauli, 1988) and display native molecular masses comprised between 200 to 800-kDa. In contrast, during heat shock or other forms of stress these proteins concentrate as insoluble structures inside the nucleus (Arrigo, 1980; Vincent and Tanguay, 1982). A similar phenomenon has been observed in all eukaryotic cells tested so far (Vincent and Tanguay, 1979; Collier and Schlesinger, 1986; Arrigo and Welch, 1987; Collier *et al.*, 1988; Arrigo *et al.*, 1988; Beaulieu *et al.*, 1989; Nover *et al.*, 1989; Rossi and Lindquist, 1989; Arrigo, 1990b; Mehlen *et al.*, 1993) and appears to be modulated by the metabolic state (Rossi and Lindquist, 1989) and the degree of thermotolerance (Arrigo, 1987; Arrigo *et al.*, 1988) of the cells. The biochemical significance of this particular behaviour of the small hsps is unknown.

7.4.2 Protective function(s) of the small hsps, and in particular hsp27 from *Drosophila*

Some of the heat shock proteins, particularly hsp70, appear to play a major role in the resistance and the recovery to heat shock (Arrigo, 1980; Pelham, 1984; Riabowol *et al.*, 1988). However, the function of the small hsps has remained obscure for a long time. In *Drosophila* no mutant of these proteins have been isolated, in spite of extensive mutagenesis in the region bearing the small hsps genes (Leicht and Bonner, 1988). In 1987, Eissenberg and Elgin have observed that a P-element insertion upstream of the hsp27 coding sequence that partially disrupts the transcription of the gene, is nevertheless viable. In other species mutagenesis studies have led to rather confusing observations. For example, Loomis and Wheeler (1982) describe a *Dictyostelium* mutant with an altered expression of the small hsps which is not viable at elevated temperature. This contrasts with the deletion analysis of the yeast hsp26 gene which does not appear to result in an altered phenotype (Petko and Lindquist, 1986). Another approach consists of the inhibition of the small hsps expression by using antisense RNA constructs (McGarry and Lindquist, 1986; Nicole and Tanguay, 1987). This technology was unsuccessful in defining the function of the small hsps. In 1983, Berger and Woodward proposed that the small hsps could play a role in the protection against heat shock. Recently, this hypothesis has been confirmed in three different studies (Landry *et al.*, 1989; Rollet *et al.*, 1992; Mehlen *et al.*, 1993). These experiments used expression vectors containing the gene encoding the small hsp under the control of a strong constitutive promotor. The DNA constructs were expressed in different types of mammalian cells and the cellular resistance to heat shock or other forms of stress was tested. The results clearly demonstrate that the small mammalian hsps as well as hsp27 from *Drosophila* are able to protect mammalian cells against heat shock. In addition, we observed that this protein also protects against oxidative stress (Mehlen *et al.*, 1993).

The role of the α-crystallin domain of hsp27 was analysed in deletion mutants in this conserved region (Figure 5). The first deletion mutant (Nco^-) covers 62 amino acids at the 5′ end of the α-crystallin domain and extends through the first 41 amino

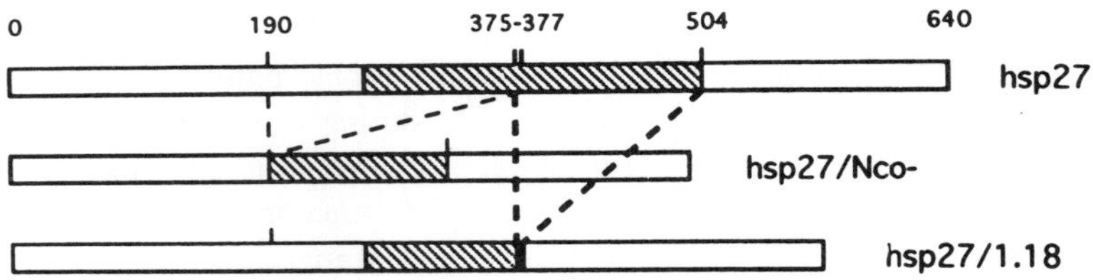

Figure 5 Schematic representation of hsp27 and deletion mutants of this protein in the conserved α-crystallin domain. The box represents the translated polypeptide and the hatched box shows the domain homologous to α-crystallin. Deletion Nco$^-$ was obtained by deleting the DNA fragment comprised between two NcoI sites. An AccI site was used to generate the 1.18 deletion by Bal 31 digestion (for details see Mehlen *et al.*, 1993).

acids of this region. Hsp27 and α-crystallin molecules have similar hydropathic profiles in their shared regions and a prominent hydrophilic peak is observed in this region which probably represents a major surface structural domain common to these proteins (Southgate *et al.*, 1983). Curiously, deletion of the first part of the α-crystallin domain of hsp27 (Nco$^-$ mutant), which partially includes the hydrophilic peak, does not appear to much affect the protective activity of this hsp. In contrast, hsp27 is no more active when the last 42 amino acids of the domain are deleted (1.18 mutant) (Mehlen *et al.*, 1993). These results suggest that the C-terminal part of the α-crystallin domain is important for the protective function of hsp27.

Hsp27, as well as the deletion mutants of this protein described above, are localized in the nucleus of unstressed COS cells (Figure 6). This confirms that the protective activity of hsp27 is localized inside the nucleus. Another interesting point is that hsp27 is a phosphoprotein (Rollet and Best-Belpomme, 1986), that appears to be heavily phosphorylated during development (Arrigo and Pauli, 1988). Thus, it is possible that the protective anti-stress activity of this protein is regulated by its level of phosphorylation.

Based on the above consideration, it is reasonable to propose that *in vivo*, hsp27 protects *Drosophila* cells against stress. Consequently, its expression during neural, germ-line and imaginal disc development can be regarded as being part of a mechanism that enhances the resistance of these sensible organs to environmental or physiological changes.

7.5 Expression of small hsps during development in species other than *Drosophila* or mammals

Few studies were performed to analyse the constitutive expression of the small hsps in species other than *Drosophila* or mammals. However, it can be mentioned that the unique small hsps of yeast is constitutively expressed during meiosis and ascospore formation (Kurtz *et al.*, 1986). In *Xenopus*, low levels of hsp30 are observed

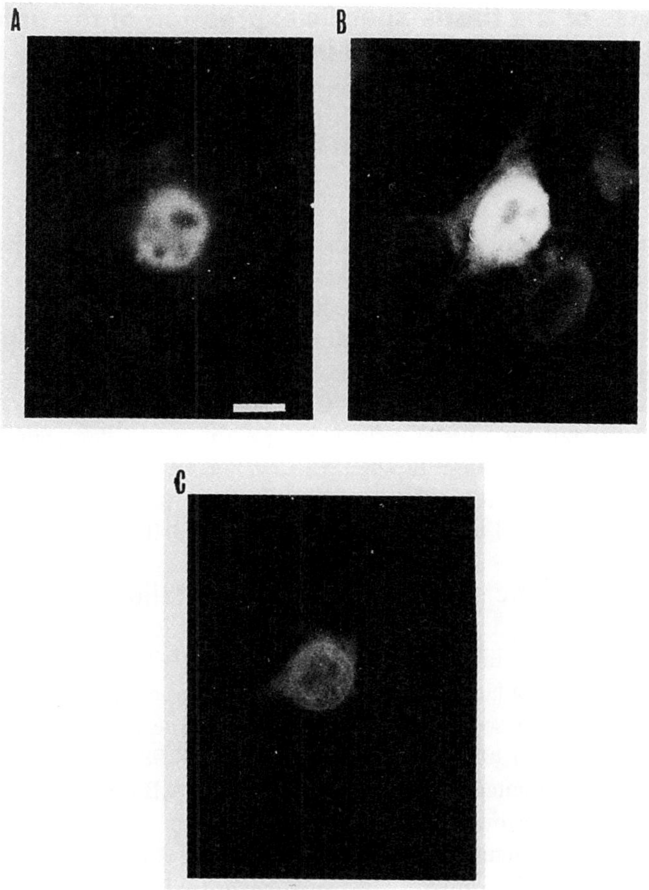

Figure 6 Immunofluorescence analysis of monkey COS cells expressing hsp27 from *Drosophila.* COS cells transiently expressing hsp27 or the different deletion mutants of this protein described in Figure 5 were fixed with cold methanol before being processed for indirect immunofluorescence analysis using an antibody specific for this hsp. (A) hsp27; (B) Mutant 1.18; (C) mutant Nco$^-$. Note the nuclear staining of hsp27 and its deleted forms. Bar: 10 μm.

in unstressed embryos (Heikkila *et al.*, 1991). Interestingly, sensory ganglia from tadpoles but not from adult bullfrogs synthesize heat shock-like proteins *in vitro* at non-heat shock temperature (Hammerschlag *et al.*, 1989). These proteins include a 25-kDa polypeptide which is probably related to the small hsps.

7.6 Analysis of the tissue-specific expression of the unique small hsp and αB-crystallin in mammalian tissues

7.6.1 Tissue-specific expression of the unique mammalian small hsp

In mammalian cells, there is only one small hsp (designated either as hsp25, hsp27 or hsp28), which is encoded by a gene family of four members (Hickey *et al.*, 1986). Similarly to the small hsps of *Drosophila*, this protein is constitutively expressed in a number of different tissues following treatment with steroid hormones (Fuqua *et al.*, 1989). Through days 13–20 of mouse development, this protein is present in neurons of the spinal cord and Purkinje cells (Gernold *et al.*, 1993). It is also heavily present in tissues containing muscles cells. During early rat development, Walsh *et al.* (1991) observed the presence of hsp28 at the level of neural tube closure. In young or adult mice, this protein is absent from the brain but it is still present in tissues that contain smooth or skeletal muscles (Klemenz *et al.*, 1993; Tanguay *et al.*, 1993).

7.6.2 Tissue-specific expression of αB-crystallin

αA,B-crystallins are two highly related polypeptides (19–20-kDa) that accumulate to high levels in the cytoplasm of eye lens cells. As described in Figure 1, these polypeptides contain a long sequence of homology with the small hsps (Ingolia and Craig, 1982b; Wistow, 1985; Wistow and Piatigorsky, 1988; de Jong *et al.*, 1989). Thus, it has been postulated that the small hsps and αA,B-crystallins share a common ancestor gene (Wistow, 1985; de Jong *et al.*, 1989). Moreover, these proteins are considered now as small hsps since their expression is stimulated by heat shock (Klemenz *et al.*, 1991). Recently, it has been shown that the A and B forms of α-crystallin differ in their tissue-specific expression. The constitutive expression of the A form of α-crystallin is restricted to lens cells, while the B form is detected not only in lens but also in different tissues (Bhat and Nagineni, 1989; Dubin *et al.*, 1989; Klemenz *et al.*, 1993). For example, in the human brain, αB-crystallin accumulates in the Rosenthal fibres of astrocytes from patients suffering from Alexander's disease (infantile neuronal degenerescence) (Iwaki *et al.*, 1989). Table 2 summarizes those tissues where elevated levels of αB-crystallin are observed in adult mice. Similarly to hsp28, αB-crystallin is strongly expressed in tissues containing muscle cells, but in contrast to this hsp it is present also in the brain. During mouse development αB-crystallin is expressed in the CNS, particularly at the level of glia cells (Gernold *et al.*, 1993).

7.6.3 Cellular locale and biochemical properties

Similarly to the other small hsps, hsp28 is an oligomeric protein with heterogeneous native molecular masses (200–800-kDa). This protein accumulates in the

Table 2 Tissue-specific expression of the unique small hsp and αB-crystallin in mice.

Tissues	Small hsp	αB-crystallin
Adult		
Skin	++	+++
Stomach	+++	++
Intestine	+	–
Colon	++	–
Lung	++	+
Brain	–	++
Liver	–	–
Spleen	–	–
Thymus	+/–	–
Salivary Gland	–	–
Bladder	+++	+
Kidney	+/–	++
Testes	++	–
Ovary	++	–
Uterus	++	–
Pancreas	–	–
Heart	++++	+++++
Skeletal Muscle	++	++
Lens	+++	++++
NIH 3T3 cells		
Unstressed	–	–
HS treated	++	+++

The levels of the mammalian small hsp and αB-crystallin were analysed in immunoblots using specific antibodies. A visual estimation of the data of Klemenz *et al.* (1992) and Tanguary *et al.* (1993) are presented. Note: –, not detectable; +/–, barely detectable. Each additional + means a two- to three-fold increased amount of proteins.

cytoplasm and is able to redistribute inside the nucleus during heat shock (Arrigo and Welch, 1987; Arrigo *et al.*, 1988; see also Section 7.4.1 and Figure 7). In the cytoplasm of human tissue culture cells, this protein is usually concentrated in a polarized perinuclear zone close to the Golgi apparatus. The native molecular mass and the cellular locale of α-crystallin resembles that of hsp28 (Seizen *et al.*, 1978; Arrigo, unpublished observations) and similarly to the small hsps, this protein also shows a redistribution inside the nucleus during heat shock (Klemenz *et al.*, 1991). An interesting property of αA,B-crystallins and hsp28 concerns their high levels of phosphorylation in different cell conditions; a property shared by the *Drosophila* hsp27. The phosphorylation of hsp28 and αB-crystallin is specific for serine residues (Voorter *et al.*, 1989; Gaestel *et al.*, 1991; Landry *et al.*, 1992). In the case of hsp28,

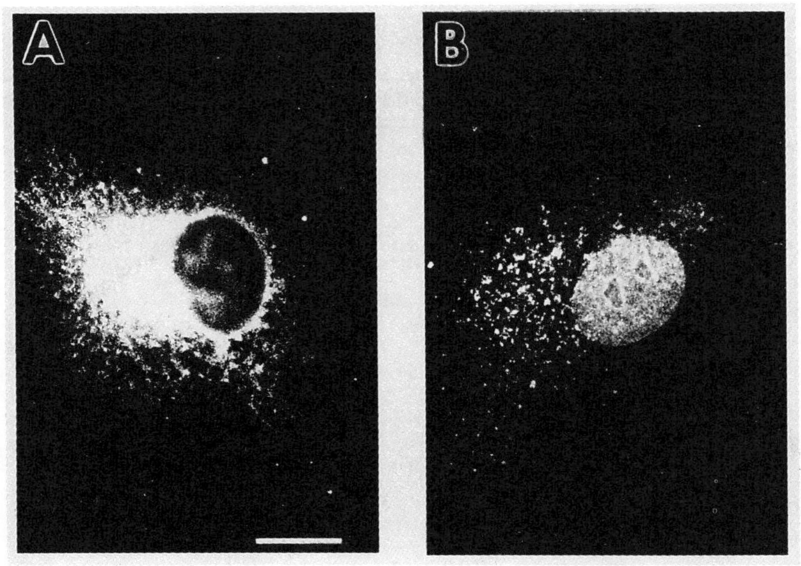

Figure 7 Cellular localization of the unique mammalian small hsp. Unstressed or heat-shock treated (43°C for 90 min) monkey COS cells growing on glass cover-slips at 37°C were fixed with cold methanol and processed for indirect immunofluorescence using an antibody specific for hsp28. (A) cells kept at normal temperature; (B) cells exposed to heat shock. Note the polarized perinuclear localization of hsp28 in unstressed cells and the redistribution of this protein in the nucleus following heat shock. Bar: 8 μm.

this phenomenon has been observed during heat shock (Arrigo and Welch, 1987), and after treatment with agents which either modulate cell division (fresh serum, mitogens), or cell differentiation (phorbol-12-myristate-13-acetate, retinoic acid) (Welch, 1985; Arrigo, 1990b). This protein modification appears of great importance since preliminary results indicate that the protective anti-stress activity of the mammalian small hsp is modulated by its level of phosphorylation (J. Landry, Winternational Symposium, Québec, Canada, March 1993).

7.6.4 The mammalian small hsp and αB-crystallin are molecular chaperones

Over-expression of the mammalian small hsp (Landry *et al.*, 1989) or αB-crystallin (Gernold *et al.*, 1993; Mehlen *et al.*, in preparation) enhances the resistance of mammalian cells to heat shock or oxidative stresses. Thus, at least three small hsps (hsp28 from mammals, αB-crystallin and *Drosophila* hsp27) are able to protect cells against stress-induced damages. Moreover, recent *in vitro* studies have demonstrated that αB-crystallin and the mammalian small hsp are able to renature denatured or mal-conformed proteins (Jakob *et al.*, 1993). This phenomenon is ATP independent.

The *Drosophila* hsp27 has not been tested yet. These observations clearly demonstrate that αB-crystallin and hsp28 can act *in vitro* as molecular chaperones.

Consequently, the expression of the small hsp and αB-crystallin in the developing brain of mammals reflects a biochemical mechanism that may protect neural cells against environmental or physiological stresses.

7.7 Possible function(s) of the small hsps in the control of cell division and differentiation

In addition to their role as molecular chaperones during stress, the small hsps may play additional role(s) in the normal unstressed cell. Such a role could be at the level of the control of cell division/differentiation. During *Drosophila* development, hsp27 expression is observed in tissues which contain either highly proliferating cells (neurocytes of the developing brain) or cells which stop dividing and differentiate (imaginal discs) (Pauli *et al.*, 1990). Similarly, the expression of hsp22 and 23 is triggered by agents (β-ecdysterone, teratogens . . .) which also inhibit cell division and stimulate cell differentiation (Buzin and Bounias-Vardiabasis, 1982). Interestingly, the unique mammalian small hsp displays a similar behaviour: following transplantation in mice, Ehrlich ascites tumour cells cease to divide concomitantly with the accumulation of hsp28 (Gaestel *et al.*, 1989). This protein also accumulates during the differentiation and growth-arrest of several human cells (Spector *et al.*, 1992; Shakoori *et al.*, 1992). Another important observation concerns the fact that a cytoplasmic turkey protein that is related to the small hsp acts *in vitro* as an inhibitor of actin polymerization (Miron *et al.*, 1991). Hsp28 from mammals appears to share this particular property (Lavoie *et al.*, 1993). These proteins may then play a role in the control of cell division by regulating the polymerization of actin microfilaments which are necessary during the cytokinesis step of the cell cycle. In this respect, the fact that some cells over-expressing hsp28 have a reduced rate of cell division is very intriguing (Knauf *et al.*, 1992). High levels of constitutively expressed small hsps may then be required to control the rate of cell division in some cells with a high mitotic index (such as those of the developing nervous system) or when cells need to stop dividing before they enter the differentiation pathway. Whether hsp23 and hsp27 from *Drosophila* or αB-crystallin, which are highly expressed during the development of the central nervous system, share this function remains to be shown. However, since hsp27 is nuclear in unstressed cells (see Section 7.4), the mode of action of this *Drosophila* protein may differ from that of the other cytoplasmic small hsps (Mehlen *et al.*, 1993).

7.8 Concluding remarks

During development, the small hsps are expressed in a tissue- and stage-specific manner. They are particularly abundant in the central nervous system of different organisms and least abundant in the developing eye of *Drosophila*. At the cellular level, the small hsps display different intracellular locale, being either cytoplasmic or nuclear. Recently, several studies have provided decisive hints concerning the function of these fascinating proteins. *In vivo*, a modulation of the rate of cell division and a protection against heat shock or oxidative stress have been observed in cells over-expressing these proteins. Moreover, *in vitro* biochemical analysis has demonstrated their ability to renature mal-folded and denatured proteins. Based on the above considerations, it is reasonable to propose that the small hsps probably have two roles: the first, at the level of the control of cell division/differentiation in unstressed cells, and the second as molecular chaperones during stress. The presence of these proteins in the developing central nervous system may then represent a molecular mechanism that controls the growth of this tissue and protects it against environmental or physiological stress. Future studies should bring us further understanding on the mechanisms regulating the tissue-specific expression and the role of these hsps in the developing central nervous system.

Acknowledgements

P.M. was supported by a doctoral training fund from the Ecole Normale Supérieure (ENS). This work was supported by grants 6011 from the Association pour la Recherche sur le Cancer and 91.C.388 from the Ministère de la Recherche et de la Technologie (to A.-P.A.).

References

Aoyama, A., Fröhli, E., Schafer, R. & Klemenz, R. (1993) αB-crystallin expression in mouse NIH 3T3 fibroblasts: glucocorticoid responsiveness and involvement in thermal protection. *Mol. Cell Biol.* **13**, 1824–1835.

Arrigo, A.-P. (1980) Investigation of the function of the heat shock proteins in *Drosophila melanogaster* tissue culture cells. *Mol. Gen. Genet.* **178**, 517–524.

Arrigo, A.-P. (1987) Cellular localization of hsp23 during *Drosophila* development and subsequent heat shock. *Dev. Biol.* **122**, 39–48.

Arrigo, A.-P. (1990a) The monovalent ionophore *monensin* maintains the nuclear localization of the human stress protein hsp28 during heat shock recovery. *J. Cell Science.* **96**, 419–427.

Arrigo, A.-P. (1990b) Tumour necrosis factor α induces the rapid phosphorylation of the mammalian low molecular weight heat shock protein hsp28. *Mol. Cell Biol.* **10**, 1276–1280.

Arrigo, A.-P. & Ahmad-Zadeh, C. (1981) Immunofluorescence localization of a small heat shock protein (hsp23) in *Drosophila melanogaster. Mol. Gen. Genet.* **184**, 73–79.

Arrigo, A.-P. & Welch, W.J. (1987) Characterization and purification of the mammalian 28,000 dalton heat shock protein. *J. Biol. Chem.* **262**, 15 359–15 369.

Arrigo, A.-P. & Pauli, D. (1988) Characterization of hsp27 and of the three immunologically related polypeptides during *Drosophila* development. *Exp. Cell. Res.* **175**, 169–183.

Arrigo, A.-P. & Tanguay, R.M. (1991) Expression of heat shock proteins during development in *Drosophila*. In *Heat Shock and Development* (eds Hightower, L. & Nover, L.), pp 106–119. Berlin, Springer-Verlag.

Arrigo, A.-P., Fakan, S. & Tissières, A. (1980) Localization of the heat shock induced proteins in *Drosophila melanogaster* tissue culture cells. *Dev. Biol.* **78**, 86–103.

Arrigo, A.-P., Suhan, J. & Welch, W.J. (1988) Dynamic changes in the structure and locale of the mammalian low molecular weight heat shock protein. *Mol. Cell. Biol.* **8**, 5059–5071.

Arrigo, A.-P., Darlix, J.-D., Khandjian, E.W. & Spahr, P.-F. (1985) Characterization of the prosome from *Drosophila* and its similarity to the cytoplasmic structure formed by the low molecular weight heat shock proteins. *EMBO J.* **4**, 2942–2954.

Ayme, A. & Tissières, A. (1985) Locus 67B of *Drosophila melanogaster* contains seven, not four, closely related heat shock genes. *EMBO J.* **4**, 2949–2954.

Beaulieu, J.F., Arrigo, A.-P. & Tanguay, R.M. (1989) Interaction of *Drosophila* 27-kDa heat shock protein with the nucleus of heat shocked and ecdysterone-stimulated cultured cells. *J. Cell. Science* **92**, 29–36.

Berger, E.M. & Woodward, M.P. (1983) Small heat shock proteins of *Drosophila* may confer thermal tolerance. *Exp. Cell. Res.* **147**, 437–442.

Bhat, S.P. & Nagineni, C.N. (1988) αB-subunit of lens-specific αB-crystallin is present in other ocular and non-ocular tissues. *Biochem. Biophys. Res. Comm.* **158**, 319–325.

Bond, U. & Schlesinger, M.J. (1987) Heat shock proteins and development. *Adv. Genet.* **24**, 1–29.

Buzin, C.H. & Bournias-Vardiabasis, N. (1982) The induction of a subset of heat shock proteins by drugs that inhibit differentiation in *Drosophila* embryonic cell cultures. In *Heat Shock from Bacteria to Man* (eds Schlesinger, M.J., Ashburner, M., Tissières, A.), pp 387–394. New York, Cold Spring Harbor Press.

Cheney, C.M. & Shearn, A. (1983) Developmental regulation of *Drosophila* imaginal discs proteins: Synthesis of a heat shock protein under non heat shock conditions. *Dev. Biol.* **95**, 325–330.

Chomyn, A. & Mitchell, H.K. (1982) Synthesis of the 84 000 dalton protein in normal and heat shocked *Drosophila melanogaster* cells as detected by specific antibody. *Insect. Biochem.* **12**, 105–104.

Cohen, R.S. & Meselson, M. (1985) Separate regulatory element for the heat-inducible and ovarian expression of the *Drosophila* hsp26 gene. *Cell* **43**, 737–746.

Collier, N.C. & Schlesinger, M. (1986) The dynamic state of heat shock proteins in chicken embryo fibroblasts. *J. Cell. Biol.* **103**, 1495–1507.

Collier, N.C., Heuser, J., Aach-Levy, M. & Schlesinger, M. (1988) Ultrastructural and biochemical analysis of the stress granules in chicken embryo fibroblasts. *J. Cell Biol.* **106**, 1131–1139.

Corces, V., Holmgren, R., Freund, R., Morimoto, R. & Meselson, M. (1980) Four heat shock proteins of *Drosophila melanogaster* coded within a 12-kb region in chromosome subdivision 67B. *Proc. Natl Acad. Sci. USA.* **77**, 5390–5393.

Craig, E.A. & McCarthy, B.J. (1980) Four *Drosophila* heat shock genes at 67B: Characterization of recombinant plasmids. *Nucleic Acids. Res.* **8**, 4441–4457.

Craig, E.A., Ingolia, T.D. & Manseau, L.J. (1983) Expression of *Drosophila* cognate genes during heat shock and development. *Dev. Biol.* **99**, 418–426.

de Jong, W.W., Leeunissen, J.A.M., Leenen, P.J.M., Zweers, A. & Versteeg, M. (1989) Dogfish α-crystallin sequences. *J. Biol. Chem.* **263**, 5141–5149.

Duband, J.-L., Lettre, F., Arrigo, A.-P. & Tanguay, R. (1986) Expression and cellular localization of HSP23 in unstressed and heat shocked *Drosophila* culture cells. *Canad. J. Genet. Cytol.* **28**, 1088–1092.

Dubin, R.A., Wawrousek, E.F. & Piatigorsky, J. (1989) Expression of the murine αB-crystallin gene is not restricted to the lens. *Mol. Cell Biol.* **9**, 1083–1091.

Dubrovsky, E.B. & Zhimulev, I.F. (1988) Trans-regulation of ecdysterone-induced protein synthesis in *Drosophila melanogaster* salivary glands. *Dev. Biol.* **127**, 33–44.

Eissenberg, J.C. & Elgin, S.C.R. (1987) HSP28^{stl}: A P-element insertion mutation that alters the expression of a heat shock gene in *Drosophila melanogaster. Genetics* **115**, 333–340.

Fuqua, S.A.W., Blum-Salingaros, M. & McGuire, W.L. (1989) Induction of the estrogen-regulated '24-kDa' protein by heat shock. *Cancer Res.* **49**, 4126–4129.

Gaestel, M., Schröder, W., Benndorf, R., Lippman, C., Buchner, K., Huchot, F., Ermann, V.A. & Bielka, H. (1991) Identification of the phosphorylation sites of the murine small heat shock protein 25. *J. Biol. Chem.* **266**, 14 721–14 724.

Gaestel, M., Gross, B., Benndorf, R., Strauss, M., Schunk, W.-H., Kraft, R., Otto, A., Böhm, H., Stahl, J., Drabsch, H. & Bielka, H. (1989) Molecular cloning, sequencing and expression in *Escherichia coli* of the 25-kDa growth-related protein of Ehrlich ascites tumour and its homology to mammalian stress proteins. *Eur. J. Biochem.* **179**, 209–213.

Gernold, M., Knauf, U., Gaestel, M., Stahl, J. & Kloetzel, P.-M. (1993) Development and tissue-specific distribution of mouse small heat shock protein hsp25. *Devel. Gen.* **14**, 103–111.

Glaser, R.L. & Lis, J.T. (1990) Multiple, compensatory regulatory elements specify spermatocyte-specific expression of the *Drosophila melanogaster* hsp26 gene. *Mol. Cell Biol.* **10**, 131–137.

Glaser, R.L., Wolfner, M.F. & Lis, J.T. (1986) Spatial and temporal pattern of hsp26 expression during normal development. *EMBO J.* **5**, 747–754.

Haass, C., Klein, U. and Kloetzel, P.M. (1990) Developmental expression of *Drosophila melanogaster* small heat shock proteins. *J. Cell. Sci.* **96**, 413–418.

Hammerschlag, R., Maines, S. & Ando, M. (1989) Sensory glia from tadpoles but not adult bullfrogs synthesize heat shock-like proteins *in vitro* at non heat shock temperature. *J. Neurosci. Res.* **23**, 416–424.

Handler, A.M. (1982) Ecdysteroid titres during pupal and adult development in *Drosophila melanogaster. Dev. Biol.* **93**, 73–82.

Heikkila, J.J., Krone, P.H. & Ovseneck, N. (1991) Regulation of heat shock gene expression during *Xenopus* development. In *Heat Shock and Development* (eds Hightower, L. & Nover, L.), pp 120–137. Berlin, Springer-Verlag.

Hofman, E.S., Gerring, S.L. & Corces, K.J. (1987) The ovarian, ecdysterone and heat shock-responsive promoters of *Drosophila melanogaster* hsp27 gene react differently to perturbation of DNA sequence. *Mol. Cell Biol.* **7**, 973–981.

Hickey, E., Brandon, S.E., Potter, R., Stein, G., Stein, J. & Weber, L.A. (1986) Sequence and organization of genes encoding the human 27-kDa heat shock protein. *Nucleic Acids Res.* **14**, 4127–4144.

Ingolia, T.D. & Craig, E. (1982a) *Drosophila* gene related to the major heat shock-induced gene is transcribed at normal temperature and not induced by heat shock. *Proc. Natl Acad. Sci. USA.* **79**, 525–529.

Ingolia, T.D. & Craig, E. (1982b) Four small *Drosophila* heat shock proteins are related to each other and to mammalian α-crystallin. *Proc. Natl Acad. Sci. USA* **79**, 2360–2364.

Ireland, R.C. & Berger, E.M. (1982) Synthesis of the low molecular weight heat shock proteins stimulated by ecdysterone in a cultured *Drosophila* cell line. *Proc. Natl Acad. Sci. USA* **79**, 855–859.

Ireland, R.C., Berger, E.M., Sirotkin, K., Yaud, M.A., Osterburg, D. & Fristom, J. (1982) Ecdysterone induces the transcription of four heat shock genes in *Drosophila* S3 cells and imaginal discs. *Dev. Biol.* **93**, 498–507.

Iwaki, T., Kume-Iwaki, A., Liem, R.K.H. & Golman, J.E. (1989) αB-crystallin is expressed in non-lenticular tissues and accumulates in Alexander's disease brain cells. *Cell* **57**, 71–78.

Jakob, U., Gaestel, M., Engel, K. & Buchner, J. (1993) Small heat shock proteins are molecular chaperones. *J. Biol. Chem.* **268**, 1517–1520.

Knauf, U., Bielka, H. & Gaestel, M. (1992) Over-expression of the small heat shock protein, hsp25, inhibits growth of Ehrlich ascites tumour cells. *FEBS Lett.* **308**, 297–302.

Klemenz, R. & Gehring, W.J. (1986) Sequence requirement for expression of the *Drosophila melanogaster* heat shock protein hsp22 gene during heat shock and normal development. *Mol. Cell Biol.* **6**, 2011–2019.

Klemenz, R., Andres, A.-C., Froehli, E., Schaefer, R. & Aoyama, A. (1993) Expression of the murine small heat shock protein hsp25 and αB-crystallin in the absence of stress. *J. Cell Biol.* **120**, 639–645.

Klemenz, R., Froehli, E., Steiger, R.H., Schaefer, R. & Aoyama, A. (1991) α-crystallin is a small heat shock protein. *Proc. Natl Acad. Sci. USA* **88**, 3652–3656.

Kurtz, S., Rossi, J., Petko, J. & Lindquist, S. (1986) An ancient developmental induction: Heat shock protein induced in sporulation and oogenesis. *Science* **231**, 1154–1157.

Landry, J., Chretien, P., Lambert, H., Hickey, E. & Weber, L.A. (1989) Heat shock resistance conferred by expression of the human HSP27 gene in rodent cells. *J. Cell Biol.* **109**, 7–15.

Landry, J., Lambert, H., Zhou, M., Lavoie, J.N., Hickey, E., Weber, L. & Anderson, C. (1992) Human hsp27 is phosphorylated at serine 78 and 82 by heat shock and mitogen-activated kinases that recognize the same amino acid motif as S6 kinase II. *J. Cell. Biol.* **267**, 794–803.

Lavoie, J.N., Gingras-Breton, G., Tanguay, R.M. & Landry, J. (1993) Induction of Chinese hamster hsp27 gene expression in mouse cells confers resistance to heat shock. *J. Biol. Chem.* **268**, 3420–3429.

Leicht, B.G. & Bonner, J.J. (1988) Genetic analysis of chromosomal region 67 A-D of *Drosophila melanogaster. Genetics* **11**, 579–593.

Loomis, W.F. & Wheeler, S. (1980) Heat shock response in *Dictyostelium. Dev. Biol.* **79**, 399–408.

Marin, R., Valet, J.-P. & Tanguay, R. (1993) Hsp23 and hsp26 exhibit distinct spatial and temporal patterns of constitutive expression in *Drosophila* adults. *Devel. Gen.* **14**, 112–118.

Mason, P.J., Hall, L.M.C. & Gausz, J. (1984) The expression of heat shock genes during normal development in *Drosophila melanogaster. Mol. Gen. Genet.* **194**, 73–78.

McGarry, T.J. & Lindquist, S. (1986) Inhibition of heat shock protein synthesis by heat inducible antisense RNA. *Proc. Natl Acad. Sci. USA* **83**, 399–403.

Mckenzie, S.L., Henikoff, S. & Meselson, M. (1975) Localization of RNA from heat-induced polysomes at puff sites in *Drosophila melanogaster. Proc. Natl Acad. Sci. USA* **72**, 1117–1121.

Mehlen, P., Briolay, J., Smith, L., Diaz-Latoud, C., Fabre, N., Pauli, D. & Arrigo, A.-P. (1993) Analysis of the resistance to heat and hydrogen peroxide stresses in COS cells transiently expressing wild type or deletion mutants of the *Drosophila* 27-kDa heat shock protein. *Eur. J. Biochem.* **215**, 277–284.

Mestril, R., Shiller, P., Amin, J., Klapper, H., Jayakumar, A. & Voellmy, R. (1986) Heat shock and ecdysterone activation of *Drosophila melanogaster* hsp23 gene; a sequence element implied in developmental regulation. *EMBO J.* **5**, 1667–1673.

Mirault, M.-E., Goldschmidt-Clermont, M., Moran, L., Arrigo, A.-P. and Tissières, A. (1978) The effect of heat shock on gene expression in *Drosophila melanogaster. Cold Spring Harbor Symp. Quant. Biol.* **42**, 819–827.

Miron, T., Vancompernolle, K., Vanderkerckhove, J., Wilcheck, M. & Geiger, B. (1991) A 25-kDa inhibitor of actin polymerization is a low molecular weight heat shock protein. *J. Cell. Biol.* **114**, 255–261.

Morimoto, R.I., Tissières, A. & Georgopoulos, C. (1990) Stress proteins in biology and medicine. New York, Cold Spring Harbor Laboratory Press.

Nicole, L. & Tanguay, R.M. (1987) On the specificity of antisense RNA to arrest *in vitro* translation of mRNA coding for *Drosophila* hsp23. *Biosci. Rep.* **7**, 239–246.

Nover, L., Sharf, K.D. & Neuman, D. (1989) Cytoplasmic heat shock granules are formed from precursor particles and are associated with a specific set of mRNAs. *Mol. Cell. Biol.* **9**, 1298–1308.

Pauli, D. & Tissières, A. (1990) Developmental expression of the heat shock genes in *Drosophila melanogaster*. In *Stress proteins in Biology and Medicine* (eds Morimoto, R., Tissiöres, A. & Georgopoulos, C.), pp 361–378 New York, Cold Spring Harbor Press.
Pauli, D., Arrigo, A.-P. & Tissières, A. (1992) Heat shock response in *Drosophila. Experientia* **48**, 623–629.
Pauli, D., Tonka, C.-H., Tissières, A. & Arrigo, A.-P. (1990) Tissue-specific expression of the heat shock protein hsp27 during *Drosophila melanogaster* development. *J. Cell Biol.* **111**, 817–828.
Pauli, D., Arrigo, A.-P., Vasquez, J., Tonka, C.-H. & Tissières, A. (1989) Expression of the small heat shock genes during *Drosophila* development: Comparison of the accumulation of HSP23 and hsp27 mRNAs and polypeptides. *Genome.* **31**, 671–676.
Pelham, H.R.B. (1984) HSP70 accelerates the recovery of nucleolar morphology after heat shock. *EMBO J.* **3**, 3095–3100.
Petersen, N.S., Moeller, G. & Mitchell, H.K. (1979) Genetic mapping of the coding regions for three heat shock proteins in *Drosophila melanogaster. Genetics* **92**, 891–902.
Petko, L. & Lindquist, S. (1986) Hsp26 is not required for growth at high temperature, nor for thermotolerance, spore development or germination. *Cell* **45**, 885–894.
Riabowol, K.T., Mizzen, L.A. & Welch, W.J. (1988) Heat shock is lethal to fibroblasts microinjected with antibodies against hsp70. *Science* **242**, 433–436.
Riddlehough, G. & Pelham, H.R.B. (1986a) Activation of the *Drosophila* hsp27 promoter by ecdysones involves independent and remote regulatory sequences. *EMBO J.* **5**, 1653–1658.
Riddihough, G. & Pelham, H.R.B. (1986b) An ecdysone response element in the *Drosophila* hsp27 promoter. *EMBO J.* **6**, 3729–3734.
Rollet, E. & Best-Belpomme, M. (1986) Hsp26 and 27 are phosphorylated in response to heat and ecdysterone in *Drosophila melanogaster* cells. *Biochem. Biophys. Res. Comm.* **141**, 426–433.
Rollet, E., Lavoie, J.N., Landry, J. & Tanguay, R.M. (1992) Expression of *Drosophila*'s 27-kDa heat shock protein into rodent cells confers thermal resistance. *Biochem. Biophys. Res. Comm.* **185**, 116–120.
Rossi, J. & Lindquist, S. (1989) The intracellular location of yeast heat shock protein 26 varies with metabolism. *J. Cell. Biol.* **108**, 425–439.
Seizen, R.J., Bindel, J.G. & Hoenders, H.J. (1978) The quaternary structure of bovine α-crystallin. *Eur. J. Biochem.* **91**, 387–396.
Shakoori, A.R., Oberdorf, A.M., Owen, T.A., Weber, L.A., Hickey, E., Stein, J., Lian, J.B. & Stein, G. (1992) Expression of heat shock genes during differentiation of mammalian osteoblasts and promyelocytic leukemia cells. *J. Cell. Biochem.* **48**, 277–287.
Sirotkin, K. & Davidson, N. (1982) Developmentally regulated transcription from *Drosophila melanogaster* site 67B. *Dev. Biol.* **89**, 196–210.
Southgate, R., Ayme, A. & Voellmy, R. (1983) Nucleotide sequence analysis of the *Drosophila* small heat shock gene cluster at *locus* 67B. *J. Mol. Biol.* **165**, 35–57.
Southgate, R., Mirault, M.-E., Ayme, A. & Tissières, A. (1985) Organization, sequences and induction of heat shock genes. In *Changes in Eukaryotic Gene Expression in Response to Environmental Stress* (eds Atkinson, B.G. & Walden, D.B.), pp 3–30. New York, Academic Press.
Spector, N.L., Samson, W., Ryan, C., Gribben, J., Urba, W., Welch, W.J. & Nadler, L.M. (1992) Growth arrest of human B-lymphocytes is accompanied by induction of the low molecular weight mammalian heat shock protein (hsp28). *J. Immunol.* **148**, 1668–1673.
Tanguay, R.M. (1989) Localized expression of a small heat shock protein, sp23, in specific cells of the central nervous system during early embryogenesis in *Drosophila. J. Cell. Biol.* **109**, 155a.
Tanguay, R.M., Wu, Y. & Khandjian, E.W. (1993) Tissue-specific expression of heat shock proteins of the mouse in the absence of stress. *Devel. Gen.* **14**, 112–118.

Tanguay, R.M., Duband, J.-L., Lettre, F., Valet, J.-P., Arrigo, A.-P. & Nicole, L. (1985) Biochemical and immunocytochemical localization of heat shock proteins *Drosophila* culture cells. International Conference on Intermediate Filaments. *Ann. N.Y. Acad. Sci.* **455**, 711–714.

Thomas, S.R. & Lengyel, J.A. (1986) Ecdysteroid-regulated heat shock gene expression during *Drosophila melanogaster* development. *Dev. Biol.* **115**, 434–438.

Voellmy, R., Goldschmidt-Clermont, M., Southgate, R., Tissières, A., Levis, R. & Gehring, W.J. (1981) A DNA segment isolated from chromosomal site 67B in *Drosophila melanogaster* contains four closely linked heat shock genes. *Cell* **45**, 185–193.

Vincent, M. & Tanguay, R. (1979) Heat shock induced proteins present in the cell nucleus of *Chironomus tentans* salivary gland. *Nature (London)*, **281**, 501–503.

Vincent, M. & Tanguay, R. (1982) Different intracellular distribution of heat shock and arsenite induced proteins in *Drosphila* KC cells. *J. Mol. Biol.* **162** 365–378.

Vitek, M. & Berger, E. (1984) Steroid and high temperature induction of the small protein genes in *Drosophila. J. Mol. Biol.* **178**, 173–189.

Voorter, C.E.M., de Haard-Hoekman, W.A., Roersma, E.S., Meyer, H.E., Bloemendal, H. & de Jong, W.E. (1989) The *in vivo* phosphorylation sites of bovine αB-crystallin. *FEBS Lett* **259**, 50–52.

Wadsworth, S., Craig, E.A. & McCarthy, B.J. (1980) Genes for three *Drosophila* heat shock induced proteins at a single *locus. Proc. Natl Acad. Sci. USA* **77**, 2134–2137.

Walsh, D., Li, K., Crowther, C., Marsh, D. & Edwards, M. (1991) Thermotolerance and heat shock response during early development of the mammalian embryo. In *Heat Shock and Development* (eds Hightower, L. & Nover, L.), pp 58–70. Berlin, Springer-Verlag.

Welch, W.J. (1985) Phorbol ester, calcium ionophore, or serum added to quiescent rat embryo fibroblast cells result in the elevated phosphorylation of two 28 000 dalton mammalian stress proteins. *J. Biol. Chem.* **260**, 3058–3062.

Wistow, G. (1985) Domain structure and evolution in α-crystallins and small heat shock proteins. *FEBS Lett.* **181**, 1–6.

Wistow, G. & Piatigorsky, J. (1988) Lens crystallins: The evolution and expression of proteins for a highly specialized tissue. *Ann. Rev. Biochem.* **57**, 479–504.

Zimmerman, J.L., Petri, W. & Meselson, M. (1983) Accumulation of a specific subset of *Drosophila melanogaster* mRNAs in normal development without heat shock. *Cell* **32**, 1161–1170.

CHAPTER 8

NEURONAL HOMOLOGUES OF THE BACTERIAL HEAT SHOCK PROTEIN DnaJ

Michael E. Cheetham*, Jean-Pierre Brion†, and Brian H. Anderton*

Department of Neuroscience, Institute of Psychiatry, De Crespigny Park, Denmark Hill, London, SE5 8AF. UK and †Laborotorie d'Anatomie Pathologique, Universite Libre de Bruxelles, 808 Route de Lennik, 1070 Brussels, Belgium*

Table of Contents

8.1 Introduction

Heat shock proteins are a family of functionally related polypeptides that are amongst the most highly conserved proteins throughout evolution. This conservation is demonstrated by the major mammalian heat shock protein family, the hsp70 family, which shows approximately 50% identity at the amino acid level to the bacterial form of hsp70, called DnaK. As such, it was surprising that until 1989 no eukaryotic equivalents of another bacterial heat shock protein, DnaJ, had been identified, especially as the *DnaJ* and *DnaK* genes form an operon in *E. coli*.

In 1989, Sadler *et al.*, identified *Sec63*, a yeast gene involved in protein trans-

Heat Shock Proteins in the Nervous System
ISBN 0-12-480960-X

location into the nucleus and endoplasmic reticulum. Sec63 contains a small domain of approximately 70 amino acids that is homologous to the N-terminal domain of DnaJ. Subsequently, 1991 and 1992 have seen an explosion in eukaryotic DnaJ homologues displaying a wide range of similarity to DnaJ. This new 'DnaJ family' can be sub-classified on the basis of their similarity to the full protein sequence of DnaJ, but it is the similarity to the N-terminal domain of DnaJ (which has been termed the 'J' domain), which is obligate for membership of the family.

We identified a human member of this 'DnaJ family' in an immunoscreen searching for components of the neurofibrillary tangle (NFT) of Alzheimer's disease. NFT are one of the classical neuropathological hallmarks of this devastating neurodegenerative disease. Under the electron microscope NFT can be seen to be composed of 10 nm filaments that are twisted about each other in pairs and are referred to as paired helical filaments (PHF). NFT are insoluble in a wide range of strong solvents, including SDS and while this property has made conventional biochemical analysis difficult, it has been exploited to prepare enriched fractions of PHF from Alzheimer brain.

We used a polyclonal antiserum raised to an enriched preparation of PHF (Brion *et al.*, 1985) to screen a cDNA expression library derived from Alzheimer brain frontal cortex and isolated a positive clone, HSJ1′. However, we have been unable to confirm the association of HSJ1′ with NFT using independent antisera raised to HSJ1′-β-galactosidase fusion protein or synthetic peptides derived from the predicted HSJ1′ amino acid sequence, suggesting that the isolation of HSJ1′ may be serendipitous.

Re-screening of another brain cDNA library with HSJ1′ led to the isolation of full-length clones which correspond to alternatively spliced transcripts (HSJ1a and b) of a single copy gene, *HSJ1*. The N-termini of these cDNA transcripts contain the 'J' domain, identifying these proteins as neuronal members of the 'DnaJ family' (Cheetham *et al.*, 1992).

8.2 The DnaJ family

8.2.1 DnaJ structural elements and the 'J' motif

E. coli DnaJ is a protein of M_r 43-kDa that appears to act as a dimer in solution (Zylicz *et al.*, 1985). The amino acid sequence of DnaJ can be divided into four putative domains, the 'J' domain, the 'G/F' domain and the cysteine repeat domain, the remaining C-terminal part of the protein has no strong sequence motifs (Figure 1). Although, as yet, these regions have not been identified as distinct structural or functional domains, the conservation of these domains varies within the DnaJ family suggesting a functional significance (Table 1).

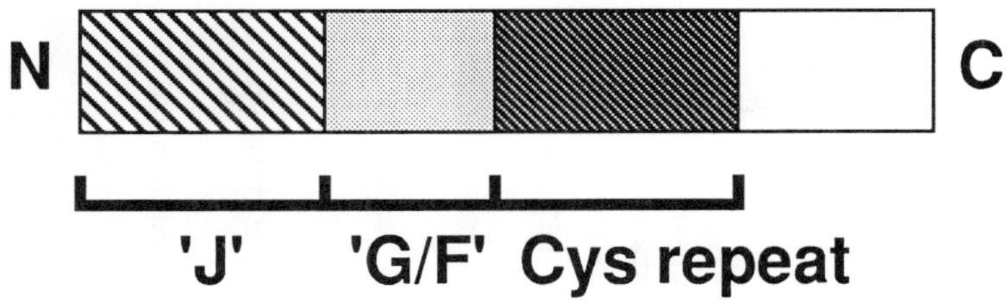

Figure 1 Schematic representation of the domains of *E. coli* DnaJ; **'J'** – the 'J' domain; **'G/F'** – the glycine phenylalanine rich region; **Cys repeat** – the cysteine repeat domain.

The 'J' domain is 70 amino acids located at the N-terminus of DnaJ. Although the sequence contains no striking structural or signal elements, computer-based secondary structure predictions suggest that the domain may form two α-helices separated by a flexible domain, a distinct structural conformation with functional implications. It is within this 'J' domain that the DnaJ family shows the highest degree of conservation and is the only domain present in all family members. This 'J' domain evolutionary conservation can be as high as 50% amino acid identity between *E. coli* DnaJ and human HSJ1 proteins and many other residues display semi-conservative changes. All DnaJ family members show high levels of amino acid identity and semi-conservation within the 'J' motif across all phylogeny and within species suggesting a strong evolutionary pressure to maintain the function of this domain. The most probable function of this domain is the interaction with the hsp70 family of proteins, an interaction that has been demonstrated *in vitro* for DnaJ (Liberek *et al.*, 1991) and YDJ1p (Cyr *et al.*, 1992).

The 'J' motif is separated from the rest of the DnaJ protein by the 'G/F' domain. The 'G/F' region is rich in both glycine and phenylalanine residues that have little secondary structural character, which suggest this region may act as a spacer separating the 'J' domain from the rest of the polypeptide. However, this domain may have a more important role in stabilizing the 'J' domain in 'cytosolic' forms of the family. Sec63, a membrane bound protein has a 'J' motif but no 'G/F' domain, the 'J' domain is between two membrane spanning-domains which may act to stabilize the 'J' domain. This hypothesis is supported by deletion analysis of DnaJ, that suggests the 'J' and 'G/F' domains are required for ATPase stimulation of DnaK (Karazi and McMaken, personal communication in Caplan *et al.*, 1993).

The middle of DnaJ contains a region of cysteine repeats. Although this domain is similar to those seen in 'zinc-finger' proteins, DnaJ is not thought to be a 'zinc-finger' protein. Rather this domain may represent a zinc binding motif (Berg, 1990). The cysteine repeat domain is conserved in some DnaJ family members, which also possess the 'J' motif and the 'G/F' region, these may be 'true homologues' of the progenitor DnaJ. These 'true homologues' may display a tight conservation of DnaJ properties, as opposed to other family members which may only display limited functional conservation.

Table 1 The DnaJ Family (June 1993).

Organism	Protein	Conserved domains 'J'	'G/F'	Cys	C-term	Comments	References
E. coli	DnaJ	+	+	+	+	Progenitor protein, regulates DnaK ATP hydrolysis.	Bardwell *et al.*, 1986, Ohki *et al.*, 1986; Liberek *et al.*, 1991.
R. freddi	NolC	+	+	–	+	Required for nodulation specificity.	Krishnan and Pueppke, 1991.
S. cerevisiae	YDJ1/MAS5	+	+	+	+	Required for protein transport, farnesylated at C-terminus, regulates hsp70.	Caplan and Douglas, 1991; Atencio and Yaffe, 1992; Caplan *et al.*, 1992b.
	SCJ1	+	+	+	+	Appears to be mitochondrial.	Blumberg and Silver, 1991.
	SIS1	+	+	–	+	Required for translation initiation.	Luke *et al.*, 1991; Zhong and Arndt, 1993.
	Sec63	+	–	–	–	Part of ER translocon.	Sadler *et al.*, 1989; Feldheim *et al.*, 1992.
	Zuotin	+	–	–	–	Binds Z-DNA.	Zhang *et al.*, 1992.
A. nummulairia	ANJ1	+	+	+	+	Plant YDJ1-like.	Zhu *et al.*, 1993.
D. melanogaster	CSP	+	–	–	–	Brain and retina specific, restricted to pre-synaptic membranes.	Zinsmaier *et al.*, 1990.
T. californica	$CCCS_1$	+	–	–	–	Candidate calcium channel subunit.	Gundersen and Umbach, 1992.
Human	HSJ1a and b	+	+	–	–	Two neurone-specific isoforms.	Cheetham *et al.*, 1992.
	HDJ1	+	–	–	+		Raabe and Manley, 1991.
	Hsp40	+	?	?	?	Highly inducible, co-localizes with hsc70.	Hattori *et al.*, 1992; 1993.

8.2.2 Bacterial DnaJ

The *DnaK* and *DnaJ* genes were discovered because mutations in them were shown to block bacteriophage replication (review: Georgopoulos *et al.*, 1990) and indeed, these genes appear to act together in many cellular processes. The *DnaK* gene forms an operon with *DnaJ* (Saito and Uchida, 1978), with the order being promoter-*DnaK-DnaJ*. The *DnaK* gene-product has been shown to be approximately 50% identical, at the amino acid sequence level, to the eukaryotic hsp70 family of proteins (Bardwell and Craig, 1984; Hunt and Morimoto, 1985).

Mutations in the *DnaJ* gene in *E. coli* result in pleiotropic effects on cell growth and metabolism that are similar in many ways to mutations in the *DnaK* gene. Similar to DnaK and another protein GrpE, DnaJ appears to be involved in the negative regulation of σ32 synthesis and stability, and thereby regulating its own synthesis and the synthesis of other hsps (Gross *et al.*, 1990). Mutations in the *DnaJ* gene also appear to block both host DNA and RNA syntheses at non-permissive temperatures (Saito and Uchida, 1978; Wada *et al.*, 1982), and arrest the cell cycle at a defined stage, prior to the synthesis of a specific subset of proteins (Ohki *et al.*, 1987).

The DnaJ protein has been purified to homogeneity and has been shown to be a basic, dimeric protein (Zylicz *et al.*, 1985). DnaJ protein appears to be mostly localized at the bacterial membrane and binds both single- and double-stranded DNA with no sequence specificity (Zylicz *et al.*, 1985). DnaJ protein appears to be able to act as a molecular chaperone, as it can bind substrates independently of DnaK. DnaJ interacts specifically with DnaB helicase and weakly with P-protein in bacteriophage replication (Alfano and McMaken, 1989; Liberek *et al.*, 1990). Also the DnaJ dimer forms a stable protein complex with a dimer of the plasmid P1 RepA initiator protein (Wickner, 1990), via this complex DnaJ and DnaK act as chaperones to mediate an alteration in RepA, rendering it much more active for DNA binding to oriP1, the P1 replication origin (Wickner *et al.*, 1991).

The DnaJ and DnaK proteins act together in the above situations, and probably act together in other cellular processes as judged by the similar phenotypes displayed by mutations in the two genes. This interaction is also supported by the observation that DnaJ binds to a DnaK-affinity column (Liberek *et al.*, 1990). Although the precise nature of this interaction is not yet defined the presence of DnaJ and GrpE can lead to a 50-fold increase in the ATPase activity of DnaK, DnaJ alone leading to an increase in the rate of ATP hydrolysis, and GrpE alone increasing the rate of release of bound ATP or ADP (Liberek *et al.*, 1991a). This stimulation of the ATPase activity of DnaK may contribute to its more efficient recycling and help 'target' proteins for DnaK action. The hsp70 family of proteins are highly conserved at the amino terminus, including homology with DnaK, and a 44-kDa amino terminal fragment has the ability to bind ATP. Several reports have shown that the hydrolysis of ATP by the hsp70 family is required for substrate release (Lewis and Pelham, 1985; Munro and Pelham, 1986; Clarke *et al.*, 1988; Liberek *et al.*, 1991b).

In *E. coli* the *DnaK/DnaJ* gene-products may also interact with other molecular chaperones, including the chaperonins (hsp60, GroEL/ES). In a reconstituted folding pathway *in vitro*, DnaJ and DnaK co-operate with GroEL to act catalytically in the production of native state proteins (Langer *et al.*, 1992). This interaction can also be seen in rpoH mutants, that lack the heat shock σ subunit and are deficient in hsps, resulting in the aggregation of newly synthesized proteins. Re-introduction of DnaJ/DnaK or GroEL/ES can reverse this aggregation, all four proteins together are effective at physiological concentrations (Gragerov *et al.*, 1992), suggesting these proteins have complementary functions in the folding and assembly of a wide range of proteins.

8.2.3 Eukaryotic DnaJ family members

The number of eukaryotic DnaJ family members has increased noticeably in the last two years, and these newly discovered proteins are as diverse as they are numerous. These homologues display a range of similarity to DnaJ (see Table 1) and do not merely reflect the conservation of DnaJ, but also suggest a development and diversification of DnaJ function throughout evolution.

8.2.3.1 Yeast DnaJ homologues

In the yeast, *Saccharomyces cerevisiae*, five homologues of DnaJ have been found at the time of writing. Although all these polypeptides possess the 'J' domain, they have different proposed sub-cellular localizations and different proposed functions. The following is a brief description of their known functions and localizations.

As mentioned above, Sec63 was the first described eukaryotic DnaJ family member and has subsequently been shown to be part of the ER translocon, a translocation machinery comprised of a complex of polypeptides spanning the ER membrane (Deshaies *et al.*, 1991). The 'J' domain of Sec63 projects into the lumen of the ER (Feldheim *et al.*, 1992) and genetic studies indicate that Sec63 interacts with Kar2 (BiP, the ER lumenal form of hsp70) (Vogel and Rose, personal communication in Gething and Sambrook, 1992; Scidmore and Rose, personal communication in Caplan *et al.*, 1993). Although details of this interaction are not known at the moment, both proteins are required together for pre-protein translocation.

The importance of yeast DnaJ family members in protein transport and translocation is also supported by two other polypeptides SCJ1 and YDJ1. *SCJ1* was identified as over-expression of SCJ1 protein resulted in the missorting of nuclear targeted chimeric protein to mitochondria. SCJ1 has a putative N-terminal mitochondrial signal sequence and appears to be localized in mitochondria (Blumberg and Silver, 1991).

YDJ1 although initially isolated in an immunoscreen for components of the nuclear matrix laminar pore complex appears to be cytoplasmic in distribution, but is also partially associated with membranes (Caplan and Douglas, 1991). YDJ1 has a C-terminus isoprenylation motif. The addition of this lipid moiety affects the association of YDJ1 with membranes and the ability of the protein to function at high temperatures (Caplan *et al.*, 1992a). The MAS5 gene, identified as a temperature-sensitive mutation, resulting in defects in mitochondrial protein import at elevated temperatures, was found to be identical to YDJ1 (Atencio and Yaffe, 1992).

The conservation of the bacterial DnaK/DnaJ interaction *in vitro* has been demonstrated for hsp70(SSA1)/YDJ1 (Cyr *et al.*, 1992). Like hsp70 proteins, YDJ1 appears to have a post-translational function in mitochondrial import and possibly post-translational translocation into the ER (Caplan *et al.*, 1992b). This transport function appears to be evolutionarily conserved since *E. coli* DnaJ and SIS1 can suppress, in transfected mutant yeast cells, the transport defects seen in a YDJ1 mutant strain (Caplan *et al.*, 1992b).

The *SIS1* gene is a high-copy number suppressor of the slow growth phenotype of strains containing mutations in the *SIT4* gene, which encodes a predicted serine/threonine phosphatase (Luke *et al.*, 1991). SIS1 protein is found in the cytoplasm and inside the nucleus and stably interacts with another protein termed p40 in a large macromolecular complex (Luke *et al.*, 1991). Recent studies have shown that SIS1 is associated with 40S ribosomal subunits and polysomes and is required for the normal initiation of translation (Zhong and Arndt, 1993). The requirement for SIS1 in this process may reflect a need for the dissociation of a translation machinery protein complex, prior to the initiation of translation.

The most recently discovered yeast DnaJ-like protein, Zuotin, is also perhaps the most enigmatic. Zuotin is a putative nuclear Z-DNA binding protein as it binds to oligonucleotides that adopt a putative Z-DNA conformation. Zuotin (ZUO1) possess only the 'J' domain but also has a sequence motif similar to histone H1 (Zhang *et al.*, 1992).

8.2.3.2 Viral and parasite DnaJ-like motifs

As eukaryotic DnaJ proteins were being discovered, several related sequence motifs were highlighted in polyoma and papovavirus large and small T-antigens (Cheetham *et al.*, 1992) and *Plasmodium falciparum* ring-infected erythrocyte surface antigen (RESA) protein (Bork *et al.*, 1992). These motifs are only similar to the N-terminal part of the 'J' domain and the level of similarity can be as high as 44% identity in a 48 amino acid overlap (budgerigar-fledgling-disease virus, BVFD, large and small T-antigen compared to HSJ1). Although the interaction of SV40 and polyoma T-antigens with hsp70 is well documented (Walter *et al.*, 1987; Sawai and Butel, 1989; May *et al.*, 1991), whether these proteins are true DnaJ family members as demonstrated by some functional conservation remains to be seen. It is possible that the presence of these motifs in these parasitic organisms represents a form of molecular mimicry, designed to hijack the normal host substrates of DnaJ-like proteins.

8.2.3.3 Higher organism DnaJ homologues

The conservation of DnaJ-like proteins in multicellular organisms has been shown in both the plant and animal kingdoms. A plant homologue of DnaJ, ANJ1 has been described in the higher plant *Atriplex nummularia*. ANJ1 protein has an isoprenylation motif similar to YDJ1, and indeed can fully suppress YDJ1 mutations (Zhu *et al.*, 1993). Another protein, LDJ1, that is highly related to ANJ1 has been described in the leek (Bessoule, 1993). The conservation of yeast DnaJ-like proteins has been suggested by Zhu *et al.*, as they have found *A. nummularia* proteins that are antigenically related to SCJ1and Sec63. The possible conservation of yeast DnaJ-like proteins in higher eukaryotes is supported by data from the *Caenorhabditis elegans* genome mapping project. ESTs (Expressed Sequence Tags) which show homology to YDJ1 and SCJ1 have been sequenced (McCombie *et al.*, 1992; Waterston *et al.*, 1992).

As yet, no homologues of the yeast DnaJ-like proteins have been described in mammals, but three distinct DnaJ-like proteins have been discovered in humans. HSJ1 is the subject of this chapter and will be discussed at length later. HDJ1 was discovered in an immunoscreen for the poly-A-polymerase in a HeLa cell cDNA expression library (Raabe and Manley, 1991) and N-terminal sequence analysis of hsp40 has revealed the presence of a 'J' domain (Hattori *et al.*, 1992).

The tissue distribution of these two proteins is not known but a *Drosophila* protein, cysteine string-protein (CSP), has a 'J'-like domain and is expressed in brain and retina at pre-synaptic membranes (Zinsmaier *et al.*, 1990; Caplan *et al.*, 1993). A homologue of CSP has recently been identified in the electric lobe of *Torpedo californica*, this protein $CCCS_1$ is a candidate calcium channel subunit (Gundersen and Umbach, 1992). Although the role of the highly conserved 'J' domains of CSP and $CCCS_1$ proteins in this putative function is unclear at the moment, it is possible that the 'J' domain is modulating calcium channel function by mediating protein complex assembly and disassembly, or other protein/protein interactions.

The brain-specific expression of these higher organism DnaJ-like proteins, HSJ1, CSP and $CCCS_1$ encourages speculation that higher organisms may have a large number of DnaJ-like proteins, with conservation of the yeast DnaJ family and tissue specific forms with functions specialized to a particular cell type.

8.2.4 The DnaJ family and the stress response

The expression of *E. coli* DnaJ is under the control of the same promoter as DnaK(hsp70) and, as such, shows a rapid induction in response to stress. However, the steady-state levels of DnaJ are 10-fold lower than DnaK, suggesting post-transcriptional control of the levels of DnaJ expression. The induction of the stress response in *E. coli* is mediated by the heat shock transcription factor σ32. DnaK and DnaJ can bind to σ32 independently, suggesting that the induction of heat shock proteins can be regulated by distinct mechanisms involving DnaK and DnaJ (Gamer *et al.*, 1992).

Although no regulatory role has been defined for the eukaryotic DnaJ family, the family does display a wide range of responses to stress. Although the majority of family members discovered at this moment are constitutively expressed, many of the genes contain a heat shock consensus element in their promoter region and display a rapid increase in transcription in the response to stress (Luke *et al.*, 1991; Atencio and Yaffe, 1992; Caplan *et al.*, 1992a).

However, two DnaJ family members, ANJ1 and hsp40, are particularly interesting in the context of their response to stress. Although ANJ1 appears to be a plant analogue of YDJ1, the stress response appears to be finely tuned with respect to ANJ1. *A. nummularia* is a halophyte that is native to the Negev desert, and while ANJ1 and hsp70 show a co-ordinate response to heat shock, only ANJ1, and not hsp70, is induced by NaCl shock in plant cells which have been de-adapted for NaCl (Zhu *et al.*, 1993). This non-coordinate stress response suggests that hsp70 and ANJ1 can be induced by different mechanisms in *A. nummularia*, and that the ANJ1 response to NaCl shock may represent a specific adaptive response to the high concentrations of NaCl found in the native habitat.

Human hsp40 was identified as a novel heat shock protein (Ohtsuka *et al.*, 1990) and N-terminal amino acid sequence analysis has revealed the 'J' motif (Hattori *et al.*, 1992). In HeLa cells, the steady state levels of hsp40 are very low, however a heat shock of 15 min at 45°C increases the levels of hsp40 dramatically. The kinetics of this induction are similar to those for hsp72 (the highly inducible hsp70) and the newly synthesized hsp40 co-localizes with hsp73 (the hsp70 cognate hsc70) in the nucleolus (Hattori *et al.*, 1993), a localization that has also been demonstrated for hsp72 (review: Pelham, 1990). This co-ordinate induction and localization suggests a co-ordinate role in the stress response and the recovery from stress for DnaJ family members and the hsp70 family.

8.3 Characterization of HSJ1

HSJ1 was initially isolated in an immunoscreen for components of the Alzheimer NFT. Although we have been unable, as yet, to confirm this association, the identification of HSJ1 proteins raises the possibility for the study of molecular chaperones and the stress response in neurones.

8.3.1 HSJ1 isoforms and alternative splicing

In our immunoscreen for PHF components, the initial anti-PHF serum immunopositive clone, HSJ1′, was terminated at the 5′ end by an internal *Eco*RI site and did not contain an initiation codon. This clone was used to isolate longer cDNA clones by screening another brain frontal cortex cDNA library. Two clones encoded putative full length open reading frames which differed by the presence of a 1.1 kb insert. The presence of the insert results in a C-terminal amino acid sequence alternative

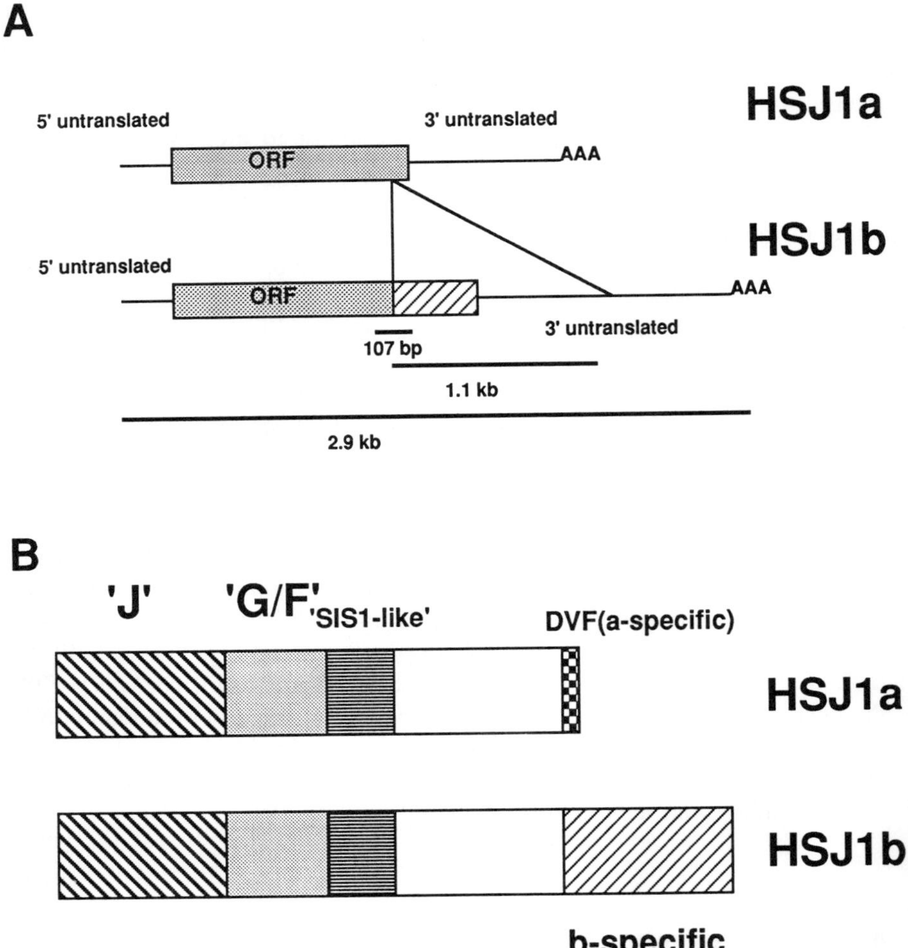

Figure 2 (A) Schematic representation of the cDNA transcripts corresponding to HSJ1a and HSJ1b. Note the 1.1 kb insert in HSJ1b. The probe of 107 bp seen in Figure 3 is marked: **AAA** – poly-A-tail; **ORF** – open reading frame. (B) Schematic representation of the domain structure of **HSJ1a** and **HSJ1b; 'J' 'G/F'** as in Figure 1; **'SIS1-like'** – small region of similarity to SIS1; **DVF** – HSJ1a specific C-terminal amino acid residues.

to that predicted for the shorter form, that is longer by 74 amino acids (Figure 2). We have designated these two isoforms of *HSJ1*, HSJ1a and HSJ1b.

The possibility existed that the larger isoform, HSJ1b was not a real mRNA species but an unspliced hnRNA. The extra 1.1 kb of sequence present in HSJ1b could possibly start with a 'GT' and end with an 'AG'. These are the highly conserved intron splice donor and acceptor sites. In addition, seven of the bases in the sequence surrounding the 'GT' match the nine-base intron splice consensus

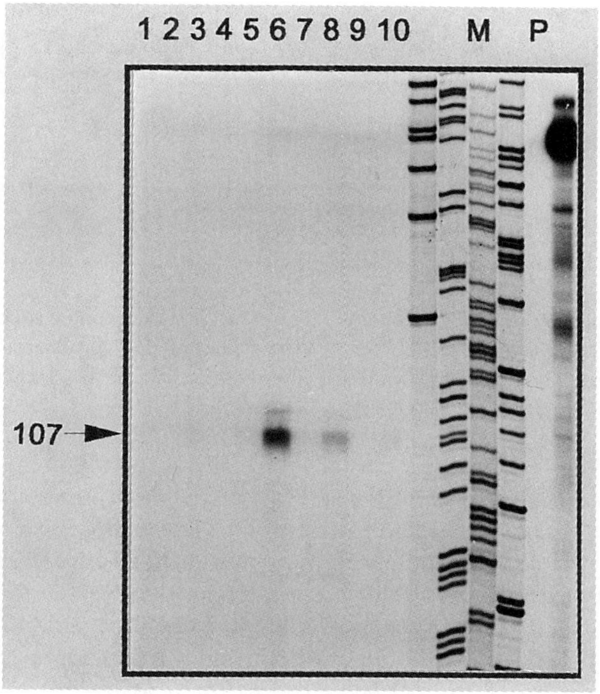

Figure 3 RNase protection analysis of HSJ1 expression. 40 μg of total RNA from; (1) liver; (2) kidney; (3) skeletal muscle; (4) testis; (5) spleen; (6) brain frontal cortex; (7) brain cerebellum; (8) brain hippocampus; (9) neuroblastoma cell line SH-SY5Y; (10) neuroblastoma cell line SH-SY5Y after 1-h recovery from a 15-min heat shock of 42°C; **M** sequencing ladder; **P** no RNase control. The band at 107 bp is specific for HSJ1b, a smaller band specific for HSJ1a is not shown.

sequence, and the sequence immediately preceding the 'AG' is very pyrimidine-rich, again typical of a splice acceptor site. However, the existence of HSJ1b as a genuine mRNA species was supported by the presence of two species of approximately the correct size detected by Northern blotting (see Figure 7). RNase protection analysis confirmed that HSJ1b was a distinct mRNA species present in human brain (Figure 3). The band at 107 bp is specific for HSJ1b (a smaller band specific for HSJ1a is not shown).

Polyclonal antisera raised to the HSJ1′-βgalactosidase fusion protein and to synthetic peptides derived from the HSJ1′ sequence detect two bands of approximately 36- and 42-kDa on Western blots of human brain homogenate. The difference in size between these two bands corresponds to the difference in the predicted molecular weight of HSJ1a and HSJ1b, suggesting that they correspond to the translated products of the two mRNA species. This has been confirmed by the use of monospecific polyclonal antisera raised to C-terminal peptides derived from

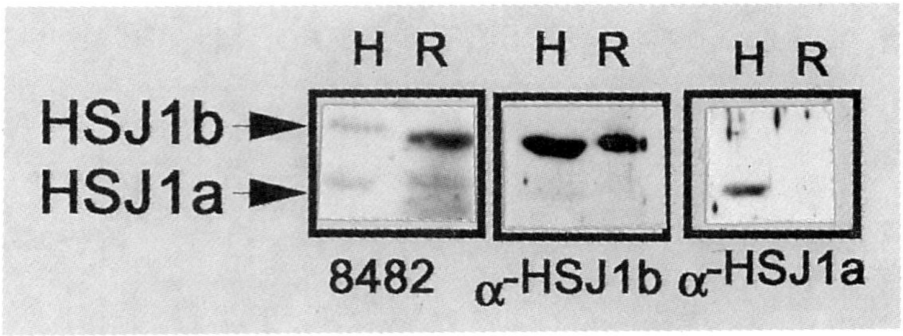

Figure 4 Western blot analysis of human (H) and rat (R) brain homogenate. Bands corresponding to HSJ1a and HSJ1b are arrowed. 8482 is an anti-HSJ1′βgalactosidase fusion protein serum; **α-HSJ1b** and **α-HSJ1a** are raised to C-terminal specific synthetic peptides. All sera diluted 1:1000, detection Amersham ECL.

HSJ1a and HSJ1b (Figure 4), despite only three HSJ1a specific amino acid residues. These antisera recognize only HSJ1a and HSJ1b and demonstrate that larger transcript is translated into the larger protein, HSJ1b. The antisera also recognize antigenetically-related proteins present in rat brain, and it would appear that the C-terminus of HSJ1b is conserved in the rat, whereas the C-terminus of HSJ1a appears not to be. These mono-specific antisera should enable us to study the localizations of HSJ1a and HSJ1b in the brain and within neurones, and aid in the identification of any isoform specific functions.

8.3.2 Sequence conservation of HSJ1

The predicted amino acid sequences of the N-termini of HSJ1 proteins possess a 'J' domain of approximately 70 residues and no leader or signal peptide sequence. The amino acid identity with *E. coli* DnaJ in this region is over 50% and many changes are semi-conservative (Figure 5). Comparison of this domain within the DnaJ family reveals that although some residues are conserved in all family members, for example the HPD motif (the proline and aspartate residues are necessary for Sec63 function – Feldheim *et al.*, 1992), the general pattern is one of a spread of identity complemented by semi-conservative change. Throughout, the 'J' domain residues are conserved between one or two family members and the overall impression of comparing family members is that the domain conservation is extended. For example, in an alignment of HSJ1 proteins with *E. coli* DnaJ and four yeast homologues, over 75% of HSJ1a and b amino acid residues are identical to residues in at least one of the other five proteins (Cheetham *et al.*, 1992).

Adjacent to the 'J' domain is a short region which is rich in glycine residues (9 of 24 residues), followed by a high density of phenylalanine residues. This 'G/F' region is present in DnaJ and many other homologues and separates the 'J' domain from the distal C-terminal half of the protein. Unlike DnaJ, the phenylalanine-rich region extends to residue 174 in HSJ1a and b and is interspersed with serine residues, forming three repeated motifs of FSF (the number of serine residues varies

```
1        M A S Y Y E I L D V P R S A S A D D I K K A Y R R K A L Q W H P D K      HSJ1a & b
1  M A K Q D Y Y E I L G V S K T A E E R E I R K A Y K R L A M K Y H P D R  E. coli DnaJ

35 N P D N K E F A E K K F K E V A E A Y E V L S D K H K R E I Y D R Y G      HSJ1a & b
37 N Q G D K E - A E A K F K E I S E A Y E V L T D S Q K R A A Y D Q Y G  E. coli DnaJ
```

Figure 5 Alignment of the 'J' domain of HSJ1a and b and *E. coli* DnaJ, identical residues are boxed. Amino acid residues are numbered on the left, one space (-) has been introduced to optimize alignment.

between one and four). This serine-rich region shows similarity to a great number of polypeptides that have a high serine content that probably has little functional significance. However, more interestingly, this region does display a limited similarity to SIS1 (Cheetham *et al.*, 1992). This similarity does not include the methionine/glycine repeat present in SIS1 (Luke *et al.*, 1991), but is concentrated on phenylalanine, serine and threonine residues surrounding this repeat. The functional significance of this similarity is unclear at the moment, though the level of sequence similarity does not suggest that HSJ1 proteins represent human forms of SIS1. Rather, the domain structure conservation combined with this low level of similarity suggests that HSJ1 and SIS1 may be members of a subset of DnaJ-like proteins with some common properties.

The predicted C-termini of HSJ1a and HSJ1b distal to this region do not display any amino acid identity to any other proteins currently in the database.

8.3.3 DnaJ family functional conservation

The key question relating to the structural conservation seen within the DnaJ family is whether it corresponds to a conservation of function. This question has been addressed by examining the complementation of yeast strains which have mutations in DnaJ-like proteins. The benchmark for this functional conservation has become the complementation of *Δydj1* strains of yeast. These strains which lack YDJ1 have a slow-growth phenotype and are temperature sensitive. The over-expression of yeast SIS1 in *Δydj1* cells results in a partial restoration of wild-type phenotype in these cells (Caplan and Douglas, 1991). However, this relationship is not reciprocal, *SIS1* is an essential gene, and yeast lacking *SIS1* are inviable, the over-expression of YDJ1 cannot restore viability to *Δsis1* null mutants (Luke *et al.*, 1991).

The expression of bacterial DnaJ in yeast can also partially suppress the slow growth phenotype of a *Δydj1* null mutant and, indeed, can suppress the protein transport defects seen in *ydj1-151*^{TS}, a temperature sensitive mutant of *YDJ1* (Caplan *et al.*, 1992b). This is perhaps not surprising as YDJ1 and DnaJ exhibit domain structure conservation. However, the expression of HSJ1b under the control of the Gal1 promoter in *Δydj1* yeast can also partially suppress the slow-growth phenotype, despite only 22% amino acid identity when the full-length protein sequences are compared (Figure 6 and Caplan *et al.*, 1992b). This functional conservation

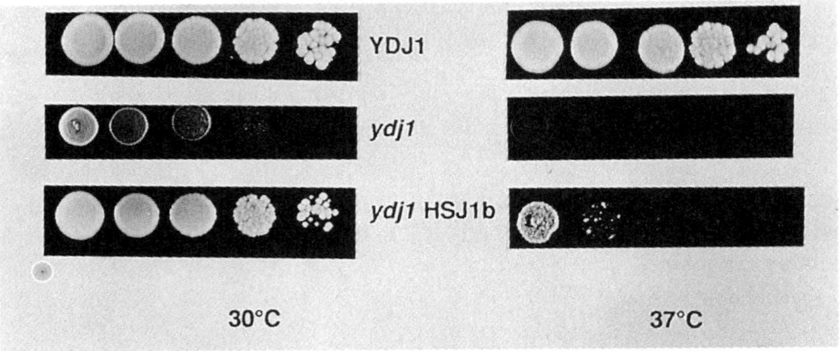

Figure 6 Partial complementation of a yeast *Δydj1* null slow growth phenotype by HSJ1b: **YDJ1** wild-type, strain W303 MATα adel-2 leu2-3.112 hsi3-11.15 trpl-1 ura 3-1 canl-100, YDJ1; **ydj1** YDJ1 null mutant, strain as W303 but no YDJ1; **ydj1 HSJ1b**, ACY143 MATα ade 2-1 leu2-3.112 his3-11.15 trpl-1 ura 3-1 canl-100 ydl1-1::HIS3, pGALHSJ1b (HSJ1b, URA3); note the temperature sensitivity of *ydj1* null mutants is only slightly complemented by HSJ1b. Figure courtesy of Dr A. Caplan.

between YDJ1 and HSJ1b suggests that the 'J' domain (YDJ1 and HSJ1 proteins have over 50% amino acid identity in the 'J' domain) is a distinct functional unit that has a highly conserved function and that it is the 'J' domain that is mediating the partial suppression of *Δydj1* by DnaJ and SIS1.

8.3.4 Neuronal specificity of HSJ1

Analysis of the expression of HSJ1 mRNAs suggests that *HSJ1* displays a tissue-dependent pattern of expression limited almost exclusively to brain. Northern blot analysis detects two messenger RNA species of approximately two and three kb in human frontal cortex polyadenylated RNA (Figure 7). These species can also be detected in hippocampus-derived RNA. No signal can be detected in RNA derived from liver, kidney or skeletal muscle tissue. Two mRNA species of two and three kb can also be detected at high stringency by Northern blot analysis of rat brain RNA, confirming the presence of HSJ1-like gene transcripts in the rat. This brain specific expression of HSJ1 has been confirmed by RNase protection analysis of total RNA derived from a variety of human tissues *post mortem* (Figure 3). The level of expression of HSJ1 mRNAs was much higher in the brain frontal cortex than in the hippocampus or cerebellum and no signal was detected in liver, kidney or testis and only an extremely weak signal can be detected in spleen and skeletal muscle.

We have examined the expression of *HSJ1* within the hippocampus by *in situ* hybridization. The highest levels of HSJ1a and HSJ1b are in the neuronal layers of the hippocampus and temporal cortex, where HSJ1a and HSJ1b show lamination (Figure 8). The intensity of hybridization showed discrete regional variations. The signal was stronger in the dentate gyrus, the entorhinal cortex and the temporal

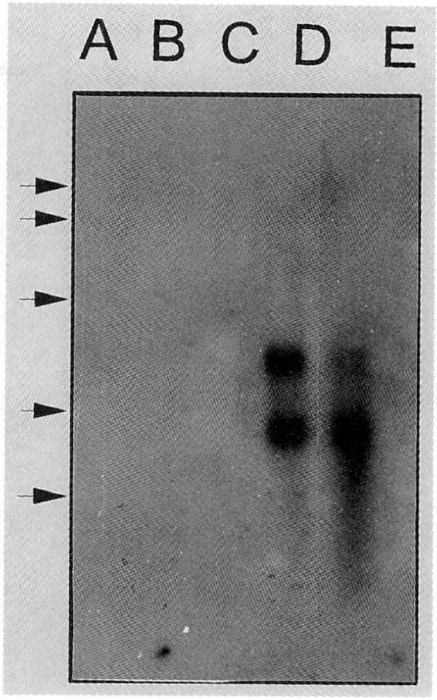

Figure 7 Northern blot analysis of HSJ1 expression: 5 μg of polyadenylated RNA extracted from human tissues; (A) liver; (B) kidney; (C) skeletal muscle; (D) brain frontal cortex; (E) brain hippocampus. Molecular weight markers are arrowed on the left (from the top) 9.5, 7.5, 4.4, 2.4, 1.4 kb.

cortex than in the CA1 sector of Ammon's horn. However, due to the weakness of the signal with ^{35}S labelled probes, single cell localization has not yet been possible. Notwithstanding this, the relative intensity of the neuronal layers strongly suggests that HSJ1a and HSJ1b are expressed at much higher levels in some CNS neurones than in any other cell type.

8.4 Concluding remarks

The DnaJ family of proteins have been implicated as being important in several fundamental cellular biological processes. Although little is known about the functions of HSJ1a and HSJ1b at the moment, we can extrapolate some of the known conserved properties of the family to the possible functions of HSJ1 proteins.

The most striking conserved function of the DnaJ family of proteins is the interaction with the hsp70 family. This interaction has been demonstrated for several

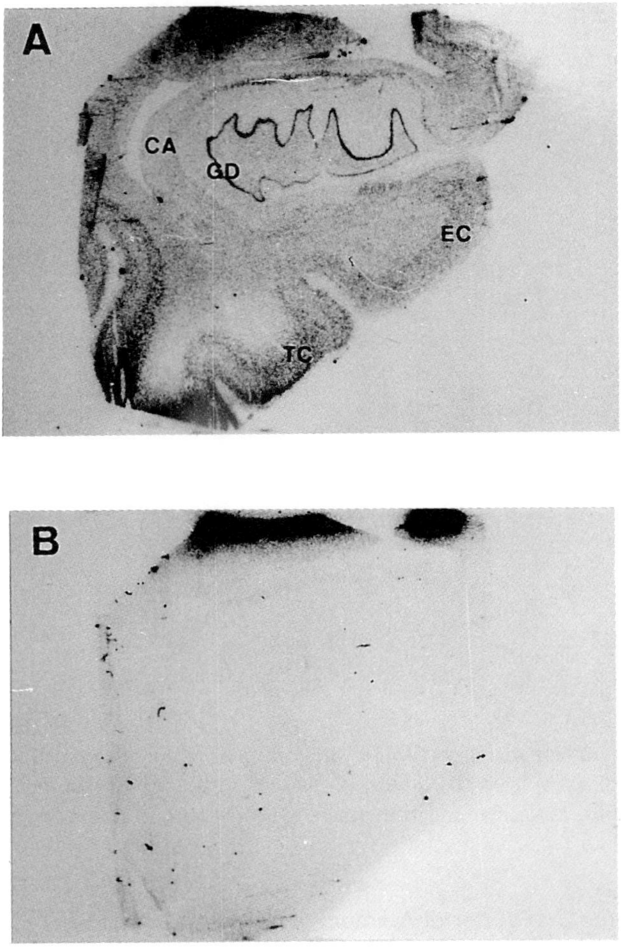

Figure 8 *In situ* hybridization histochemistry of human hippocampus and the adjacent temporal cortex: (A) antisense ^{32}P labelled riboprobe derived from HSJ1a; (B) sense ^{32}P labelled riboprobe; (DG) dentate gyrus; (CA) pyramidal layers of Ammon's horn; (EC) entorhinal cortex; (TC) temporal cortex.

DnaJ-like proteins, including DnaJ, YDJ1, Sec63 and hsp40. The interaction of Sec63 with hsp70 proteins appears to be mediated solely by the 'J' domain. The hypothesis that the 'J' domain is sufficient for the interaction with hsp70 is strengthened by the complementation of *Δydj1* null mutants by HSJ1b. This partial complementation of yeast lacking YDJ1 by human neuronal DnaJ-like proteins strongly implicates HSJ1 proteins in the regulation of hsp70 proteins by a highly conserved mechanism.

In vitro studies using purified DnaJ (Liberek *et al.*, 1991) and YDJ1 (Cyr *et al.*,

1992) have demonstrated the ability of DnaJ-like proteins to stimulate the ATPase activity of hsp70 proteins, an activity which is closely linked to hsp70/substrate interactions. The stimulation of the DnaK ATPase activity is mediated by DnaJ and another heat shock protein, GrpE. Whether eukaryotic homologues of GrpE exist remains to be determined, though the degree to which this accessory protein is conserved is much lower than DnaJ in prokaryotes. So eukaryotic homologues of GrpE may have diverged beyond recognition. Although the precise physiological significance of the DnaJ family mediated stimulation of hsp70 ATP hydrolysis is not yet understood, this interaction may have a variety of effects *in vivo*. While the interaction of YDJ1 can stimulate the release of carboxy-methylated lactalbumin from a normally stable complex with hsp70 (Cyr *et al.*, 1992); DnaJ can also stabilize hsp70:substrate interactions (Langer *et al.*, 1992). Although this may seem paradoxical, a discrimination of chaperone:substrate release may be necessary *in vivo*.

The compartmentalization and multiplicity of members of the DnaJ and hsp70 families raises the possibility of specific interactions between members of each family. Such that one DnaJ-like protein may target a particular hsp70 family member and no other, for example the interaction between Sec63 and the ER lumenal form of hsp70, BiP. Whether this specificity of DnaJ/hsp70 interaction is a general feature of DnaJ-like proteins remains to be determined. Investigation of the interaction of HSJ1 proteins with different hsp70 family members may help to elucidate this question.

The constitutively expressed form of hsp70, hsc70, is expressed at high levels in the mammalian brain and has been implicated as the clathrin-uncoating ATPase (Chappell *et al.*, 1987). This proposed role for hsc70, one which it fulfils *in vitro*, is a controversial one and the co-operation of a DnaJ-like protein in hsc70 mediated clathrin-uncoating is an attractive possibility. HSJ1 proteins, due to their neuronal expression, are excellent candidates for an involvement in this process and we shall be investigating the involvement of HSJ1 proteins in clathrin-uncoating.

DnaJ appears to act not only as an accessory protein to hsp70 but also appears to be a molecular chaperone in its own right. DnaJ can bind independently of hsp70 to a number of substrates (see Wickner *et al.*, 1991; Gamer *et al.*, 1992; Langer *et al.*, 1992). It can be reasoned that if the 'J' domain is responsible for the interaction with hsp70 proteins then the remaining domains of DnaJ-like proteins mediate specific polypeptide associations. We believe that HSJ1 proteins are molecular chaperones with functions specialized to neurones. The alternative splicing of *HSJ1* into two isoforms with different C-termini suggests that the two isoforms may perform different functions within neurones, potentially targeting different substrates.

HSJ1 proteins may interact with other molecular chaperones in a eukaryotic equivalent of the 'hand off' of unfolded proteins from one molecular chaperone to another that can be reconstituted *in vitro* for a prokaryotic protein folding pathway (Langer *et al.*, 1992). The eukaryotic cytosol contains an equivalent of hsp60, the TCP-1 complex. TCP-1 complex polypeptides have been shown to be important factors in actin and tubulin biogenesis (Gao *et al.*, 1992; Yaff *et al.*, 1992). The neuronal cytoskeleton is a complex and dynamic structure that contains many neuronespecific proteins, including the microtubule-associated proteins, τ and MAP2. These microtubule associated proteins have been demonstrated to be associated with PHF in Alzheimer's disease (Goedert *et al.*, 1988; Brion *et al.*, 1990). A eukaryotic *in vivo* folding pathway that involves successive molecular chaperone action, may include HSJ1 proteins in the biogenesis of such neurone-specific proteins.

The alternative splicing of *HSJ1* into HSJ1a and HSJ1b represents an evolutionary diversification from bacterial DnaJ. This diversification of the DnaJ family can be seen in many of the DnaJ-like proteins so far identified and suggests that there may be a great number of specialized DnaJ-like proteins in higher organisms. Amongst the specialized functions that these proteins may fulfil is the protection of the cell from stress. Hsp40 is a highly inducible human heat shock protein and shows an induction pattern very similar to hsp70 in the response to stress, however, the stress required to induce hsp40 is much greater than that required to induce hsp70 (Hattori *et al.*, 1993). The difference in the response to stress of hsp40 and hsp70 may reflect a more stringent requirement for DnaJ-like proteins in the stress response. ANJ1 appears to have evolved an adaptive response to osmotic shock (Zhu *et al.*, 1993), suggesting a possible refined stress response in DnaJ-like proteins. The response of HSJ1 to stress is not known, although preliminary results suggest that a small induction of HSJ1a and b mRNA may be seen in SH-SY5Y neuroblastoma cells following a 15-min heat shock of 42°C (Figure 3). We are currently investigating the expression of HSJ1 and related proteins in response to a variety of neuronal insults.

HSJ1 was initially identified by an antiserum raised to PHF of Alzheimer's disease, as yet we have been unable to confirm this association, so at the present time the isolation of HSJ1 must be regarded as serendipitous. However, the possible involvement of DnaJ-like proteins in the pathogenesis of neurodegenerative diseases cannot be discounted. Hsp70 proteins have been demonstrated to be associated with NFT in Alzheimer's disease (Hamos *et al.*, 1991) and ubiquitin is a common component of many pathological inclusions present in neurodegenerative diseases (see Chapters 9, 10 and 11). The interaction of DnaJ-like proteins with hsp70 and other stress proteins is an interesting aspect of the possible involvement of stress proteins and molecular chaperones in the events preceding the deposition of the protein aggregates that are characteristic of neurodegenerative diseases.

Acknowledgements

The authors would like to thank Dr Avrom Caplan for useful discussion during the preparation of this manuscript. This work was supported by the Wellcome Trust.

References

Alfano, C. & McMacken, R. (1989) Ordered assembly of nucleoprotein structures at the bacteriophage λ replication origin during the initiation of DNA replication. *J. Biol. Chem.* **264**, 10 699–10 708.

Antencio, D.P. & Yaff, M.P. (1992) MAS5, a yeast homologue of DnaJ involved in mitochondrial protein import. *Mol. Cell Biol.* **12**, 283–291.

Bardwell, J.C.A. & Craig, E. (1984) Major heat shock gene of *Drosophila* and the *Escherichia coli* heat-inducible DnaK gene are homologous. *Proc. Natl Acad. Sci. USA* **81**, 848–852.

Bardwell, J.C.A., Tilly, K., Craig, E., King, J., Zylicz, M. & Georgopoulos, C. (1986) The nucleotide sequence of the *Escherichia coli* K12 DnaJ+ gene. *J. Biol. Chem.* **261**, 1782–1785.

Berg, J.M. (1990) Zinc Fingers and other metal-binding domains. *J. Biol. Chem.* **265**, 6513–6516.

Bessoule, J.-J. (1993) Occurrence and sequence of a DnaJ protein in plant (*Allium porrum*) epidermal cells. *FEBS Lett.* **323**, 51–54.

Blumberg, H. & Silver, P.A. (1991) A homologue of the bacterial heat shock gene DnaJ that alters protein sorting in yeast. *Nature (Lond.).* **349**, 627–630.

Bork, P., Sander, C., Valencia, A. & Bakau, B. (1992) A module of the DnaJ heat shock proteins found in malaria parasites. *TIBS.* **17**, 129.

Brion, J.P., Couck, A.M., Passareiro, E. & Flament-Durand, J. (1985) Neurofibrillary tangles of Alzheimer's disease: An immunohistological study. *J. Submicrosc. Cytol.* **17**, 89–96.

Brion, J.P., Cheetham, M.E., Couck, A.M., Flament-Durand, J., Hanger, D.P. & Anderton, B.H. (1990) Characterization of a partial cDNA specific for the high molecular weight microtubule-associated protein MAP2 that encodes epitopes shared with paired helical filaments of Alzheimer's disease. *Dementia* **1**, 304–315.

Caplan, A.J. & Douglas, M.G. (1991) Characterization of YDJ1: A yeast homologue of the *E. coli* DnaJ gene. *J. Cell Biol.* **114**, 609–622.

Caplan, A.J., Tsai, J., Casey, P.J. & Douglas, M.G. (1992a) Farnesylation of the YDJ1p is required for function at elevated temperatures in *S. cerevisiae. J. Biol. Chem.* **267**, (18)890–18 895.

Caplan, A.J., Cyr, D.M. & Douglas, M.G. (1992b) YDJ1p facilitates polypeptide translocation across different intracellular membranes by a conserved mechanism. *Cell* **71**, 1143–1155.

Caplan, A.J., Cyr, D.M. & Douglas, M.G. (1993) Eukaryotic homologues of *Escherichia coli* DnaJ: A diverse protein family that functions with hsp70 stress proteins. *Mol. Biol. Cell.* **4**, 555–563.

Chappell, T.G., Konforti, B.B., Schmid, S.L. & Rothman, J.E. (1987) The ATPase core of a clathrin-uncoating protein. *J. Biol. Chem.* **262**, 746–751.

Cheetham, M.E, Brion, J.-P. & Anderton, B.H. (1992) Human homologues of the bacterial heat shock protein DnaJ are preferentially expressed in neurons. *Biochem. J.* **284**, 469–476.

Clarke, C.F., Cheng, K., Frey, A.B., Stein, R., Hinds, P.W. & Levine, A.J. (1988) Purification of complexes of nuclear oncogene p53 with rat and *Escherichia coli* heat shock proteins: *In vitro* dissociation of hsc70 and DnaK from murine p53 by ATP. *Mol. Cell Biol.* **8**, 1206–1215.

Cyr, D.M., Lu, X. & Douglas, M.G. (1992) Regulation of eukaryotic hsp70 function by a DnaJ homologue. *J. Biol. Chem.* **267**, 20 927–20 931.

Deshaies, R.J., Sanders, S.L., Feldheim, D.A. & Schekman, R. (1991) Assembly of yeast sec proteins involved in translocation into the endoplasmic reticulum into a membrane-bound multi-subunit complex. *Nature (Lond.)* **349**, 806–809.

Feldheim, D., Rothblatt, J. & Schekman, R. (1992) Topology and functional domains of SEC63p, an endoplasmic reticulum membrane protein required for secretory protein translocation. *Mol. Cell Biol.* **12**, 3288–3296.

Gamer, J., Bujard, H. & Bakau, B. (1992) Physical interactions between heat shock proteins DnaK, DnaJ and GrpE and the bacterial heat shock transcription factor σ32. *Cell* **69**, 833–842.

Gao, Y., Thomas, J.O., Chow, R.L., Lee, G.H. & Cowan, N.J. (1992) A cytoplasmic chaperonin that catalyzes β-actin folding. *Cell* **69**, 1043–1050.

Georgopoulos, C., Ang, D., Liberek, K. & Zylicz, M. (1990) Properties of the *Escherichia coli* heat shock proteins and their role in bacteriophage λ-growth. In *Stress Proteins in Biology and Medicine* (eds Morimoti, R.I., Tissieres, A. & Georgopoulos, C.), pp 191–221. New York, Cold Spring Harbor Laboratory Press.

Gething, M.-J. & Sambrook, J. (1992) Protein folding in the cell. *Nature (Lond.)* **335**, 33–45.

Goedert, M., Wischik, C.M., Crowther, R.A., Walker, J.E. & Klug, A. (1988) Cloning and sequencing of the cDNA encoding a core protein of the paired helical filament of Alzheimer disease: Identification as the microtubule-associated protein tau. *Proc. Natl Acad. Sci. USA* **85**, 4051–4055.

Gragerov, A., Nudler, E., Komissarova, N., Gaitanaris, G.A., Gottesman, M.E. & Nikiforov, V. (1992) Co-operation of GroEL/GroES and DnaK/DnaJ heat shock proteins in preventing protein misfolding in *Escherichia coli. Proc. Natl Acad. Sci. USA* **89**, 10 341–10 344.

Gross, C.A., Strauss, D.B., Erickson, J.W. & Yura, T. (1990) The function and regulation of heat shock proteins in *Escherichia coli*. In *Stress Proteins in Biology and Medicine* (eds Morimoto, R.I., Tissieres, A. & Georgopoulos, C.), pp 167–190. New York, Cold Spring Harbor Laboratory Press.

Gundersen, C.B. & Umback, J.A. (1992) Suppression cloning of the cDNA for a candidate subunit of a presynaptic calcium channel. *Neuron* **9**, 527–537.

Hamos, J.E., Oblas, B., Pulaski-Salo, D., Welch, W.J., Bole, D.G. & Drachman, D.A. (1991) Expression of heat shock proteins in Alzheimer's disease. *Neurology* **41**, 345–350.

Hattori, H., Lui, Y-C., Tohnai, I., Ueda, M., Kaneda, T., Kobayashi, T., Tanabe, K. & Ohtsuka, K. (1992) Intracellular localization and partial amino acid sequence of a hsp-inducible 40-kDa protein in HeLa cells. *Cell Struct. & Function.* **17**, 77–86.

Hattori, H., Kaneda, T., Lokeshwar, B., Laszlo, A. & Ohtsuka, K. (1993) A stress-inducible 40-kDa protein (hsp40): Purification by a modified two-dimensional gel electrophoresis and co-localization with hsc70(p73) in heat shocked HeLa cells. *J. Cell Sci.* **104**, 629–638.

Hunt, C. & Morimoto, R.I. (1985). Conserved features of eukaryotic hsp70 genes revealed by comparison with the nucleotide sequence of human hsp70. *Proc. Natl Acad Sci. USA* **82**, 6455–6459.

Krishnan, H.B. & Pueppke, S.G. (1991) Nolc, a *Rhizobium fredii* gene involved in cultivar specific nodulation of soybean, shares homology with a heat shock gene. *Mol. Micro.* **5**, 737–745.

Langer, T., Lu, C., Echols, H., Flanagan, J., Hayer, M.K. & Hartl, F.-U. (1992) Successive action of DnaK, DnaJ and GroEL along the pathway of chaperone-mediated protein folding. *Nature (Lond.)* **356**, 683–689.

Lewis, M.J. & Pelham, H.R.B. (1985) Involvement of ATP in the nuclear and nucleolar functions of the 70-kDa heat shock protein. *EMBO J.* **4**, 3137–3143.

Liberek, K., Osipiuk, J., Zylicz, M., Ang, D., Skorko, J. & Georgopoulos, C. (1990) Physical interactions between bacteriophage and *Escherichia coli* proteins required for initiation of λ-replication. *J. Biol. Chem.* **265**, 3022–3029.

Liberek, K., Marszalek, J., Ang, D., Georgopoulos, C. & Zylicz, M. (1991a) *Escherichia coli* DnaJ and GrpE heat shock proteins jointly stimulate ATpase activity of DnaK. *Proc. Natl Acad. Sci. USA* **88**, 2874–2878.

Liberek, K., Skowyra, D., Zylicz, M., Johnson, C. & Georgopoulos, C. (1991b) The *Escherichia coli* DnaK chaperone, the 70-kDa heat shock protein eukaryotic equivalent, changes conformation upon ATP hydrolysis, thus triggering its dissociation from bound target protein. *J. Biol. Chem.* **266**, 14 491–14 496.

Luke, M.M., Sutton, A. & Arndt, K.A. (1991) Characterization of SIS1, a *Saccharomyces cerevisiae* homologue of bacterial DnaJ proteins. *J. Cell Biol.* **114**, 623–638.
McCombie, W.R., Adams, M.D., Kelley, J.M., Fitzgerald, M.G., Utterback, T.R., Khan, M., Dubnick, M., Kerlavage, A.R., Venter, J.C. & Fields, C. (1992) *Caenorhabditis elegans* expressed sequence tags, identify gene families and potential disease gene homologues. *Nature Genetics* **1**, 124–131.
May, E., Breugnot, C., Duthu, A. & May, P. (1991) Immunological evidence for the association between simian virus 40 115 super T-antigen and hsp70 proteins in rat, monkey, and human cells, *Virology* **180**, 285–293.
Munro, S. & Pelham, H.R.B. (1986) An hsp70 like protein in the ER: Identity of the 78-kDa glucose regulated protein and immunoglobulin heavy-chain binding protein. *Cell* **46**, 291–300.
Ohki, M., Tamura, F., Nishimura, S. & Uchida, H. (1986) Nucleotide sequence of the *Escherichia coli* DnaJ gene and purification of the gene product. *J. Biol. Chem.* **261**, 1778–1781.
Ohki, M., Uchida, H., Tamura, F., Ohki, R. & Nishimura, S. (1987) The *Escherichia coli* DnaJ mutation affects biosynthesis of specific proteins, including those of the lac operon. *J. Bacteriol.* **169**, 1917–1922.
Ohtsuka, K., Masuda, A., Nakai, A. & Nagata, K. (1990) A novel 40-kDa protein induced by heat shock and other stresses in mammalian and avian cells. *Biochem. Biophys. Res. Commun.* **166**, 642–647.
Pelham, H.R.B. (1990) Functions of the hsp70 protein family: An overview. In *Stress Proteins in Biology and Medicine* (eds Morimoto, R.I., Tissieres, A. & Georgopoulos, C.), pp 287–300. New York, Cold Spring Harbor Laboratory Press.
Raabe, T. & Manley, J.L. (1991) A human homologue of the *Escherichia coli* DnaJ heat shock protein. *Nucleic Acids Res.* **19**, 6645.
Sadler, I., Chiang, A., Kurihara, T., Rothblatt, J., Way, J. & Silver, P. (1989) A yeast gene important for protein assembly into the endoplasmic reticulum and the nucleus has homology to DnaJ, an *Escherichia coli* heat shock protein. *J. Cell Biol.* **109**, 2665–2675.
Saito, H. & Uchida, H. (1978) Organization and expression of the DnaJ and DnaK genes of *Escherichia coli* K12. *Mol. Gen. Genet.* **164**, 1–8.
Sawai, E.T. & Butel, J.S. (1989) Association of a cellular heat shock protein with simian virus 40 large T-antigen in transformed cells. *J. Virol.* **63**, 3961–3973.
Wada, M., Kadokami, K. & Itikawa, H. (1982) Thermosensitive synthesis of DNA and RNA in DnaJ mutants of *Escherichia coli* K-12. *Jpn. J. Genet.* **57**, 407–413.
Walter, G., Carbone, A. & Welch, W.J. (1987) Medium tumour antigen of polyomavirus transformation-defective mutant NG59 is associated with 73-kDa heat shock protein. *J. Virol.* **61**, 405–410.
Waterston, R., Martin, C., Craxton, M., Huynh, C., Coulson, A., Hillier, L., Durbin, R., Green, P., Showkeen, R., Halloran, N., Metzstein, M., Hawkins, T., Wilson, R., Berks, M., Du, Z., Thomas, K., Thierry-Mieg, J. & Sulston, J. (1992) A survey of expressed genes in *Caenorhabditis elegans*. *Nature Genetics* **1**, 114–123.
Wickner, S.H. (1990) Three *Escherichia coli* heat shock proteins are required for P1 plasmid DNA replication: Formation of an active complex between *E. coli* DnaJ protein and the P1 initiator protein. *Proc. Natl Acad. Sci. USA* **87**, 2690–2694.
Wickner, S.H., Hoskins, J. & McKenney, (1991) Monomerization of RepA dimers by heat shock proteins, activates binding to DNA replication origin. *Proc. Natl Acad. Sci. USA* **88**, 7903–7907.
Yaffe, M.B., Farr, G.W., Miklos, D., Horwich, A.L., Sternlicht, M.L. & Sternlicht, H. (1992) TCP1 complex is a molecular chaperone in tubulin biogenesis. *Nature (Lond.)* **358**, 245–248.
Zhang, S., Lockshin, C., Herbert, A., Winter, E. & Rich, A. (1992) Zuotin, a putative Z-DNA binding protein in *Saccharomyces cerevisiae*. *EMBO J.* **11**, 3787–3796.

Zhong, T. & Arndt, K.T. (1993) The yeast SIS1 protein, a DnaJ homologue, is required for the initiation of translation. *Cell* **73**, 1175–1186.

Zhu, J-K., Shi, J., Bressan, R.A. & Hasegawa, P.M. (1993) Expression of an *Atriplex nummularia* gene encoding a protein homologous to the bacterial molecular chaperone DnaJ. *The Plant Cell* **5**, 341–349.

Zinsmaier, K.E., Hofbauer, A., Heimbeck, G., Pflugfelder, G.O., Buchner, S. & Buchner, E. (1990) A cysteine-string protein is expressed in retina and brain of *Drosophila*. *J. Neurogenetics* **7**, 15–29.

Zylicz, M., Yamamoto, T., McKittrick, N., Sell, S. & Georgopoulos C (1985) Purification and properties of the DnaJ replication protein of *Escherichia coli*. *J. Biol. Chem.* **260**, 7591–7598.

CHAPTER 9

CELLULAR ROLES OF UBIQUITIN

Keith D. Wilkinson

Department of Biochemistry
4017 Rollins Research Building, Emory University School of Medicine, Atlanta, GA 30322, USA

Table of Contents

Heat Shock Proteins in the Nervous System
ISBN 0-12-480960-X

Preface

As the other chapters in this book amply demonstrate, the cellular responses to stress are many-fold. This may be particularly important in the nervous system where damage to a small number of cells can result in profound consequences. In general, it appears that many cells deal with cellular damage by one of three protective mechanisms: repair or refolding of damaged proteins catalysed by chaperonins and other of the stress-induced gene products; isolation and sequesteration of the damage, particularly as observed with neural inclusions and lesions; and frank degradation of the damaged proteins and cellular components. It is the latter process, at least as it involves ubiquitin, that is the subject of this chapter. Several recent reviews have highlighted different aspects of the system and the reader is referred to these for more detailed information (Hershko and Ciechanover, 1992; Jentsch *et al.*, 1991; Jentsch, 1992; Varshavsky, 1992; Finley and Chau, 1991; Rechsteiner, 1991; Hershko, 1991).

Ubiquitin is a small protein, highly conserved, and present universally in eukaryotic cells. In all known cases it functions by being covalently attached to target proteins in order to target those proteins for some cellular fate. The linkage is an isopeptide bond between the C-terminal glycine of ubiquitin and side-chain amino groups of lysine on the target protein. It is useful to think of ubiquitin as a post-translational signal-sequence which is attached to a variety of cellular proteins and which can respond in context to the environment. Therefore, the ultimate consequences of ubiquitination depend on the protein to which it is attached, as well as the localization of the conjugate and complex enzymatic specificities. Only a few of those consequences are understood, and only protein degradation has been studied in any detail.

At first glance, the machinery for accomplishing ubiquitin-dependent proteolysis seems numbingly complicated. There are well over 50 gene products involved. It is this complexity however, which gives this system the flexibility to be both a 'garbage collection' system for damaged proteins in the cell and a pathway for the degradation of short-lived regulatory proteins. In general, every step in the pathway is catalysed by a family of isozymes. In some cases, the specificity of individual isozymes is known to vary and the expression of different isozymes is spatially and developmentally regulated. This complexity allows the system to accomplish a variety of roles in different tissues at different stages of development, differentiation, and adaptation.

9.1 Introduction

Ubiquitin is of interest in the current context because it participates in the degradation of damaged proteins, because it is a stress-induced protein, and because it has been recognized as a marker of neural damage in a variety of clinical and pathological situations.

9.1.1 Degradation of regulatory and damaged proteins

Ubiquitin was first shown to be involved in ATP-dependent protein degradation catalysed by rabbit reticulocyte extracts (Wilkinson *et al.*, 1980). This system has been extensively characterized with exogenous substrates and the best substrates are damaged or denatured proteins (Evans and Wilkinson, 1985; Chin *et al.*, 1982; Hershko *et al.*, 1982). It is now widely accepted that ubiquitin-dependent protein degradation is also involved in the turnover of short-lived proteins such as the cyclins (Glotzer *et al.*, 1991; Hershko *et al.*, 1991), nuclear oncogenes (Ciechanover *et al.*, 1991a; Ciechanover *et al.*, 1991b; Ishida *et al.*, 1993) and p53 (Scheffner *et al.*, 1990), as well as yeast mating factors (Hochstrasser *et al.*, 1991) and plant phytochromes (Shanklin *et al.*, 1987). Clearly this system is of widespread importance and a detailed discussion of substrate specificity is given below.

Figure 1 presents an outline of ubiquitin-dependent protein degradation. The overall process can be thought of as involving four separate reactions: activation of ubiquitin requiring the expenditure of one ATP molecule; conjugation of ubiquitin to a variety of cellular proteins; proof-reading of the conjugates to either generate the target protein by removal of ubiquitin or commit the target protein to proteolysis by adding more ubiquitin molecules; and finally proteolysis of the conjugates by the proteasome or the lysosome.

Through these reactions, proteins are marked by ubiquitination and targeted for degradation in the cell. The energy contributed by ATP is utilized to exert specificity on the proteolysis step by specifically marking only certain proteins with ubiquitin. Many of the unanswered questions about this system revolve around how such proteins are recognized and ubiquitinated.

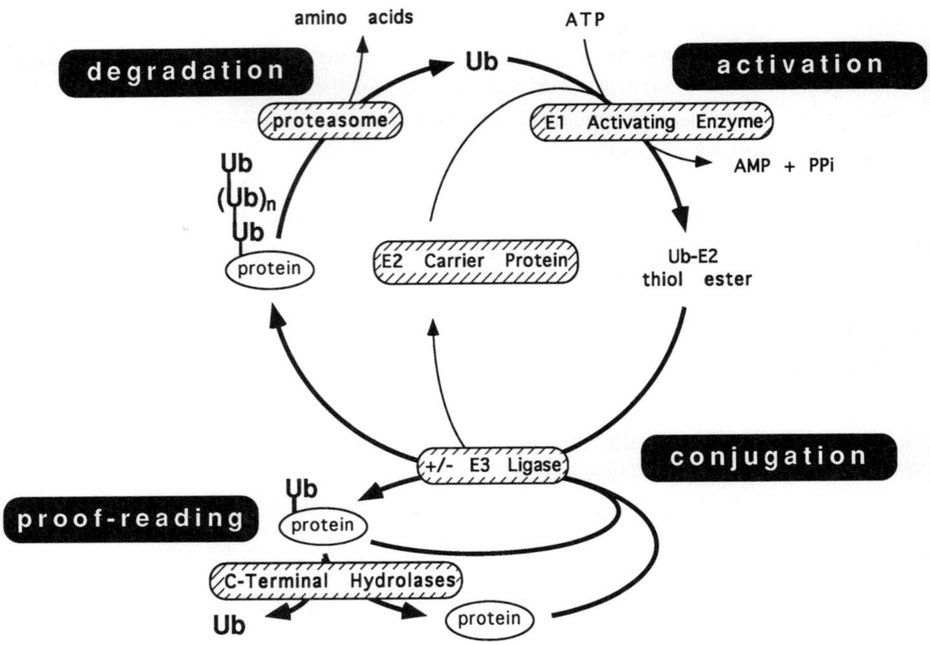

Figure 1 Ubiquitin-dependent proteolysis. The conceptual categories include Activation, Conjugation, Proof-reading and Degradation.

9.1.2 Ubiquitin metabolism in neural tissues

It has been known for over 15 years that ubiquitin is a major protein in brain and pituitary tumors (Scherrer *et al.*, 1978). Its concentration is approximately 20 μM in reticulocytes (Haas and Bright, 1985), and probably higher in brain. It is particularly rich in the post-synaptic domain of human neuromuscular junctions (Serdaroglu *et al.*, 1992). Antibodies to ubiquitin inhibit the sodium-dependent uptake of choline, γ-aminobutyric acid, glutamate, norepinephrine, aspartate, and serotonin in rat cerebral cortical synaptosomes (Meyer *et al.*, 1987). These immunoreactive species appear to consist of several proteins found in synaptic membranes (Chapman *et al.*, 1992). Finally, two ganglioside-binding proteins from brain have been shown to have ubiquitin-like sequences (Zdebska *et al.*, 1992).

All of the enzymes of the ubiquitin system are also found in neural tissues (Handley *et al.*, 1991; Kwak *et al.*, 1991; Azaryan *et al.*, 1989; Doran *et al.*, 1983). In particular, we have found that an abundant neuron-specific protein, PGP9.5 (Doran *et al.*, 1983), is a ubiquitin carboxyl-terminal hydrolase (Wilkinson *et al.*, 1989). This protein can reach 1–2% of the soluble protein in brain and it is selectively accumulated in various neural lesions (Lowe *et al.*, 1990). These findings point to an active ubiquitin metabolism in normal neural tissues.

9.1.3 Stress induces the ubiquitin system

Several of the components of the ubiquitin system are markedly elevated under stress (Finley and Chau, 1991) and a variety of pathological conditions. One of the ubiquitin genes contains a consensus heat shock promoter (Finley *et al.*, 1987; Bond and Schlesinger, 1986), and ubiquitin is a heat shock protein (Bond and Schlesinger, 1985). Increases in ubiquitin or ubiquitinated proteins in brain tissues are observed in the presence of protease inhibitors and upon ageing (Ivy *et al.*, 1989), in Alzheimer's disease (Wang *et al.*, 1991; Ali Khan *et al.*, 1992), in Scrapie-infected brain (Lowe *et al.*, 1992), and during stress such as aluminum intoxication (Takeda *et al.*, 1991) or salt-loading (Schueler and Elde, 1988). Cells in culture respond in a similar way to ischaemia (Hayashi *et al.*, 1991), hypoglycaemia (Cheng and Mattson, 1992), or treatment with serum from Alzheimer's disease patients (Brewer and Ashford, 1992). Ubiquitin mRNA is increased during hyperthermia or ischaemia (Nowak *et al.*, 1990) and upon oxidative insult caused by iron loading (Uney *et al.*, 1993).

The proteasome complex (or multicatalytic proteinase) thought to be responsible for a significant amount of ubiquitin-dependent proteolysis is present in brain tissue (Azaryan *et al.*, 1989) and elevated in Lewy bodies (Kwak *et al.*, 1991). Similarly, lysosomes are thought to participate in the degradation of some ubiquitinated proteins, are rich in neural tissues and induced by some types of stress (Mayer *et al.*, 1992).

The neuronal-specific ubiquitin carboxyl-terminal hydrolase, PGP 9.5, is present in high amounts in brain (Wilson *et al.*, 1988), is selectively accumulated in a variety of neurodegenerative lesions (Lowe *et al.*, 1990), and is induced by trimethyltin intoxication (Anderson *et al.*, 1992).

Since the stress response in neural tissues includes increased protein degradation, it will probably be important to understand the role of ubiquitin-dependent proteolysis in this process. The remainder of this chapter describes what is known about the cellular roles of ubiquitin in a variety of tissues.

9.2 Genetics and structural characteristics

In the last ten years, we have learned a great deal about the genetics and structure of the components of the ubiquitin-dependent proteolysis system. The following section takes a conventional approach, describing the factors and enzymes involved in approximately the order in which they act. This order is also approximately chronological, and will hopefully assist the reader in appreciating the logic and current understanding of the system.

9.2.1 Structure and chemistry of ubiquitin

Ubiquitin is a highly conserved 76 residue protein that is universally present in eukaryotic cells (Wilkinson, 1988). The X-ray crystal structures of three different

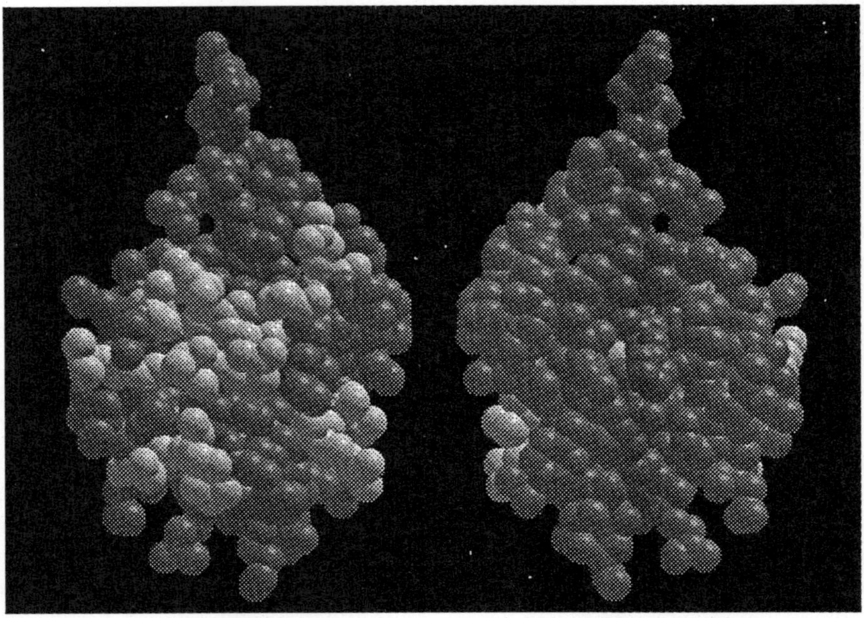

Figure 2 Observed substitutions of the ubiquitin sequence cluster on one face of the molecule. The left image shows the face containing all known substitutions of eukaryotic ubiquitin. The right image is rotated 180 degrees about the *y*-axis and demonstrates the conservation of this face of the molecule.

ubiquitins have been determined: human (Vijay-Kumar *et al.*, 1985), yeast and oat (Vijay-Kumar *et al.*, 1987). These variants all have similar physical properties, activities in supporting protein degradation (Wilkinson *et al.*, 1986b; Vierstra *et al.*, 1986) and, not surprisingly, three-dimensional structure. Figure 2 shows a space-filling representation of ubiquitin and the positions where naturally-occurring substitutions have been noted. Ubiquitin has been sequenced from 20 different organisms and only 13 of the 76 residues have been found to vary (Wostmann *et al.*, 1992). The three-dimensional surface of ubiquitin has patches of acidic residues (8 of the 11 on one face of the molecule), basic residues (7 of the 11 residues on another face), and a marked patch of hydrophobic residues on yet a third face of the molecule (Wilkinson, 1988). These landmarks are likely recognized by the different proteins that must bind to ubiquitin. Further, the location of the naturally occurring substitutions is confined to the face of the molecule containing the α-helix (Figure 2). This has suggested that this face of the molecule is under less selective pressure than the rest of the molecule (Wilkinson, 1988).

This highly conserved protein sequence has suggested to many workers that selective pressure must result from multiple protein interactions with the surfaces of ubiquitin, and that conformational changes may also be important in the mode of action (Cox *et al.*, 1986; Wilkinson and Mayer, 1986). The structural properties of ubiquitin have been described previously and only a few points will be elaborated here. First, ubiquitin is a very stable protein and one which refolds rapidly (Briggs and Roder, 1992). It is also stable to most proteases at neutral pH. These properties are probably advantageous for a protein which participates in the delivery of substrate proteins to proteases. Secondly, ubiquitin undergoes a well characterized conformational change in a hydrophobic environment (Wilkinson and Mayer, 1986; Pan and Briggs, 1992). By circular dichroism, it appears that moderate concentrations of alcohol cause a conformational change characterized by an apparent increase in the helical content of ubiquitin (Wilkinson and Mayer, 1986). By NMR, it appears that most of the native structural elements are still present in the alcohol-induced form (Pan and Briggs, 1992), but that the turn-rich region surrounding the 3_{10} helix is disturbed (Harding *et al.*, 1991). Thus, the changes are probably localized around the 3_{10} helix and spanning residues 50 to 65. This would explain the apparent exposure of the methionine sulphur which is normally hydrogen-bonded to the backbone nitrogen of residue 63, and the increase in the fluorescence of tyrosine 59 which is normally quenched by its proximity to glutamate 51. As discussed below, this conformational change may be important in the 'proofreading' of protein structure.

9.2.2 Structure of ubiquitin conjugates

The chemistry of ubiquitin is dominated by the formation and cleavage of peptide bonds at its carboxyl-terminus, gly-76. Two types of amide bonds at this position are important (Figure 3). The primary gene products of ubiquitin genes are all fusion proteins (Jentsch *et al.*, 1991) with an amino-terminal copy of ubiquitin and a C-terminal extension consisting of either more copies of ubiquitin (pro-ubiquitin), or unrelated sequences (zinc-finger ribosomal proteins). Thus, these gene products consist of ubiquitin linked via an α-amide bond to a C-terminal polypeptide. These fusion proteins must be proteolytically processed to obtain mature ubiquitin.

A second type of C-terminal derivative is formed when multiple ubiquitins are added to target proteins. The best characterized of these is formed by the conjugation of the C-terminus of ubiquitin to the side-chain amino group of lys-48 of another ubiquitin attached to the target protein (Chau *et al.*, 1989; Gregori *et al.*, 1990). Long chains of this type can be built up and must be disassembled during or after proteolysis of the target protein. These polymeric ubiquitin molecules are linked by ε-amino amide bonds and have been called multi-ubiquitin chains or trees. A more appropriate nomenclature would probably refer to this structure as poly-ubiquitin (c.f. polyadenylate and poly-ADP ribose). This structure is apparently necessary to target proteins for degradation, since replacement of lys 48 with arg prevents this poly-ubiquitination and also degradation (Gregori *et al.*, 1990). In some cases, other lysines on ubiquitin may be involved in poly-ubiquitination, although the significance and consequences of these alternative structures are not understood.

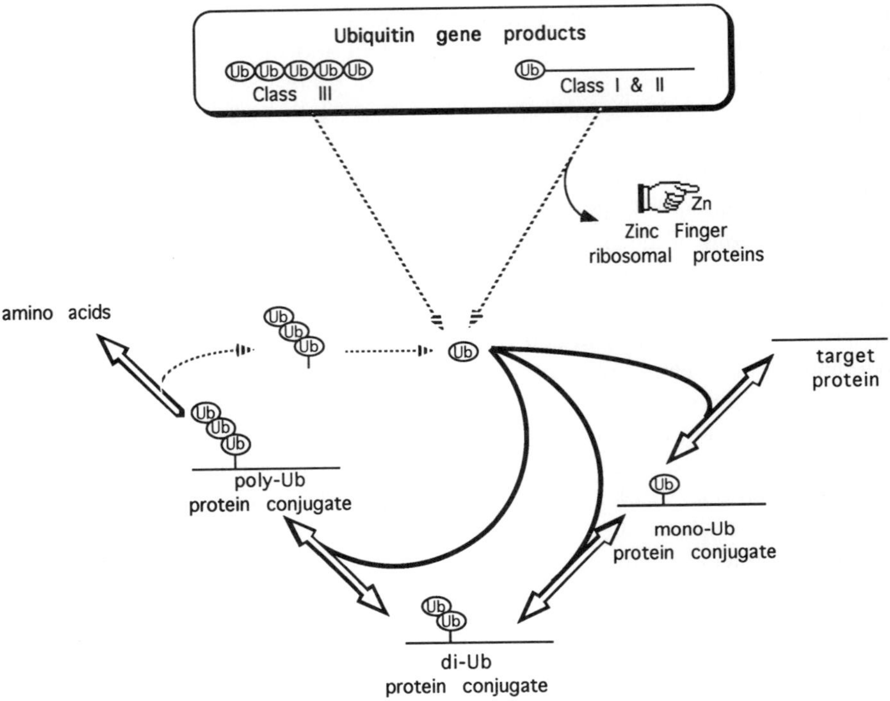

Figure 3 Metabolism of ubiquitin. (Dashed arrows) indicate processing reactions catalysed by ubiquitin carboxyl-terminal hydrolases or isoamidases. (Solid arrows) represent reactions of conjugation. (Open arrows) represent the path for targeting and degrading cellular proteins.

9.2.3 Ubiquitin genes

Ubiquitin is coded for by three classes of genes in all species studies. Class I and II genes encode two proteins, each consisting of a single copy of ubiquitin fused to a different zinc finger protein. These carboxyl extension proteins are processed to yield ubiquitin and the zinc finger proteins (Figure 3). The latter have been shown to be ribosomal proteins and are required for efficient ribosome biogenesis. Thus, transcription from Class I and II genes results in fusion proteins which are processed into two halves; one half of which is required for ubiquitin-dependent protein degradation and the other for protein synthesis. These genes are expressed constitutively in rapidly growing cells (Ozkaynak *et al.*, 1987; Taccioli *et al.*, 1991; Cabrera *et al.*, 1992) and this arrangement could be considered a mechanism to co-ordinate rates of protein synthesis and protein degradation.

The class III gene encodes a ubiquitin precursor consisting of head-to-tail repeats of the ubiquitin protein sequence (Ozkaynak *et al.*, 1987). The number of repeats varies with the organism. This transcript is strongly induced by stress and contains a consensus heat shock promoter (Bond and Schlesinger, 1986; Nowak *et al.*, 1990; Ozkaynak *et al.*, 1987). This is the transcript that is relevant to the stress response and is presumably induced in response to the increased need for ubiquitin during stress.

In addition, there are several genes encoding ubiquitin-related proteins (Toniolo *et al.*, 1988; Haas *et al.*, 1987; Guarino, 1990; Banerji *et al.*, 1990; Meyers *et al.*, 1991) whose functions are less clear. Two RNA viruses have acquired a ubiquitin-like gene which encodes a protein fusion with viral protein sequences (Guarino, 1990; Meyers *et al.*, 1991). It is not known if these are proteolytically processed, but cytopathogenicity appears correlated with the presence of this ubiquitin-like sequence (Meyers *et al.*, 1991). This suggests that the ubiquitin-like protein may function in the pathogenicity of the virus. In bacculovirus, a DNA virus, a single copy of ubiquitin is encoded with only a single tyrosine as a C-terminal extension. This viral protein is processed to an active ubiquitin, although conjugates of this viral ubiquitin are formed more slowly than with eukaryotic ubiquitin (A. Haas and L. Guardino, privileged communication). A final sequence related to viral infection is the human interferon-induced 15-kDa protein which appears to have arisen by gene duplication. This protein consists of an evolutionarily diverged N-terminal copy of ubiquitin, a more highly conserved C-terminal copy of ubiquitin, and a nine amino acid extension (Haas *et al.*, 1987; Knight *et al.*, 1988). The nonapeptide is proteolytically cleaved to produce the 15-kDa protein which has been shown to be conjugated to cellular proteins and to be present at low levels in normal cells (Loeb and Haas, 1992). Thus, under the stress of a viral infection, these ubiquitin-like sequences are expressed and may be involved in viral pathogenicity or in cellular response to interferons (see below).

9.2.4 Enzymes of ubiquitin conjugation

Ubiquitin is covalently attached to proteins to target them for proteolysis (Figure 3). Proteins can be mono-, di- or poly-ubiquitinated. In all cases, the bond is between an amino group of the target protein (usually the side-chain of lysine) and the carboxyl-terminus of ubiquitin. The specificity of ubiquitination is not understood, but involves a complex series of reactions catalysed by the first three enzymes of the system.

9.2.4.1 E1- the ubiquitin activating enzyme

Undoubtedly the most well-characterized enzyme in this pathway is the first; the ubiquitin activating enzyme (Ciechanover *et al.*, 1982). It has a molecular weight of approximately 100 000 daltons and is active as the dimer. This enzyme catalyses the formation of a thiol ester between the C-terminus of ubiquitin and a cysteine residue of the ubiquitin carrier proteins (Figure 1). The mechanism of this enzyme has been well characterized (Haas *et al.*, 1982; Haas *et al.*, 1983) and involves: (1) ordered binding of *ubiquitin and then ATP*; (2) formation of an acyl adenylate between

AMP and the C-terminus of ubiquitin with release of pyrophosphate; (3) release of AMP upon attack by a cysteine of E1 to form a thiol ester; (4) repetition of steps 1 and 2; (5) binding of an E2 ubiquitin carrier protein; (6) a trans-thiolesterification with the active site cysteine on E2 forming a thiol ester between E2 and the C-terminus of ubiquitin; and (7) repeat of steps 3–7. Inhibitors of this enzyme include a non-hydrolyzable ubiquitin adenylate analogue (Wilkinson *et al.*, 1990) and cis-platin (Isoe *et al.*, 1992).

Antibodies to purified ubiquitin activating enzyme have been reported to detect two forms of the enzyme in yeast and humans (Hoefer and Cook, 1991; Cook and Chock, 1992). Using such antibodies, E1 has been variously reported to be bound to condensed chromatin during mitosis (Cook and Chock, 1991b), found in the nucleus (Cook and Chock, 1991a), distributed widely in the cytoplasm and other organelles (Schwartz *et al.*, 1992), and bound to all three components of the cytoskeleton (Trausch *et al.*, 1993). These observations raise the possibility that ubiquitin metabolism could be important in a variety of cellular sites and stages of the cell cycle.

There is genetic evidence for one yeast enzyme (McGrath *et al.*, 1991), one human enzyme (Handley *et al.*, 1991; Zacksenhaus and Sheinin, 1990), two mouse isozymes (Kay *et al.*, 1991; Mitchell *et al.*, 1991), and three wheat activating enzymes (Hatfield and Vierstra, 1992). Note however, that there is also evidence for more than one form of the activating enzyme in yeast and humans (Deveraux *et al.*, 1990; Hoefer and Cook, 1991; Cook and Chock, 1992) and this suggests that there may be more genes which are spatially or developmentally regulated. One example of such regulated expression is known, that of the mouse gene Sby (A1s9Y-1). This gene encodes a ubiquitin-activating enzyme and is located on the short arm of the mouse Y-chromosome, in the sex-reversed (Sxr) region. A translocation of this region to the X-chromosome causes chromosomally female mice to develop as sterile males. Sby is expressed in the testis and is a candidate for the mouse testis-determining gene. Mouse ubiquitin-activating enzyme is also constitutively expressed from the A1s9X gene present between the Otc and Hprt *loci* in the proximal region of the X chromosome (Kay *et al.*, 1991). The mRNA for this X-linked enzyme is expressed throughout the cell cycle (Zacksenhaus and Sheinin, 1990).

Much of what is known about the function and biological roles of ubiquitin-dependent processes has been discerned from the phenotype of conditional lethal mutations in the ubiquitin activating enzyme. A surprising proportion of conditional lethals with defects in cell cycle progression are ts mutants in this enzyme (Jentsch *et al.*, 1991; Hershko and Ciechanover, 1992). Many investigators have utilized ts85 cells, a recessive conditional lethal derivative of the mouse mammary carcinoma line FM3A (Finley *et al.*, 1984). Synchronized cultures of these cells arrest in S and S/G2 upon shifting to the non-permissive temperature. They fail to activate ubiquitin, show reduced degradation of abnormal and short-lived proteins, and fail to condense chromatin (Finley *et al.*, 1984). Several other cell lines exist which belong to the same complementation group and/or have been shown to be rescued by ubiquitin-activating enzyme (Ohtsubo and Nishimoto, 1988; Kulka *et al.*, 1988;

Nishitani *et al.*, 1992; Ayusawa *et al.*, 1992; Zacksenhaus and Sheinin, 1990; Eki *et al.*, 1990). The specific phenotypes of these cell lines will be discussed below.

9.2.4.2 E2- ubiquitin carrier proteins

After ubiquitin is activated by thiol ester formation with E1, it undergoes a trans-thiol esterification with a cysteine residue on one of the E2 proteins (Hershko *et al.*, 1983). E2 genes have now been characterized in many organisms (Jentsch *et al.*, 1991), including at least 10 separate genes in yeast (Jentsch, 1992). This large gene family consists of three classes of genes. All share a central core of about 150 residues which are 35% identical and which contain the active site thiol. Class I enzymes consist of the core only. These 14- to 17-kDa proteins require E3 (ligase) to conjugate ubiquitin to target proteins (Figure 4A). Class II enzymes have C-terminal extensions, which are often highly charged. These extensions are required for E3-independent conjugation and are thought to be involved in substrate recognition (Figure 4B). Class III enzymes have both C- and N-terminal extensions, perhaps involved in membrane anchoring or interactions with other types of ligases.

In the case of the small E2s, the activated ubiquitin is delivered as the E2-thiol ester to the ligase-protein complex for conjugation to the target protein (Pickart and Rose, 1985). These include UBC4 and UBC5 in yeast (Jentsch, 1992), as well as the E2 (14k) from reticulocytes (Pickart and Rose, 1985), and a 15-kDa wheat E2 (Girod and Vierstra, 1993). The reticulocyte and the wheat proteins appear to exist as a dimer *in vivo*. The E3 ligase shows a significant affinity for protein, apparently binding substrate proteins more avidly than other proteins (Hershko *et al.*, 1986). The ligase must also bind the E2 ubiquitin thiol ester in order to catalyze conjugation. The specificity of this pathway depends on the specificity of E3 ligase(s) as discussed below, and probably to a smaller extent on the identity of the E2. These E2 proteins have been called ubiquitin carrier proteins because of the mechanistic similarity to fatty acid activation.

Some of the larger E2s however, can conjugate ubiquitin directly to target proteins (Pickart and Rose, 1985), or to itself (Chen and Pickart, 1990; Van Nocker and Vierstra, 1991) and have thus been referred to as ubiquitin-conjugating enzymes (Jentsch *et al.*, 1987). It is thought that the C-terminal domains contain important specificity determinants which participate in the selection of substrates for ubiquitination. Reticulocyte E2 (20k) and E2 (32k) both catalyse conjugation of ubiquitin to histones and cytochromes in a saturable fashion (Haas *et al.*, 1988), as do the yeast UBC2 (RAD6) gene product and the wheat 23-kDa UBC4 gene products. The latter two proteins have acidic C-terminal domains which are necessary and sufficient for recognition of histones (Sullivan and Vierstra, 1991; Silver *et al.*, 1992; Kolman *et al.*, 1992). Fusion of the acidic tail to the catalytic core of other E2s converts them to enzymes which are able to efficiently ubiquitinate histones. Little is known about the specific substrates recognized by the C-terminal domains of other E2s, but it is inferred that they are involved in the other non-E3 mediated ubiquitination events.

As discussed below, there is considerable genetic evidence that several of these enzymes have different substrate specificity. It is clear that the expression of E2s is tissue specific, but no comprehensive studies on the spatial and temporal patterns of expression have been reported. In light of the differing substrate specificity of these carrier proteins, it would be interesting to examine the expression during stress and other physiological perturbations.

9.2.4.3 E3- ubiquitin protein ligase(s)

One of the seminal unanswered questions about intracellular proteolysis is how proteins are chosen for degradation. One of the clearest examples is afforded by the E3 ligase-dependent ubiquitination of proteins. This enzyme was first recognized as a large molecular weight component of the ubiquitin pathway which was necessary for multiple ubiquitination and subsequent degradation of proteins in the reticulocyte extracts (Hershko *et al.*, 1983). Further investigation showed that E3 has several binding sites (Figure 4A). First, E3 can bind proteins, and appears to prefer proteins which are good substrates for the degradative pathway (Hershko *et al.*, 1986). It must also bind the E2 (14K) dimer, at least transiently, in order to transfer the ubiquitin from this carrier protein to the target protein. Finally, genetic and biochemical evidence (Hershko and Ciechanover, 1992; Varshavsky, 1992) suggests that it also is specific for the nature of the amino-terminus of the protein. There is a site which is specific for a basic residue at the N-terminus and a second site which is specific for a large hydrophobic residue at the N-terminus of the target protein. Both sites are apparently present on the same polypeptide (called E3α) and useful inhibitors of this enzyme are dipeptide esters. Proteins with basic amino termini bind to the basic site, as shown in Figure 4A, and this binding is inhibited by basic dipeptide esters. Conversely, proteins with large hydrophobic amino termini bind in the other site and are prevented from binding by the corresponding dipeptide esters. This specificity is apparently the basis of the finding that the nature of the amino-terminal residue in a protein sequence can determine the half-life of the protein (Bachmair *et al.*, 1986; Bachmair and Varshavsky, 1989; Varshavsky, 1992). Thus, it was observed that the half-life of β-galactosidase constructs could be varied from 2 min to more than 20 h simply by varying the amino-terminal residue. This was termed the 'N-end Rule' and it predicted that there would be specific enzymes which recognized these termini, particularly basic and hydrophobic termini, and targeted these proteins for degradation. It now seems clear the 'N-end rule' is simply the manifestation of the enzymatic specificity of E3α.

The yeast homologue of this enzyme, UBR1, has been cloned and sequenced (Bartel *et al.*, 1990). It encodes a protein of 1950 residues, a slightly acidic isoelectric point, and a molecular weight of approximately 225 000 daltons. It does not appear to be similar to other known proteins, and there are no significant domains or profiles detected. Null mutants of UBR1 are viable, do not degrade N-end rule substrates, and show partial defects in sporulation and growth.

The fact that UBR1 mutants are not lethal clearly suggests that there must be

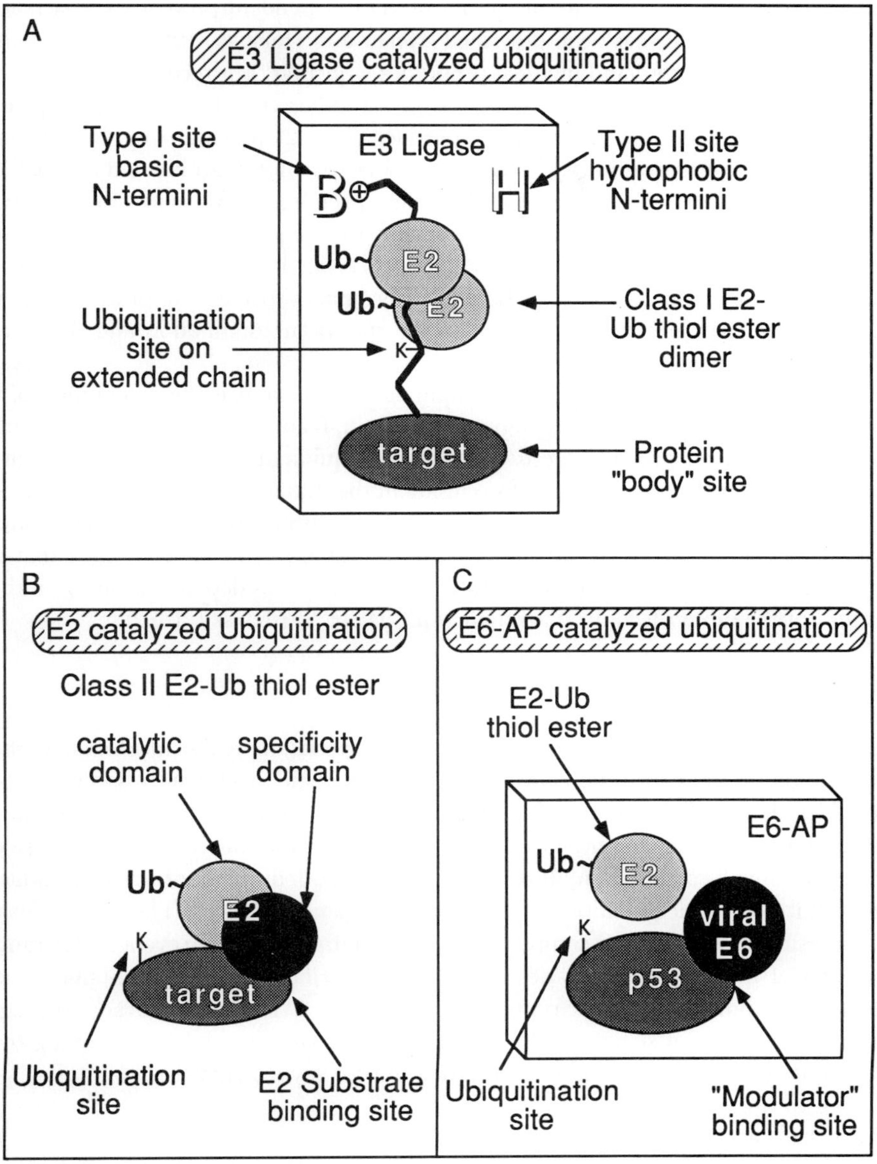

Figure 4 Enzymes involved in conjugation of ubiquitin to cellular proteins. (A) the E3 ubiquitin protein ligase responsible for conjugation to the N-end rule substrates utilizing class I E2 carrier proteins. (B) class II E2 carrier protein catalysed conjugation to basic proteins. (C) the E6-AP/E6 catalysed ubiquitination of p53. This is an example of other possible conjugation routes.

other systems which conjugate ubiquitin for ATP-dependent proteolysis. A second reticulocyte E3, E3β, has been described (Heller and Hershko, 1990). This protein shares many functional and physical properties with E3α, but recognizes substrates with A, S, or T as the amino-terminal residue. Another yeast factor (E3-R) appears to co-operate with the RAD6 gene product (UBC2) in catalysing the mono-ubiquitination of several proteins, which are otherwise not ubiquitinated by this E2 (Sharon *et al.*, 1991). Finally, a recently cloned protein, E6-AP, binds to the complex of papilloma virus E6 and p53 causing its ubiquitination and subsequent degradation (Huibregtse *et al.*, 1991; Huibregtse *et al.*, 1993) (Figure 4C). It is useful to think of E6-AP as a type of E3 which recognizes specific complexes between viral and cellular proteins, and results in the specific ubiquitination and degradation of the conjugate.

Thus, a common theme in these ubiquitination reactions is the recognition of a particular protein or structure by a binding protein, followed by the association with a ubiquitin carrier protein core domain, and ubiquitination of the target protein. Much of the specificity must therefore reside in the initial binding of protein to E3-like molecules. It is the recognition of the amino termini by the E3 enzymes and their efficiency at ubiquitination which is responsible for the 'N-end rule'; that is the observation that the degradation rate of model proteins depends strongly on the nature of the amino termini (Bachmair *et al.*, 1986; Varshavsky, 1992).

9.2.5 Proteolysis of ubiquitin conjugates

Once the target proteins have been marked for degradation by the attachment of the poly-ubiquitin, they must be delivered to the proteolytic machinery responsible for the degradation. The experimental characterization of this machinery has recently been reviewed (Hershko and Ciechanover, 1992) and will only be briefly discussed here. The involvement of a multi-enzyme complex and/or the lysosome emphasizes the fact that these degradation processes are segregated in the cell. There are at least two possible reasons for this complexity; (1) by requiring a marking event and a translocation of the substrate to a degradative site, specificity can be maintained and degradation limited to specific protein substrates; and (2) release of degradative fragments (perhaps with significant biological activities) is prevented by catalysing the degradation in a localized environment rich in proteolytic activities.

9.2.5.1 The proteasome

In the early 1980s, work from Rechsteiner's and Hershko's laboratories demonstrated that the specific proteases of the ubiquitin-dependent system in reticulocytes were large molecular weight complexes which apparently required ATP for both assembly and degradation (review: Hershko and Ciechanover, 1992). The sedimentation coefficient of this active protease was 26S and it exhibited an ATP-dependance and specificity for ubiquitinated substrates (Hough *et al.*, 1987). In addition, they noted a second form of proteolytic activity which sedimented at 20S and contained many of the same polypeptides. The peptidase activity of this complex was activated by detergents and was not ATP-dependent or selective for ubiquitinated proteins. In this chapter, the term prosome is used to refer to the 20S multicatalytic proteinase. The term proteasome is reserved for the 26S particle responsible for ubiquitin-dependent degradation. As will be seen, these are probably related particles.

The 20S prosome: The 20S complex described above has many of the same properties as a multicatalytic proteinase (Wilk and Orlowski, 1983) first isolated from pituitary tissue. The morphology and enzymatic activities of these particles also suggested a similarity to the prosome, a 19S ring-shaped particle present in all nucleated cells (Falkenburg and Kloetzel, 1989; Arrigo *et al.*, 1988). This particle contains a characteristic set of polypeptides from 20- to 34-kDa and assembles into a stack of 4 rings containing a six to eight-fold symmetry in the individual rings. The resultant barrel-shaped structure exhibits several proteolytic activities on artificial substrates and various proteins; including a trypsin-like, a chymotrypsin-like and a peptidylglutamyl hydrolyzing activity. Antibodies to these subunits do not inhibit the proteolytic activity, perhaps because the proteolytic sites are located on the inner surface of the barrel.

Many of the subunits of the 20S complex have been cloned and sequenced (Tanaka *et al.*, 1992). There is considerable sequence similarity between subunits and across species. The subunits are encoded by two gene families, homologous to the corresponding α and β subunits of a similar complex in archaebacteria (Dahlmann *et al.*, 1991; Dahlmann *et al.*, 1989). The β subunits probably occupy the central two disks of the four-disk structure and contain the active sites necessary for proteolytic activity. There is no obvious sequence similarity to other proteases and the assignment of proteolytic activities to individual subunits is just beginning using antibodies, chemical modification and genetic methods (Richter-Ruoff *et al.*, 1992; Heinemeyer *et al.*, 1993; Hilt *et al.*, 1993). Deletion mutants in yeast are lethal, suggesting an important role for this particle. To limit the polymerization of the complex to a stack of four rings, the outer subunits must block further polymerization. One way for this to be accomplished is for the α subunits to occupy the outer two disks of the 20S protease. These then would presumably be the site of binding of any accessory proteins observed in the electron microscope (Ikai *et al.*, 1991) and involved in the other activities of the prosome (see below).

The subunit compositions and intracellular localization of the 20S complex appears to vary with the developmental stage in *Drosophila* (Klein *et al.*, 1990; Haass and Kloetzel, 1989) or the proliferative rate in kidney cancer (Kanayama *et al.*, 1991). The differential subunit composition allows the chromatographic separation of different populations of prosomes (Falkenburg and Kloetzel, 1989) although it is not clear what the functional differences are. One 31-kDa subunit has been reported to be localized in prosomes present in the apical domain of hepatocytes (Briane *et al.*, 1992). This suggests some regio-specific function for a subpopulation of prosomes. It has been postulated that expression of specific isoforms may be a mechanism to alter the specificity of the 20S complex in certain tissues or developmental states (Heinemeyer *et al.*, 1993). Two messages encoding Hs PROS-30, differing only in the 5′ untranslated region and the first six amino acids encoded, have been described (Silva-Pereira *et al.*, 1992). This result suggests that at least one of the prosome subunits can be regulated by alternative splicing, a common regulatory mechanism involved in tissue specific expression. Post-translational mechanisms may also be involved in regulating the activity or half-lives of subunits. At least two of the subunits have potential phosphorylation sites that could be involved in regulation, and phosphorylation of a ~30-kDa subunit has been noted (Hershko and Ciechanover, 1992). In addition, a ubiquitinated protein is a component of and inhibitor of the 26S proteasome described below (Li and Etlinger, 1992).

The variability in subunit composition is probably involved in altering the activities, specificity or localization of the 20S complex. This complex has been reported to be associated with polysomes or as free messenger ribonucleoprotein particles (Kloetzel *et al.*, 1987), to contain a 90 nucleotide RNA (Falkenburg and Kloetzel, 1989; Skilton *et al.*, 1991) and/or a pre-tRNA 5′- processing endonuclease activity (Castano *et al.*, 1986). The 20S prosome has also been implicated in regulation of translation by the finding that it is inhibitory in reticulocyte extracts (Kuehn *et al.*, 1990). These results point out that there may be multiple functions of this 'core protease', that they could be regulated, and that the assembly of the active proteasome may be similarly regulated. A variety of endogenous inhibitors and specificity factors have been described and implicated in the assembly and regulation of the active 26S proteasome (Li *et al.*, 1991; Chu Ping *et al.*, 1992; Driscoll *et al.*, 1992; Li and Etlinger, 1992). However, to date, no individual factors have been sequenced or mechanistically characterized.

The 26S proteasome: Although it is still controversial (Seelig *et al.*, 1991), a unifying hypothesis holds that the 20S multicatalytic proteinase associates with regulatory proteins in an ATP-dependent manner to yield the active 26S proteasome. It is thought that this complex specifically binds to ubiquitinated substrates and degrades the target protein to amino acids. This would be a specific case of the general hypothesis (Hershko and Ciechanover, 1992) that the prosome is a degradative particle whose specificity is controlled by its localization, the composition of its variable subunits, and binding of accessory proteins. One of the definitive tests of this hypothesis will be to define the accessory proteins which must 'decorate' the prosome to convert it to a proteasome, and to define the mechanisms and regulation of such an interconversion.

A number of lines of evidence support the interconversion of the 20S prosome and the 26S proteasome. First, several workers have noted the co-purification of 20S prosome subunits with the 26S proteasome (Hough *et al.*, 1987; Eytan *et al.*, 1989; Driscoll and Goldberg, 1990; Kanayama *et al.*, 1992). Antibodies raised against the 20S prosome have been reported to detect similar subunits in the 26S proteasome and to immunoprecipitate the 26S proteasome, thereby interfering with ubiquitin-dependent proteolysis (McGuire *et al.*, 1988; McGuire and DeMartino, 1989; Tanaka and Ichihara, 1988; Matthews *et al.*, 1989; DeMartino *et al.*, 1991;

Rivett and Sweeney, 1991; Orino *et al.*, 1991). Several groups have demonstrated a reversible, ATP-dependent interconversion of 20S and 26S proteolytic activity (Eytan *et al.*, 1989; Ishiura *et al.*, 1989; Driscoll and Goldberg, 1990; Orino *et al.*, 1991), assumed to be the interconversion of the prosome and the proteasome. Finally, there is significant genetic evidence that the chymotryptic-like activity of the prosome is important for degradation of abnormal proteins and during stress (Heinemeyer *et al.*, 1991). Similar mutations fall into two complementation groups and reveal that the prosome is required for the degradation of ubiquitin-dependent substrates (Seufert and Jentsch, 1992; Richter-Ruoff *et al.*, 1992). Mutants defective in the peptidylglutamyl hydrolysing activity show a less striking phenotype, but show enhanced levels of ubiquitin-protein conjugates (Hilt *et al.*, 1993). Deletion of either of these activities is lethal, reinforcing the importance of the ubiquitin-dependent proteolytic system to vital cellular processes.

It may be useful to speculate on the role of the accessory proteins which bind to the prosome. Ganoth and colleagues, first noted that three protein fractions and ATP must be recombined to form the 26S proteasome and that there was a time lag in the assembly (Ganoth *et al.*, 1988). These were named CF-1, CF-2 and CF-3. CF-1 and CF-2 were shown to have ATP-binding sites and CF-3 was later shown to be the 20S prosome (Eytan *et al.*, 1989; Driscoll and Goldberg, 1990). Association of CF-1 and CF-2 with the 20S prosome is required to confer ATP-dependence and specificity for ubiquitinated proteins. CF-2 binding alone results in the inhibition of the peptidase and proteinase activities of the 20S prosome (Driscoll *et al.*, 1992). Electron microscopy of the 26S proteasome reveals additional electron density at the ends of the stacked rings characteristic of the 20S prosome (Ikai *et al.*, 1991). Thus, it seems reasonable that these accessory proteins bind to the ends of the barrel blocking access of substrates to the catalytic sites which may line the inside. Specific functions of these proteins may be: the specific binding of the polyubiquitin chain to bring the substrate into the vicinity of the degradative machinery; an ATP-dependent translocator which would require energy to unfold the protein and translocate it to the site of proteolysis; and a ubiquitin-specific protease which may disassemble the polyubiquitin chain.

9.2.5.2 The lysosome

It should be noted that the lysosome is another compartmentalized degradative organelle, and in fact has been known to be involved in protein degradation for much longer. It has usually been implicated in the enhanced protein degradation seen in starvation or serum deprivation (Dice and Walker, 1979; Dice *et al.*, 1978) and in non-specific proteolysis. The uptake of such proteins has usually been attributed to microautophagy and no specific transport or protein uptake systems were recognized. Recently, a protein uptake system has been described which shows selectivity for proteins containing a KFERQ sequence motif (Dice and Chiang, 1989; Chiang and Dice, 1988). Uptake of proteins containing these sequences is mediated by binding to hsc73, the 73-kDa heat shock cognate protein (Terlecky *et al.*, 1991; Dice, 1992). Up to 30% of the cytoplasmic protein can be subject to this enhanced degradation rate. While it is likely that this is a general non-specific stress response, it points out that there can be significant selectivity to lysosomal degradation.

It is easy to see the parallels between proteasomal and lysosomal degradation. A signal sequence (ubiquitin or the KFERQ domain) is recognized by a specific binding protein and the bound protein is delivered to a degradative environment for proteolysis. In spite of this, it was several years before it was recognized that ubiquitin (Schwartz *et al.*, 1988) could be found localized in the lysosome. It was soon demonstrated that ubiquitin-protein conjugates accumulated in the lysosome under conditions where lysosomal degradation was inhibited (Doherty *et al.*, 1989b; Doherty *et al.*, 1989a) and were enriched about 12-fold in the presence or absence of lysosomal inhibitors (Laszlo *et al.*, 1990). A second line of evidence supporting the role of the lysosome in degradation of ubiquitinated proteins is the finding that cells with a thermolabile ubiquitin activating enzyme fail to increase lysosomal proteolysis in response to heat shock (Gropper *et al.*, 1991). It has been suggested that there is some ubiquitin-dependent step in the maturation of autophagic vacuoles, but not in their formation (Lenk *et al.*, 1992; Schwartz *et al.*, 1992).

Several investigators feel that the degradation of ubiquitinated proteins can be accomplished by both soluble (26S proteasomes) and vesicular (lysosomal) proteases. Perhaps the latter is involved when the ubiquitinated protein is insoluble, membrane bound, or associated in some way with the cytoskeleton (Mayer *et al.*, 1991). Other investigators have criticized the interpretation of these results. First, it was shown that in the presence of leupeptin, the enrichment of ubiquitinated conjugates was similar to the enrichment of general markers of autophagy (Ueno and Kominami, 1991). Thus ubiquitinated conjugates may not be specifically internalized. Second, there is no apparent increase in conjugates in a variety of lysosomal storage diseases (Zhan *et al.*, 1992). Finally, it should be noted that the specific localization of these conjugates to the lysosome and their quantitation depends on immunological techniques. The anti-ubiquitin antibodies used in these studies react with the poly-ubiquitin chain better than free ubiquitin, and with denatured poly-ubiquitin chains even better (Swerdlow *et al.*, 1986). Thus, it remains a formal possibility that the apparent enrichment is due to selective exposure of the epitope in the acidic environment of the lysosome. Finally, it has been shown that peptides and proteins can survive the lysosomal environment (Isenman and Dice, 1989) and it is known that ubiquitin is stable to proteolysis (Wilkinson, 1988). It may be that the accumulation of immunoreactivity is due to normal autophagy of ubiquitinated proteins and the stability of the poly-ubiquitin chains in the lysosome. Whatever the interpretation, it is clear that at least some of the ubiquitinated proteins in the cell can be internalized and degraded therein.

9.2.6 Deconjugation of ubiquitin: proof-reading, proprotein processing and disassembly of poly-ubiquitin chains

To this point we have considered the formation and degradation of ubiquitinated proteins. We must also consider the cleavage of ubiquitin from a variety of cellular forms.

In the ubiquitin-dependent proteolysis reaction the action of these enzymes could inhibit protein degradation. Inhibition could result from the deconjugation of ubiquitin from either mono- or poly-ubiquitinated protein substrates. These reactions would reverse the conjugation of ubiquitin and define a 'futile cycle'. We have postulated that this futile cycle is a proof-reading mechanism designed to assure that only damaged proteins become marked and degraded by this system (Cox *et al.*, 1986; Wilkinson and Mayer, 1986). This hypothesis suggests that ubiquitin is conjugated to most proteins 'semi-randomly'. If the protein is stable, proteases are present which will recognize the native conformation of ubiquitin and remove it. If however, the protein is damaged, unstable, or denatured, the exposed hydrophobicity could cause a conformational change in ubiquitin. It is this altered conformation which is recognized by the committed step in the pathway, poly-ubiquitination.

The action of ubiquitin carboxyl-terminal hydrolases could also stimulate degradation by increasing the steady-state level of free ubiquitin upon the hydrolysis of poly-ubiquitin chains. These enzymes have also been suggested to hydrolyse adventitiously formed adducts between ubiquitin and small cellular nucleophiles. Enzymes with this catalytic activity are also required for the processing of all known ubiquitin gene products and metabolites. Clearly it is important to understand which enzymes catalyse these reactions and to determine their distribution and regulation.

Formally, this reaction requires proteases with specificity for cleavage of the amide bond involving the carboxyl group of gly-76 of ubiquitin. Two types of enzymes with this specificity have been described: the ubiquitin carboxyl-terminal hydrolases, which are thiol proteases of approximately 25 000 daltons (Wilkinson *et al.*, 1989; Wilkinson *et al.*, 1992) and; ubiquitin-specific proteases, a group of yeast thiol proteases between 100 000 and 150 000 daltons (Tobias and Varshavsky, 1991; Baker *et al.*, 1992). These activities have also been called isopeptidase (Matsui *et al.*, 1982) or A-24 lyase (Andersen *et al.*, 1981). We have adopted the suggestion of Hershko and Ciechanover (Hershko and Ciechanover, 1992) and will term enzymes with this general activity C-terminal hydrolases, and for those where it is confirmed that ε-amino amides are hydrolysed, we will use the term isopeptidase.

9.2.6.1 Ubiquitin carboxyl-terminal hydrolases

It is clear that there are a number of steps in ubiquitin metabolism which require proteolytic processing by cleavage of amide bonds at the C-terminus of ubiquitin. Ubiquitin carboxyl-terminal hydrolases (UCH) are a class of small cytoplasmic thiol proteases with specificity for cleavage of small esters and amides of the carboxyl terminal glycine of ubiquitin.

The UCH gene family: We have endeavoured to identify and characterize these enzymes using a generic substrate, ubiquitin C-terminal ethyl ester (Wilkinson *et al.*, 1986a). It was assumed that the use of the ethyl leaving group would be neutral and that all enzymes with UCH specificity would be detected by the hydrolysis of this substrate. That has turned out to be a gross over-simplification (see below). We were successful however, in using this assay to quantitate the purification and characterization of four UCH isozymes from bovine thymus (Mayer and Wilkinson, 1989). A human cDNA for the major thymus isozyme, UCH-L3 has been cloned and sequenced (Wilkinson *et al.*, 1989). The results showed that UCH-L3 is a member of a newly recognized gene family. The UCH-L3 sequence shows considerable similarity to a yeast protein (Miller *et al.*, 1989) which exhibits the same catalytic activity; i.e., cleavage of C-terminal esters and amides of ubiquitin. A second sequence that demonstrates a similar degree of identity is a newly identified *Drosophila* protein prominent in nurse cells (Zhang *et al.*, 1993). The fusion protein with glutathione S-transferase is enzymatically active, demonstrating that it probably serves as a UCH in *Drosophila.* The third sequence is that of human PGP 9.5 (UCH-L1), a neuronal-specific protein first characterized by Day, Thompson and collaborators (Day and Thompson, 1989; Wilson *et al.*, 1988; Doran *et al.*, 1983). We have shown that the bovine PGP 9.5 is catalytically active as a UCH (Wilkinson *et al.*, 1989). Finally, we have obtained partial peptide sequences from a third human isozyme, UCH-L2 isolated from bovine liver, and these sequences are extremely similar to those of UCH-L3.

A few conclusions can be made from the structure of these isozymes. First, these enzymes are thiol proteases based on the sensitivity to inhibitors (Miller *et al.*, 1989; Wilkinson *et al.*, 1989) and this implies that cysteine 100 (the only conserved cysteine in these sequences) must be the active site thiol. Second, there is a conserved histidine at position 107 which may furnish the required acid/base catalysis of tetrahedryl intermediate formation and decay. Finally, the active site sequences are unrelated to any other known thiol proteases. The sequences are conserved as strongly across species as they are among isozymes (approximately 40% identity), suggesting significant evolutionary pressure for the maintenance of these active structures throughout development.

Tissue specific expression of UCH isozymes: It was of interest to examine the distribution of these mammalian isozymes. The long-term objectives of our studies are to identify and characterize these related proteins, to determine the tissue distribution and regulation of the isozymes, to define the substrate specificity of these enzymes, and to elucidate the physiological role(s) for these enzymes.

Two independent approaches were taken to characterize the distribution and regulation of these enzymes. The first is an immunochemical approach. Polyclonal antibodies against homogeneous bovine UCH-L1 (PGP 9.5) and UCH-L3 were raised in rabbits. The antibodies were then affinity purified by adsorption to and elution from nitrocellulose strips containing electrophoretically purified UCH isozymes. These antibodies cross-react with the corresponding isozymes from rat and human tissues. Anti-L1 detects only isozyme L1 while anti-L3 detects both isozyme L2 and L3. In addition, several higher molecular weight species are detected in different tissues (Wilkinson *et al.*, 1992). Since the molecular weights of the UCH isozymes are slightly different, SDS-PAGE and Western blots of tissue and cell homogenates can define the isozyme content of these samples.

The second approach depends on the fact that the three mammalian isozymes can be separated by ion exchange chromatography (Mayer and Wilkinson, 1989). FPLC of tissue and cell homogenates and assay of the separated isozymes using ubiquitin ethyl ester allows us to characterize the amounts of each isozyme activity. Changes in isozyme profiles can be conveniently followed by this technique. Since these results agree with the immunochemical measurements, we will discuss only the latter. Two classes of protein are detected in these experiments: authentic UCH isozymes and immunoreactive bands that may or may not be related to UCH isozymes.

From the UCH isozyme patterns, it is apparent that there is considerable tissue specificity in the distribution of UCH isozymes L1 and L3. As discussed below, we see the L1 isozyme (neuronal PGP 9.5) predominantly in brain and the testis. There are trace amounts detected in heart also. Most tissues examined (except erythrocytes) contained significant amounts of L2 and we consider this a constituitive or 'housekeeping' enzyme. Isozyme L3 is found in red cells (and other haematopoetic cells), lung and brain. Its presence in the latter two tissues may be due to contaminating macrophages and blood cells in these preparations.

In tissue culture, the homogeneity of cell type makes the distribution clearer. Neuronal cell types (PC-12 and Neuro-2A) contain only UCH-L1 (PGP 9.5). HEL 299 (a human lung fibroblast) and Vero (an African green monkey kidney fibroblast) both contain small amounts of UCH-L1 (PGP 9.5). This is consistent with a recent report that MRC-5 lung fibroblasts express UCH-L1 (PGP 9.5), but that the level is down-regulated upon transformation (Honore *et al.*, 1991). Primary chicken hepatocyte cultures (cHEP) contain primarily isozyme L2, in good agreement with the data described above. LK 35.2, a hybridoma, contains both L2 (from the spleen cells?) and L3 (from the lymphoma cells?). Y1, an epithelial-like adrenal carcinoma, contains both L2 and L3 while Raji, a B-cell lymphoma, contains predominantly isozyme L3.

The generalizations that emerge are that isozyme L1 is strongly expressed in neuronal, neuroendocrine, and perhaps some foetal fibroblasts. Isozyme L3 is present mainly in hematopoetic cells. Many tissues and cells contain significant amounts of isozyme L2 which may be a constitutive isozyme.

Expression of UCH isozymes in neuronal disease: Since the original observations that Alzheimer's disease plaques show ubiquitin immunostaining (Cole and Timiras, 1987; Perry *et al.*, 1987) a large number of publications have noted the presence of ubiquitin immunostaining in diseased neural tissues (Mayer *et al.*, 1991). In general, ubiquitin immunostaining has been observed in cases involving deposition of abnormal proteins, particularly cytoskeletal and neurofilament proteins. This fact, and the fact that ubiquitin is a stress protein suggests that ubiquitin is found in these inclusions because it is intimately involved in the degradation of these abnormal protein deposits. It is not known whether ubiquitin is present at the time of deposition or only detected late in the process as a reflection of the cells attempt to deal with this pathological situation.

This question may be clarified by characterizing when the enzymes that metabolise ubiquitin become associated with these lesions. As discussed above, PGP 9.5 (UCH-L3) is a ubiquitin carboxyl-terminal hydrolase present in neural tissues. This enzyme could potentially play a role in proof-reading (Cox *et al.*, 1986) or, by removing ubiquitin from conjugates, it could slow the rates of proteolysis. It could also increase the rates of proteolysis by deconjugating ubiquitin from the multi-ubiquitin chains as the attached protein is being degraded (Hadari *et al.*, 1992).

When a variety of neurological tissues were immunostained for the presence of PGP 9.5, it was found that the enzyme is selectively accumulated in inclusion bodies characteristic of neurodegenerative diseases (Lowe *et al.*, 1990). The preferential accumulation of UCH-L1 (PGP 9.5) appears to be indicative of ubiquitin-dependent activity in that the 26S proteasome is similarly localized; ie., both are found in cortical Lewy bodies and loosely arranged globose-type neurofibrillary tangles (Kwak *et al.*, 1991). It is not yet clear whether this is an early or late event because of the uncertain temporal relationship of the various pathological lesions observed during disease progression.

We thus conclude that active ubiquitin metabolism is associated with neuronal damage and is most likely part of the stress response system. Many immunocytochemical observations have suggested an increase in conjugated ubiquitin occurs in various diseases and environmental insults. This presumably requires an increased expression of ubiquitin. Indeed, the stress-induced poly-ubiquitin mRNA is elevated during the stress response (Bond and Schlesinger, 1986; Nowak *et al.*, 1990; Ozkaynak *et al.*, 1987), in motor neurone disease (Heggie *et al.*, 1989) and in normal brain during fever or ischaemia (Nowak *et al.*, 1990).

Regulation of UCH gene expression: Since ubiquitin is a heat shock protein and the levels of UCH-L1 (PGP 9.5) are increased in neurological disease it was of interest to determine if the expression UCH isozymes were affected by stress. Cells were heat shocked or treated with sodium arsenite and the levels of UCH and HSP 70 immunoreactivity were determined by SDS-PAGE and Western blotting (Wilkinson *et al.*, 1992). There was no change in the level of any UCH isozymes in PC12, HEL299, LK 35.2, Y1, or Raji cells under conditions where HSP 70 was strongly induced. These results indicate that it is unlikely that expression of these enzymes is under regulation by conventional stress-response elements.

We also investigated the effects of serum deprivation, differentiation by nerve growth factor and seeding density on the isozymes expressed in PC12 cells. In short, none of these factors caused a change in the amounts of isozymes expressed. Treatment of Raji cells with TPA caused a two-fold increase in the low levels of UCH-L1 (PGP 9.5) seen in these cells. While this may indicate regulation of L1 expression via protein kinase C, the stimulation by TPA is not apparent in other cell types tested. Thus, we are not certain of the generality of this observation.

The expression of UCH isozymes is also developmentally regulated. The appearance of UCH-L1 (PGP 9.5) immunoreactivity in developing mouse brain correlates with the arrival of the neuronal precursor cells at the neural plate and the elaboration of neural processes (I. N. M. Day and R. Thompson, private communication). Similarly, in *Drosophila*, the mRNA for *uch* -D is strongly expressed in the nurse cell, the ovary and the testis (Zhang *et al.*, 1993). The transcripts are also easily identified for the first few hours of development. By 4–6 h of development, the levels of transcript drop markedly, consistent with the pattern usually seen for maternal transcripts.

The role of this UCH isozyme in development is highly speculative. The *uch* -D transcripts are preferentially localized to the ventral surface of both the oocyte and the nurse cell. It is known that both dorsal/ventral and anterior/posterior gradients of mRNA and protein accumulation are required for proper embryogenesis in *Drosophila*. We can suggest that one of the mechanisms for establishing this polarity may be the formation of a spatial gradient of proteolytic potential. The expression of the ubiquitin-dependent proteolysis system along a concentration gradient in the cell could contribute to the establishment of a gradient of several proteins. Relevant ubiquitin-dependent proteolytic substrates could be those involved in regulation of cell division, in signal transduction, cell–cell contact, hormone response, or gene regulation.

Physiological functions of UCH isozymes: A clear understanding of the role(s) of these isozymes awaits identification of physiological substrates. UCH-L1 (PGP 9.5) is clearly involved in some neuronal-specific function, perhaps related to stress response. It is not a conventional heat shock protein (see above). Since it is transported by the slow SCb transport pathway in neurons (Bizzi *et al.*, 1991) it is difficult to see how it could be involved in a rapid response in the nerve termini. It could however be constitutively expressed in these tissues or rapidly accumulated in the cell body. We have hypothesized that it is involved in the co-translational processing of the pro-ubiquitin gene product induced by stress response (Wilkinson *et al.*, 1992). Its accumulation in neuronal lesions may simply reflect its binding affinity for the poly-ubiquitin which accumulates at these sites.

Haemopoetic cells on the other hand, may have less need for this type of stress response but would need to process the ubiquitin-zinc finger gene products in order to rapidly synthesize the ribosomes needed for these cells to undergo rapid clonal expansion. Thus, isozyme L3 may be involved in co-translational processing of these proteins in rapidly dividing cells. Isozyme L2 is much more widely distributed and could play a role in removing ubiquitin from peptides generated as degradation intermediates or in salvaging ubiquitin which may have been adventitiously trapped by reaction with small cellular amines and thiols.

Recently, we have accumulated evidence for these speculations. Human UCH-L1 and UCH-L3 have been cloned into prokaryotic expression vectors carrying the kanamycin resistance gene and mRNA synthesis is driven by a T7 promoter. When these plasmids are used to transform *E. coli* containing the T7 RNA polymerase, high levels of enzymatically active protein are produced (C. N. Larsen and K. D. Wilkinson, unpublished). Co-transformation of these bacteria with a plasmid-encoding putative substrate molecules and carrying the ampicillin resistance gene results in expression of both the enzyme and the putative substrates in the same cell. Using these constructs we have shown that UCH-L1 (PGP 9.5) co-translationally processes the penta-ubiquitin gene product, but not the ubiquitin-zinc finger fusion proteins (UbCEP52 and UbCEP80). The converse result was also obtained, i.e. UCH-L3 is able to co-translationally process the Ub-zinc finger fusion proteins but not penta-ubiquitin. UCH catalysed cleavage of the purified fusion proteins is at least three orders of magnitude less efficient. It may be that co-translational processing is necessary to prevent the accumulation of these gene products, perhaps because they would resemble the poly-ubiquitin signal sequence and/or other ubiquitinated proteins.

We also were interested in knowing the basis of this specificity. Three model substrates were prepared, each consisting of a ubiquitin fused to the first ten amino acids of the normal ubiquitin gene products. As expected, UCH-L1 exhibits significant activity in hydrolysing the fusion containing the peptide corresponding to the N-terminus of ubiquitin, but little activity on the model zinc-finger substrates. This demonstrates that there must be a binding site for the leaving peptide in this class of enzymes. UCH-L3 cleaves all of the model substrates at similar rates and this suggests that the observed *in vitro* specificity requires interaction with either more of the nascent translation product or with the translational machinery. We might expect to see the enzymes associated with polysomes although we have not yet tested this hypothesis.

A working model for these enzymes suggests that they specifically bind ubiquitin with affinities in the micromolar range (Wilkinson *et al.*, 1989). In addition, the peptide leaving group must make contact with the enzyme in order to explain the specificity of cleavage. When the leaving group is a large, folded domain these enzymes will not hydrolyse the amide bond, perhaps because of steric exclusion from the active site. In co-translational processing, the substrate is presumably generated when the N-terminal copy of ubiquitin has emerged from the ribosome and folded, but before the subsequent peptide sequence has had a chance to fold into a stable structure.

9.2.6.2 Ubiquitin-specific proteases

Another class of enzymes with this general activity have been characterized from yeast (Tobias and Varshavsky, 1991; Baker *et al.*, 1992) based on the ability to hydrolyse large ubiquitin-protein fusions. Three genes were identified and expressed in *E. coli.* (UBP1, 809 amino acids; UBP2, 1264 amino acids; and UBP3, 912 amino acids). These proteins are only distantly related, if at all. Lysates of the transformed bacteria were then tested for activity and specificity on several natural and artificial ubiquitin fusion proteins. UBP1 and 2 were generally able to hydrolyse the amide bond between ubiquitin and β-galactosidase, dihydrofolate reductase and the natural ribosomal zinc-finger proteins. They would not hydrolyse either di-ubiquitin or penta-ubiquitin. UBP3 showed no activity at all in lysates. When these enzymes were coexpressed with substrate in bacteria, penta-ubiquitin was hydrolysed by UBP1 and UBP2, although considerable amount of the substrate escaped processing. In addition, UBP3 was able to cleave β-galactosidase and ribosomal zinc-finger fusions when it was co-expressed with these substrates. This result indicates that either UBP3 acts only co-translationally, or that it is inactive after lysing the bacterial cells.

The working model for these enzymes is similar to that presented above for UCH isozymes. The only significant difference is that these enzymes are large enough to have binding sites for more than one ubiquitin or for larger leaving groups. This makes them better suited to be involved in disassembling poly-ubiquitin chains, or in the selective de-ubiquitination of proteins.

9.2.6.3 Other hydrolases

There must be yet more of these hydrolases in yeast, since the deletion of all three UBP isozymes and the one known UCH isozyme is not lethal (Baker *et al.*, 1992). These cells are able to produce ubiquitin by processing the primary gene products and are able to hydrolyse ub-βgal fusions. They are sensitive to stress and canavanine, a phenotype consistent with a compromised ability to process the pro-ubiquitin sequence to monomeric ubiquitin.

Finally, there have been sporadic attempts to demonstrate UCH activity associated with the 26S proteasome. One such activity, catalysed by a 98-kDa protein termed isopeptidase T, has been reported to associate with the 26S proteasome and to stimulate protein degradation (Hadari *et al.*, 1992). It is assumed that this is because isopeptidase T can partially disassemble poly-ubiquitin chains, thereby increasing the supply of ubiquitin or relieving inhibition caused by the accumulation of such chains. We have recently found that this enzyme efficiently disassembles poly-ubiquitin chains by an exo-hydrolytic activity (K. D. Wilkinson and C. Pickart, unpublished). This activity absolutely requires an intact C-terminal copy of ubiquitin and deletion of gly-76 or replacement of it with an ala residue decreases the rate of reaction by at least three orders of magnitude. The activity is not precessive, but tri-ubiquitin binds more tightly than di-ubiquitin in agreement with earlier suggestions that there were multiple ubiquitin binding sites (Hadari *et al.*, 1992). These results imply that the poly-ubiquitin chain must be removed from the protein substrate, or peptidyl remnant, before disassembly. This could prevent such hydrolytic activity from removing the signal sequence prematurely. In addition, this mechanism assures that only intact ubiquitin is liberated from these chains for reutilization.

9.3 Known cellular roles of ubiquitin-dependent protein degradation

The biological roles of ubiquitin-dependent protein degradation can be classified as resulting from the degradation of short-lived regulatory proteins, or the more general phenomenon of degradation of abnormal or damaged proteins. Below, examples of both types are discussed.

9.3.1 Regulated degradation of short-lived protein

Ubiquitin has been implicated in the degradation of several important regulatory proteins. In some cases, it is not known if the effect is direct (i.e. requiring ubiquitination of the target protein) or indirect (for instance, regulating the level of a binding protein or other protease). In other cases the evidence is from *in vitro* studies only. The highest level of certainty comes from experiments demonstrating the involvement of ubiquitin *in vivo*.

9.3.1.1 p53

Many oncogenic viruses act by means of a viral-encoded protein which causes the degradation of the p53 tumour suppressor protein. The oncogenic papillomavirus E6 gene product is required for degradation of the p53 protein *in vivo* (Band *et al.*, 1991) and *in vitro* (Scheffner *et al.*, 1990). A related viral protein, E7, binds to the retinoblastoma protein and prevents it from acting as a tumour suppressor, but does not cause its degradation. In reticulocyte extracts, E6 from HPV-16 or -18 strains causes the degradation of p53 and a corresponding ubiquitination of the p53 protein. The E6 from the 'low risk' HPV-6 or -11 strains do not bind p53 or cause its degradation. E6 must bind to a cellular component involved in ubiquitination since hybrids between E6 and the N-terminal domain of papillomavirus E7 can target the retinoblastoma tumour suppressor protein for degradation in this assay (Scheffner *et al.*, 1992). The binding of E6 and the cellular protein, E6-AP has been directly demonstrated (Huibregtse *et al.*, 1991). Recently, the E6-AP protein has been purified and the cDNA sequenced (Huibregtse *et al.*, 1993).

It is interesting to speculate on how such a mechanism arose. E6 from oncogenic 'high risk' and 'low risk' isolates bind to E6-AP, although only the former binds p53 and causes its degradation. This would suggest that the interaction between E6 and E6-AP may be involved in normal viral infection or reproduction. Oncogenic potential may have been acquired by mutation of E6 to bind p53. The ubiquitination of E6 could indicate that E6-AP is an ubiquitin protein ligase (E3) like those described above, or it could exert an effect in trans by delivering E6 to another ubiquitin conjugating enzyme.

9.3.1.2 Other nuclear oncogenes: myc and fos

Other nuclear oncogenes such as N-*myc*, c-*myc*, c-*fos* and adenovirus E1A are notable in that their protein products are rapidly degraded. Ciechanover *et al.* (1991)a demonstrated that these proteins are also rapidly degraded in a reticulocyte lysate and that the degradation was dependent on activation of ubiquitin. The physiological significance of this observation is uncertain. Many proteins can be degraded in this *in vitro* system. Direct experiments, such as measuring the half-life of such proteins in conditional mutants, might yield more definitive answers.

9.3.1.3 MAT α2

The MAT α2 transcriptional repressor of yeast is a rapidly degraded protein involved in mating type selection (Hochstrasser and Varshavsky, 1990). There are two domains which, when fused to β-galactosidase, can target it for rapid degradation. Immunoprecipitation with antibodies to α2 and epitope-tagged ubiquitin demonstrated that ubiquitin is conjugated to MAT α2 *in vivo*. Furthermore, degradation was significantly inhibited by over-expressing inhibitory ubiquitin derivatives (the K48R ubiquitin mutant or the epitope-tagged ubiquitins). Finally, unpublished data is quoted suggesting that certain null mutants of the UBC conjugating enzymes (E2 carrier proteins) are defective in MAT α2 degradation (Hochstrasser *et al.*, 1991).

9.3.1.4 Cell cycle progression: cyclins and mos

The cyclins are important regulators of cell cycle progression. They associate with a protein kinase, p34CDC2, to form the heterodimeric complexes. The complex formed with cyclin B for instance is an active histone H1 kinase and its presence prevents cells from exiting from mitosis. The degradation rate of cyclins are altered during the cell cycle, with rapid degradation at exit from mitosis. During interphase the cyclins gradually accumulate again. Glotzer (Glotzer *et al.*, 1991) showed that, in *Xenopus* egg extracts, a region of the N-terminus of cyclin was necessary for this increase in degradation rate, that it could be fused to other proteins to cause them to be degraded at mitosis, and that the degradation was coincident with conjugation of ubiquitin to cyclin A or the fusion protein carrying the degradation sequence. Hershko (Hershko *et al.*, 1991) then showed that methylated ubiquitin was able to prevent poly-ubiquitin chain formation and degradation of cyclins A and B in clam oocyte extracts.

It is not known what triggers the timed degradation of the cyclins. The exit from M-phase upon fertilization of vertebrate eggs requires degradation of cyclins and is prevented by the presence of p39mos, the product of the c-mos gene (Lorca *et al.*, 1991). Upon fertilization, calcium levels are raised to micromolar levels, cyclins are degraded and mitosis occurs. This same calcium transient, when applied to xenopus oocyte lysates, allows the cyclin degradation to take place. This degradation can be inhibited by addition of a calcium–calmodulin binding peptide and restored by adding back calmodulin. These results implicate phosphorylation as a regulatory event, but there is little direct evidence.

Degradation of p39mos has been shown to involve ubiquitination (Nishizawa *et al.*, 1992) and to be catalysed by the proteasome (Ishida *et al.*, 1993). Phosphorylation of ser 3 prevents this ubiquitination and degradation. Thus, exit from mitosis may be regulated by phosphorylation of cyclin (enhancing its degradation) and phosphorylation of p39mos (preventing its degradation).

9.3.1.5 Phytochrome

A major photoreceptor in plants is phytochrome. It consists of a 124-kDa polypeptide with a linear tetrapyrrole chromophore and is involved in control of red light mediated morphogenesis. A brief pulse of light converts the red light-absorbing form (Pr) to a far-red light-absorbing form (Pfr) which then aggregates. The half-life of Pr is over 100 h, while Pfr is degraded with a half-life of about 2 h (Shanklin *et al.*, 1987). The enhanced degradation correlates with conjugation of ubiquitin (Jabben *et al.*, 1989). Peptide mapping of the ubiquitinated phytochrome suggested that it was attached near the carboxyl-terminus (Shanklin *et al.*, 1989).

The signal which causes increased ubiquitination and degradation in this system would appear to be a conformational change associated with a light induced isomerization of the chromophore. This altered conformation lead to increased sensitivity to trypsin (Shanklin *et al.*, 1989), aggregation and ubiquitination of both the soluble and the aggregated protein.

9.3.2 Degradation of abnormal proteins

Another role of the ubiquitin-dependent system is as a 'garbage collector' for damaged protein in the cell. This stress-response may work in concert with hsps and/or remove those proteins which cannot be repaired by the system.

9.3.2.1 Amino acid analogues

A variety of early studies on intracellular proteolysis demonstrated that incorporation of amino acid analogues, often canavanine, led to synthesis of short-lived abnormal proteins. This labelling procedure has been used as a model for normal short-lived proteins. Degradation of these proteins in hepatoma cells correlated with the appearance of ubiquitin conjugates of the analogue-containing proteins (Hershko *et al.*, 1982). Specific mutants have indicated that a significant fraction of the degradation of short-lived and analogue-containing proteins requires functional ubiquitin activating enzyme E1 (Ciechanover *et al.*, 1984) and E2 carrier proteins (Seufert and Jentsch, 1990).

9.3.2.2 Oxidized damaged or denatured proteins

The degradation rate of normal proteins has been correlated with size, stability and susceptibility to proteases (Goldberg and Dice, 1974). This has led to the idea that denaturation, especially that caused by damage may be an important signal for proteolysis. Chin (Chin *et al.*, 1982) showed that denaturation of haemoglobin with phenylhydrazine led to increased ubiquitination of the protein, and that the levels of ubiquitination correlated with the degradation rate. Aggregation of phytochrome Pfr (Shanklin *et al.*, 1989) or alkylated BSA (Evans and Wilkinson, 1985) apparently triggers ubiquitination also. Reduction of a single disulfide bond (Dunten *et al.*, 1991) is sufficient to cause a slight conformational change, the ubiquitination, and the degradation of lysozyme. Oxidation of methionines in ribonuclease (Dunten and Cohen, 1989) and lysozyme (Dunten *et al.*, 1991) greatly increase the rate of ubiquitination and degradation in reticulocyte system, probably by increasing the binding to the protein substrate binding site of E3 ubiquitin protein ligase (Hershko *et al.*, 1986). Oxidation of hexokinase causes it to be degraded more rapidly by the ubiquitin-dependent system (Thorburn and Beutler, 1989). Finally, three different cytochrome P450 isozymes have been reported to be rapidly degraded by the ubiquitin dependent system after mechanism-based inactivation (Tierney *et al.*, 1992; Correia *et al.*, 1992).

These data point to a mechanism which distinguishes 'normal' and 'damaged' proteins. The proposed proof-reading mechanism described above could be involved. Only if the protein were damaged (exposure of hydrophobic regions?) would it cause a conformational change in the first ubiquitin attached and be committed to degradation by poly-ubiquitination.

9.4 Other possible roles for ubiquitination

The results discussed above all deal with the complete proteolytic degradation of target proteins. There are however, several instances of partial degradation or limited ubiquitination which have been observed. Below we discuss the possible role of these more limited pathways.

9.4.1 Antigen presentation

Class I antigen presentation is responsible for self/non-self immune responses. Among the best antigens are viral-encoded nuclear proteins. There is good evidence that these proteins are degraded to peptides and the peptides are taken up into the endoplasmic reticulum by MHC-encoded peptide transporters. There they assemble with newly synthesized class I MHC molecules. After efflux of the antigen-MHC complex to the surface of the cell, these cells are destroyed by cytotoxic T-lymphocytes. Normal cellular proteins probably also follow this pathway but these peptides are not antigenic because self-reactive T-lymphocytes have been removed early in development (Driscoll and Finley, 1992).

It has been dogma and lore for some time that this proteolysis is due to the ubiquitin-dependent proteolysis system. The first experimental evidence supporting this came from the observation that adding a signal for ubiquitination to the N-terminus of nucleoprotein enhanced the efficiency of its presentation (Townsend *et al.*, 1988). Second, two MHC encoded genes, *Ring10* and *Ring12* are apparently members of the prosome β-gene family (Driscoll and Finley, 1992). A highly homologous yeast gene, *PRE2*, encodes a subunit with chymotryptic-like activity and is required for degradation of ubiquitinated proteins (Heinemeyer *et al.*, 1993). Finally it has been argued that the Ring genes are polymorphic and regulated by γ-interferon. Thus, the evidence for direct involvement of the ubiquitin system in antigen presentation is circumstantial at best. It does remain however, an intriguing speculation that deserves more detailed experimental examination.

9.4.2 Receptor structure/function

A number of investigators have reported ubiquitination of cell surface receptors, including: the lymphocyte homing receptor (Siegelman *et al.*, 1986); the platelet-derived growth factor receptor (Yarden *et al.*, 1986); the growth hormone receptor (Spencer *et al.*, 1988); the T-cell antigen receptor (Cenciarelli *et al.*, 1992); the high-affinity immunoglobulin E receptor (Paolini and Kinet, 1993); and the human tumour necrosis factor receptor (Loetscher *et al.*, 1990). The lymphocyte homing receptor is ubiquitinated early in biosynthesis, perhaps co-translationally (van de Rijn *et al.*, 1990). The T-cell receptor, PDGF receptor and IgE receptors are poly-ubiquitinated in response to binding of their respective physiological ligands (Cenciarelli *et al.*, 1992; Mori *et al.*, 1992; Mori *et al.*, 1993; Paolini and Kinet, 1993). The PDGF receptor is ubiquitinated on the carboxyl-terminal domain and is apparently degraded after poly-ubiquitination. In contrast, the IgE receptor is reversibly modified, with no evidence for its involvement in degradation.

From this small sampling, it appears that one function of ubiquitination could be in the down-regulation of receptors. It may also be involved in extracellular events, as monoclonal antibodies to ubiquitin decorate the surface of cells. Finally, the reversible ubiquitination on the cytoplasmic face of the receptor may indicate a participation in signal transduction events.

9.4.3 Chromosome structure/function

The first ubiquitinated protein discovered was A-24, a ubiquitinated histone H2a (Hunt and Dayhoff, 1977). Approximately 10% of the H2a molecules are ubiquitinated in interphase, and ubiquitin is removed from H2a during mitosis (Matsui *et al.*, 1979). This ubiquitination/de-ubiquitination is rapid and reversible. A review of the covalent modifications of chromatin has recently appeared (Bradbury, 1992). Several groups have suggested that histones of transcriptionally active chromatin are preferentially ubiquitinated (Levinger and Varshavsky, 1982; Davie and Murphy, 1992). At the non-permissive temperature, E1 conditional mutants are cell cycle arrested and fail to condense their chromatin (Matsumoto *et al.*, 1983). Since histone proteins are thought to be stable, it seems likely that this reversible modification has some structural implications for transcriptional activity or chromatin condensation.

Another function of ubiquitination appears to be in DNA repair. The yeast E2 carrier protein, UBC2 is encoded by the RAD6 locus. Mutants in this *locus* are defective in DNA repair (Jentsch *et al.*, 1987). Since repair of DNA lesions probably requires extensive alterations in chromatin structure, it is difficult to separate these two functions. Similarly, the CDC34 mutation (another E2 carrier protein) may affect chromatin structure, as it has been argued that most structural modifications of chromosomes will be involved in cell cycle control (Bradbury, 1992).

9.4.4 Vectoral transport

A considerably more speculative example of putative ubiquitin functions is that of vectoral protein transport. Such transport would be necessary to generate immunoreactive ubiquitin on the surface of cells or to target ubiquitinated proteins to the lysosome. It has also been reported that mitochondrial import of monoamine oxidase is inhibited by antibodies to ubiquitin and that this inhibition can be relieved by addition of ubiquitin (Zhaung and McCauley, 1989; Zhuang *et al.*, 1992). Interestingly, two yeast E2 carrier proteins have been implicated in similar processes (Jentsch, 1992). UBC6 is reported to be an integral membrane protein of the ER involved in secretion and UBC10 (PAS2) has been implicated in peroxisome biogenesis (Wiebel and Kunau, 1992).

It is easy to see how the machinery of this proteolysis system could accomplish vectoral transport. A ubiquitin-binding protein could deliver the conjugate to a transporter (perhaps a pore resembling a modified 20S prosome with no proteolytic activity?) The action of a ubiquitin carboxyl-terminal hydrolase could remove the ubiquitin signal coincident with or after transversal of the membrane. We have in fact, noted membrane bound forms of UCH immunoreactivity (Wilkinson *et al.*, 1992) but have not identified these to date.

9.4.5 Viral survival mechanisms

It is also clear that viruses have usurped the ubiquitin dependent system in ill defined ways to assist in their survival and replication. *Herpes* infection induces the transcription of poly-ubiquitin by the trans-activating factor ICP4 (Latchman *et al.*, 1987). TMV-virons contain on average one ubiquitin covalently attached to the coat protein (Dunigan *et al.*, 1988). Several viral isolates encode ubiquitin-like molecules (see above). In the case of the baculoviral ubiquitin, it accumulates upon infection, is conjugated to a variety of cellular proteins and can support proteolysis, although at a reduced rate (A. Haas and L. Guarino, personal communication).

Similarly, the interferon induced 15-kDa ubiquitin homologue can be conjugated to proteins, although the precise consequences of that are unknown (Loeb and Haas, 1992). Finally, an E2 carrier protein homologous to the RAD6 and CDC34 gene products is encoded by an african swine fever virus (Hingamp *et al.*, 1992).

These viral 'tricks' point out the potential for manipulating important cellular functions by modifications of ubiquitin-mediated metabolism. Perhaps future studies of these pathways will illuminate ways in which we can manipulate the pathway for treatment of disease.

9.5 Future directions

It is both risky and enlightening to speculate on the future directions a field will take. It is probably safe to say that it will continue to be important to define substrates for the system, and to define new biological implications. In addition, a better understanding of the protein–protein interactions, as elucidated by structural biology, will clarify a number of questions concerning the specificity and scope of the system.

9.5.1 Repair of cellular damage

Perhaps one of the clearest conclusions which can be drawn from the above discussion is that this proteolytic pathway is exquisitely regulated. The degradation of specific proteins depends variously on the presence or absence of a signaling sequence, the phosphorylation state of that sequence, the conformational changes induced by ligand binding, and/or a variety of damage events. Both host- and pathogen-encoded gene products are used in some circumstances, leading to a complex interplay. For instance, the interactions of this degradation system with the cellular stress-response is unclear at best, but a hypothesis can be formulated. Stress proteins are thought to be involved in repair of damage, perhaps by acting as catalysts for refolding proteins. There is good evidence that the constitutive homologues often are important for proper folding or localization of proteins. If refolding can be accomplished by the stress proteins, the cell survives. The ubiquitin system may only be invoked when that system fails, and could even involve the complex with hsp's as signals for ubiquitination. If the ubiquitin system fails, the cell dies. The presence of ubiquitin in lysosomes and in a variety of neural pathogenic lesions may be the 'scars' left by this unsuccessful attempt to repair the damage.

9.5.2 Regulation of development and differentiation

The genetics of the system in yeast has been tremendously important. It can not however, completely elucidate the complex developmental biology of vertebrates, nor the repertoire of cellular reponses of a multicellular organism. It will be important to move the analysis of these genes into more complex organisms. A start has been made in *Drosophila* and we can probably look forward to transgenic mammals soon. The incredible redundancy of the system makes it likely that there are a number of interesting and informative mutants possible.

9.5.3 Inhibitors of the system

Perhaps the most useful piece of the puzzle still missing is specific inhibitors of the process. Without these, we cannot fully exploit the physiological implications, nor develop pharmaceuticals which will be necessary to treat what we all feel will be important pathological processes involving the ubiquitin system.

References

Ali Khan, Z., Normand, J., Alizadeh Khiavi, K., Robitaille, Y. & Chronopoulos, S. (1992) Ubiquitin profile and amyloid enhancing factor activity in Alzheimer and 'normal' human brain extracts. *Neurosci. Lett.* **139**, 24–28.

Andersen, M., Raghuveera Ballal, N., Goldknopf, I.L. & Busch, H (1981) Protein A24 lyase activity in nucleoli of thioacetamide-treated rat liver releases histone 2A and ubiquitin from conjugated protein A24. *Biochemistry* **20**, 1100–1104.

Anderson, V.E., Hajimohammadreza, I., Gallo, J.M., Anderton, B.H., Uney, J., Brown, A.W., Nolan, C.C., Cavanagh, J.B. & Leigh, P.N. (1992) Ubiquitin, PGP 9.5 and dense body formation in trimethyltin intoxication: Differential neuronal responses to chemically induced cell damage. *Neuropathol. Appl. Neurobiol.* **18**, 360–375.

Arrigo, A.P., Tanaka, K., Goldberg, A.L. & Welch, W.J. (1988) Identity of the 19S 'prosome' particle with the large multifunctional protease complex of mammalian cells (the proteasome). *Nature* **331**, 192–194.

Ayusawa, D., Kaneda, S., Itoh, Y., Yasuda, H., Murakami, Y., Sugasawa, K., Hanaoka, F. & Seno, T. (1992) Complementation by a cloned human ubiquitin-activating enzyme E1 of the S-phase-arrested mouse FM3A cell mutant with thermolabile E1. *Cell Struct. Funct.* **17**, 113–122

Azaryan, A., Banay Schwartz, M. & Lajtha, A. (1989) The presence of ATP+ubiquitin-dependent proteinase and multicatalytic proteinase complex in bovine brain. *Neurochem. Res.* **14**, 995–1001.

Bachmair, A., Finley, D. & Varshavsky, A. (1986) *In vivo* half-life of a protein is a function of its amino-terminal residue. *Science* **234**, 179–186.

Bachmair, A & Varshavsky, A. (1989) The degradation signal in a short-lived protein. *Cell* **56**, 1019–1032.

Baker, R.T., Tobias, J.W. & Varshavsky, A. (1992) Ubiquitin-specific proteases of *Saccharomyces cerevisiae.* Cloning of UBP2 and UBP3, and functional analysis of the UBP gene family. *J. Biol. Chem.* **267**, 23 364–23 375.

Band, V., De Caprio, J.A., Delmolino, L., Kulesa, V. & Sager, R. (1991) Loss of p53 protein in human papillomavirus type 16 E6-immortalized human mammary epithelial cells. *J. Virol.* **65**, 6671–6676.

Banerji, J., Sands, J., Strominger, J.L. & Spies, T. (1990) A gene pair from the human major histocompatibility complex encodes large proline-rich proteins with multiple repeated motifs and a single ubiquitin-like domain. *Proc. Natl Acad. Sci. USA* **87**, 2374–2378.

Bartel, B., Wunning, I. & Varshavsky, A. (1990) The recognition component of the N-end rule pathway. *EMBO J.* **9**, 3179–3189.

Bizzi, A., Schaetzle, B., Patton, A., Gambetti, P. & Autilio Gambetti, L. (1991) Axonal transport of two major components of the ubiquitin system: Free ubiquitin and ubiquitin carboxyl-terminal hydrolase PGP 9.5. *Brain Res.* **548**, 292–299.

Bond, U. & Schlesinger, M.J. (1985) Ubiquitin is a heat shock protein in chicken embryo fibroblasts. *Mol. Cell Biol.* **5**, 949–956.

Bond, U. & Schlesinger, M.J. (1986) The chicken ubiquitin gene contains a heat shock promoter and expresses an unstable mRNA in heat shocked cells. *Mol. Cell Biol.* **6**, 4602–4610.

Bradbury, E.M. (1992) Reversible histone modifications and the chromosome cell cycle. *Bioessays* **14**, 9–16.

Brewer, G.J. & Ashford, J.W. (1992) Human serum stimulates Alzheimer markers in cultured hippocampal neurons. *J. Neurosci. Res.* **33**, 355–369.

Briane, D., Olink Coux, M., Vassy, J., Oudar, O., Huesca, M., Scherrer, K. & Foucrier, J. (1992) Immunolocalization of a specific type of prosome close to the bile canaliculi in faetal and adult rat liver. *Eur. J. Cell Biol.* **57**, 30–39.

Briggs, M.S. & Roder, H. (1992) Early hydrogen-bonding events in the folding reaction of ubiquitin. *Proc. Natl Acad. Sci. USA* **89**, 2017–2021.

Cabrera, H.L., Barrio, R. & Arribas, C. (1992) Structure and expression of the *Drosophila* ubiquitin-52-amino-acid fusion-protein gene. *Biochem. J.* **286**, 281–288.

Castano, J.G., Ornberg, R., Koster, J.G., Tobian, J.A. & Zasloff, M. (1986) Eukaryotic pre-tRNA 5′ processing nuclease: copurification with a complex cylindrical particle. *Cell* **46**, 377–385.

Cenciarelli, C., Hou, D., Hsu, K.C., Rellahan, B.L., Wiest, D.L., Smith, H.T., Fried, V.A. & Weissman, A.M. (1992) Activation-induced ubiquitination of the T-cell antigen receptor. *Science* **257**, 795–797.

Chapman, A.P., Courtney, S.C., Smith, S.J., Rider, C.C. & Beesley, P.W. (1992) Ubiquitin immunoreactivity of multiple polypeptides in rat brain synaptic membranes. *Biochem. Soc. Trans.* **20**, 155S.

Chau, V., Tobias, J.W., Bachmair, A., Marriott, D., Ecker, D.J., Gonda, D.K. & Varshavsky, A. (1989) A multiubiquitin chain is confined to specific lysine in a targeted short-lived protein. *Science* **243**, 1576–1583.

Chen, Z. & Pickart, C.M. (1990) A 25-kilodalton ubiquitin carrier protein (E2) catalyses multi-ubiquitin chain synthesis via lysine 48 of ubiquitin. *J. Biol. Chem.* **265**, 21 835–21 842.

Cheng, B. & Mattson, M.P. (1992) Glucose deprivation elicits neurofibrillary tangle-like antigenic changes in hippocampal neurons: Prevention by NGF and βFGF. *Exp. Neurol.* **117**, 114–123.

Chiang, H.L. & Dice, J.F. (1988) Peptide sequences that target proteins for enhanced degradation during serum withdrawal. *J. Biol. Chem.* **263**, 6797–6805.

Chin, D.T., Kuehl, L. & Rechsteiner, M. (1982) Conjugation of ubiquitin to denatured haemoglobin is proportional to the rate of haemoglobin degradation in HeLa cells. *Proc. Natl Acad. Sci. USA* **79**, 5857–5861.

Chu Ping, M., Slaughter, C.A. & DeMartino, G.N. (1992) Purification and characterization of a protein inhibitor of the 20S proteasome (macropain). *Biochim. Biophys. Acta* **1119**, 303–311.

Ciechanover, A., DiGiuseppe, J.A., Bercovich, B., Orian, A., Richter, J.D., Schwartz, A.L. & Brodeur, G.M. (1991a) Degradation of nuclear onco-proteins by the ubiquitin system *in vitro*. *Proc. Natl Acad. Sci. USA* **88**, 139–143.

Ciechanover, A., DiGiuseppe, J.A., Schwartz, A.L. & Brodeur, G.M. (1991b) Degradation of MYCN onco-protein by the ubiquitin system. *Prog. Clin. Biol. Res.* **366**, 37–43.

Ciechanover, A., Elias, S., Heller, H. & Hershko, A. (1982) 'Covalent affinity' purification of ubiquitin-activating enzyme. *J. Biol. Chem.* **257**, 2537–2542.

Ciechanover, A., Finley, D. & Varshavsky, A. (1984) Ubiquitin dependence of selective protein degradation demonstrated in the mammalian cell cycle mutant ts85. *Cell* **37**, 57–66.

Cole, G.M. & Timiras, P.S. (1987) Ubiquitin-protein conjugates in Alzheimer's lesions. *Neurosci. Lett.* **79**, 207–212.
Cook, J.C. & Chock, P.B. (1991a) Immunocytochemical localization of ubiquitin-activating enzyme in the cell nucleus. *Biochem. Biophys. Res. Commun.* **174**, 564–571.
Cook, J.C. & Chock, P.B. (1991b) Association of ubiquitin-activating enzyme with HeLa cell chromosomes during mitosis. *Proc. Natl Acad. Sci. USA* **88**, 11 388–11 392.
Cook, J.C. & Chock, P.B. (1992) Isoforms of mammalian ubiquitin-activating enzyme. *J. Biol. Chem.* **267**, 24 315–24 321.
Correia, M.A., Davoll, S.H., Wrighton, S.A. & Thomas, P.E. (1992) Degradation of rat liver cytochromes P450 3A after their inactivation by 3,5-dicarbethoxy-2,6-dimethyl-4-ethyl-1,4-dihydropyridine: Characterization of the proteolytic system. *Arch. Biochem. Biophys.* **297**, 228–238.
Cox, M.J., Haas, A.L. & Wilkinson, K.D. (1986) Role of ubiquitin conformations in the specificity of protein degradation: Iodinated derivatives with altered conformations and activities. *Arch. Biochem. Biophys.* **250**, 400–409.
Dahlmann, B., Kopp, F., Kuehn, L., Hegerl, R., Pfeifer, G. & Baumeister, W. (1991) The multicatalytic proteinase (prosome, proteasome): Comparison of the eukaryotic and archaebacterial enzyme. *Biomed. Biochim. Acta* **50**, 465–469.
Dahlmann, B., Kopp, F., Kuehn, L., Niedel, B., Pfeifer, G., Hegerl, R. & Baumeister, W. (1989) The multicatalytic proteinase (prosome) is ubiquitous from eukaryotes to archaebacteria. *FEBS Lett.* **251**, 125–131.
Davie, J.R. & Murphy, L.C. (1992) Level of ubiquitinated histone H2B in chromatin is coupled to ongoing transcription. *Biochemistry* **29**, 4752–4757.
Day, I.N.M. & Thompson, R.J. (1989) Molecular Cloning of cDNA coding for human PGP 9.5 protein. *FEBS Lett.* **210**, 157–160.
DeMartino, G.N., McCullough, M.L., Reckelhoff, J.F., Croall, D.E., Ciechanover, A. & McGuire, M.J. (1991) ATP-stimulated degradation of endogenous proteins in cell-free extracts of BHK 21/C13 fibroblasts. A key role for the proteinase, macropain, in the ubiquitin-dependent degradation of short-lived proteins. *Biochim. Biophys. Acta* **1073**, 299–308.
Deveraux, Q., Wells, R., and Rechsteiner, M. (1990) Ubiquitin metabolism in ts85 cells, a mouse carcinoma line that contains a thermolabile ubiquitin activating enzyme. *J. Biol. Chem.* **265**, 6323–6329.
Dice, J.F. (1992) Selective degradation of cytosolic proteins by lysosomes. *Ann. NY. Acad. Sci.* **674**, 58–64.
Dice, J.F. & Chiang, H.L. (1989) Peptide signals for protein degradation within lysosomes. *Biochem. Soc. Symp.* **55**, 45–55.
Dice, J.F. & Walker, C.D. (1979) Protein degradation in metabolic and nutritional disorders. *Ciba. Found. Symp.* P 331–50.
Dice, J.F., Walker, C.D., Byrne, B. & Cardiel, A. (1978) General characteristics of protein degradation in diabetes and starvation. *Proc. Natl Acad. Sci. USA* **75**, 2093–2097.
Doherty, F.J., Laszlo, L., Lowe, J., Lennox, G., Landon, M. & Mayer, R.J. (1989a) Ubitquitin-protein conjugates: Clinical and experimental findings. *Revis. Biol. Celular.* **20**, 255–273.
Doherty, F.J., Osborn, N.U., Wassell, J.A., Heggie, P.E., Laszlo, L. & Mayer, R.J. (1989b) Ubiquitin-protein conjugates accumulate in the lysosomal system of fibroblasts treated with cysteine proteinase inhibitors. *Biochem. J.* **263**, 47–55.
Doran, J.F., Jackson, P., Kynoch, P.A.M. & Thompson, R.J. (1983) Isolation of PGP 9.5, a new human neurone-specific protein detected by high-resolution two-dimensional electrophoresis. *J. Neurochem.* **40**, 1542–1547.
Driscoll, J. & Finley, D. (1992) A controlled breakdown: antigen processing and the turnover of viral proteins. *Cell* **68**, 823–825.
Driscoll, J., Frydman, J. & Goldberg, A.L. (1992) An ATP-stabilized inhibitor of the proteasome is a component of the 1500-kDa ubiquitin conjugate-degrading complex. *Proc. Natl Acad. Sci. USA* **89**, 4986–4990.

Driscoll, J. & Goldberg, A.L. (1990) The proteasome (multicatalytic protease) is a component of the 1500-kDa proteolytic complex which degrades ubiquitin-conjugated proteins. *J. Biol. Chem.* **265**, 4789–4792.

Dunigan, D.D., Dietzgen, R.G., Schoelz, J.E. & Zaitlin, M. (1988) Tobacco mosaic virus particles contain ubiquitinated coat protein subunits. *Virology* **165**, 310–312.

Dunten, R.L. & Cohen, R.E. (1989) Recognition of modified forms of ribonuclease A by the ubiquitin system. *J. Biol. Chem.* **264**, 16 739–16 747.

Dunten, R.L., Cohen, R.E., Gregori, L. & Chau, V. (1991) Specific disulphide cleavage is required for ubiquitin conjugation and degradation of lysozyme. *J. Biol. Chem.* **266**, 3260–3267.

Eki, T., Enomoto, T., Miyajima, A., Miyazawa, H., Murakami, Y., Hanaoka, F., Yamada, M. & Ui, M. (1990) Isolation of temperature-sensitive cell cycle mutants from mouse FM3A cells: Characterization of mutants with special reference to DNA replication. *J. Biol Chem.* **265**, 26–33.

Evans, A.C., Jr. & Wilkinson, K.D. (1985) Ubiquitin-dependent proteolysis of native and alkylated bovine serum albumin: Effects of protein structure and ATP concentration on selectivity. *Biochemistry* **24**, 2915–2923.

Eytan, E., Ganoth, D., Armon, T. & Hershko, A. (1989) ATP-dependent incorporation of 20S protease into the 26S complex that degrades proteins conjugated to ubiquitin. *Proc. Natl Acad. Sci. USA* **86**, 7751–7755.

Falkenburg, P.E. & Kloetzel, P.M. (1989) Identification and characterization of three different subpopulations of the *Drosophila* multicatalytic proteinase (proteasome). *J. Biol. Chem.* **264**, 6660–6666.

Finley, D. & Chau, V. (1991) Ubiquitination. *Annu. Rev. Cell Biol.* **7**, 25–69.

Finley, D., Ciechanover, A. & Varshavsky, A. (1984) Thermolability of ubiquitin-activating enzyme from the mammalian cell cycle mutant ts85. *Cell* **37**, 43–55.

Finley, D., Ozkaynak, E. & Varshavsky, A. (1987) The yeast poly-ubiquitin gene is essential for resistance to high temperatures, starvation, and other stresses. *Cell* **48**, 1035–1046.

Ganoth, D., Leshinsky, E., Eytan, E. & Hershko, A. (1988) A multicomponent system that degrades proteins conjugated to ubiquitin. Resolution of factors and evidence for ATP-dependent complex formation. *J. Biol. Chem.* **263**, 12 412–12 419.

Girod, P.A. & Vierstra, R.D. (1993) A major ubiquitin conjugation system in wheat germ extracts involves a 15-kDa ubiquitin-conjugating enzyme (E2) homologous to the yeast UBC4/UBC5 gene products. *J. Biol. Chem.* **268**, 955–960.

Glotzer, M., Murray, A.W. & Kirschner, M.W. (1991) Cyclin is degraded by the ubiquitin pathway. *Nature* **349**, 132–138.

Goldberg, A.L. & Dice, J.F. (1974) Intracellular protein degradation in mammalian and bacterial cells. *Annu. Rev. Biochem.* **43**, 835–869.

Gregori, L., Poosch, M.S., Cousins, G. & Chau, V. (1990) A uniform isopeptide-linked multi-ubiquitin chain is sufficient to target substrate for degradation in ubiquitin-mediated proteolysis. *J. Biol. Chem.* **265**, 8354–8357.

Gropper, R., Brandt, R.A., Elias, S., Bearer, C.F., Mayer, A., Schwartz, A.L. & Ciechanover, A. (1991) The ubiquitin-activating enzyme, E1, is required for stress-induced lysosomal degradation of cellular proteins. *J. Biol. Chem.* **266**, 3602–3610.

Guarino, L.A. (1990) Identification of a viral gene encoding a ubiquitin-like protein. *Proc. Natl Acad. Sci. USA* **87**, 409–413.

Haas, A.L., Ahrens, P., Bright, P.M. & Ankel, H. (1987) Interferon induces a 15-kilodalton protein exhibiting marked homology to ubiquitin. *J. Biol. Chem.* **262**, 11 315–11 323.

Haas, A.L. & Bright, P.M. (1985) The immunochemical detection and quantitation of intracellular ubiquitin-protein conjugates. *J. Biol. Chem.* **260**, 12 464–12 473.

Haas, A.L., Bright, P.M. & Jackson, V.E. (1988) Functional diversity among putative E2 isozymes in the mechanism of ubiquitin-histone ligation. *J. Biol. Chem.* **263**, 13 268–13 275.

Haas, A.L., Warms, J.V., Hershko, A. & Rose, I.A. (1982) Ubiquitin-activating enzyme. Mechanism and role in protein-ubiquitin conjugation. *J. Biol. Chem.* **257**, 2543–2548.

Haas, A.L., Warms, J.V. & Rose, I.A. (1983) Ubiquitin adenylate: Structure and role in ubiquitin activation. *Biochemistry* **22**, 4388–4394.

Haass, C. & Kloetzel, P.M. (1989) The *Drosophila* proteasome undergoes changes in its subunit pattern during development. *Exp. Cell Res.* **180**, 243–252.

Hadari, T., Warms, J.V., Rose, I.A. & Hershko, A. (1992) A ubiquitin C-terminal isopeptidase that acts on poly-ubiquitin chains. Role in protein degradation. *J. Biol. Chem.* **267**, 719–727.

Handley, P.M., Mueckler, M., Siegel, N.R., Ciechanover, A. & Schwartz, A.L. (1991) Molecular cloning, sequence, and tissue distribution of the human ubiquitin-activating enzyme E1. *Proc. Natl Acad. Sci. USA* **88**, 258–262.

Harding, M.M., Williams, D.H. & Woolfson, D.N. (1991) Characterization of a partially denatured state of a protein by two-dimensional NMR: Reduction of the hydrophobic interactions in ubiquitin. *Biochemistry* **30**, 3120–3128.

Hatfield, P.M. & Vierstra, R.D. (1992) Multiple forms of ubiquitin-activating enzyme E1 from wheat. Identification of an essential cysteine by *in vitro* mutagenesis. *J. Biol. Chem.* **267**, 14 799–14 803.

Hayashi, T., Takada, K. & Matsuda, M. (1991) Changes in ubiquitin and ubiquitin-protein conjugates in the CA1 neurons after transient sublethal ischaemia. *Mol. Chem. Neuropathol.* **15**, 75–82.

Heggie, P., Burdon, T., Lowe, J., Landon, M., Lennox, G., Jefferson, D. & Mayer, R.J. (1989) Ubiquitin gene expression in brain and spinal cord in motor neurone disease. *Neurosci. Lett.* **10**, 343–348.

Heinemeyer, W., Gruhler, A., Mohrle, V., Mahe, Y. & Wolf, D. H. (1993) PRE2, highly homologous to the human major histocompatibility complex-linked RING10 gene, codes for a yeast proteasome subunit necessary for chrymotryptic activity and degradation of ubiquitinated proteins. *J. Biol. Chem.* **268**, 5115–5120.

Heinemeyer, W., Kleinschmidt, J.A., Saidowsky, J., Escher, C. & Wolf, D.H. (1991) Proteinase yscE, the yeast proteasome/multicatalytic-multifunctional proteinase: Mutants unravel its function in stress-induced proteolysis and uncover its necessity for cell survival. *EMBO J.* **10**, 555–562.

Heller, H. & Hershko, A. (1990) A ubiquitin-protein ligase specific for type III protein substrates. *J. Biol. Chem.* **265**, 6532–6535.

Hershko, A. (1991) The ubiquitin pathway for protein degradation. *Trends Biochem. Sci.* **16**, 265–268.

Hershko, A. & Ciechanover, A. (1992) The ubiquitin system for protein degradation. *Ann. Rev. Biochem.* **61**, 761–807.

Hershko, A., Eytan, E., Ciechanover, A. & Haas, A.L. (1982) Immunochemical analysis of the turnover of ubiquitin-protein conjugates in intact cells. Relationship to the breakdown of abnormal proteins. *J. Biol. Chem.* **257**, 13 964–13 970.

Hershko, A., Ganoth, D., Pehrson, J., Palazzo, R.E. & Cohen, L.H. (1991) Methylated ubiquitin inhibits cyclin degradation in clam embryo extracts. *J. Biol. Chem.* **266**, 16 376–16 379.

Hershko, A., Heller, H., Elias, S. & Ciechanover, A. (1983) Components of ubiquitin-protein ligase system. Resolution, affinity purification, and role in protein breakdown. *J. Biol. Chem.* **258**, 8206–8214.

Hershko, A., Heller, H., Eytan, E. & Reiss, Y. (1986) The protein substrate binding site of the ubiquitin-protein ligase system. *J. Biol. Chem.* **261**, 11 992–11 999.

Hilt, W., Enenkel, C., Gruhler, A., Singer, T. & Wolf, D.H. (1993) The PRE4 gene codes for a subunit of the yeast proteasome necessary for peptidylglutamyl-peptide-hydrolyzing activity. Mutations link the proteasome to stress- and ubiquitin-dependent proteolysis. *J. Biol. Chem.* **268**, 3479–3486.

Hingamp, P.M., Arnold, J.E., Mayer, R.J. & Dixon, L.K. (1992) A ubiquitin conjugating enzyme encoded by African swine fever virus. *EMBO J.* **11**, 361–366.

Hochstrasser, M., Ellison, M.J., Chau, V. & Varshavsky, A. (1991) The short-lived MAT α-2 transcriptional regulator is ubiquitinated *in vivo*. *Proc. Natl Acad. Sci. USA* **88**, 4606–4610.

Hochstrasser, M. & Varshavsky, A. (1990) *In vivo* degradation of a transcriptional regulator: The yeast α-2 repressor. *Cell* **61**, 697–708.

Hoefer, M. & Cook, J.C. (1991) Purification and partial characterization of ubiquitin-activating enzyme from *Saccharomyces cerevisiae*. *FEBS Lett.* **289**, 54–58.

Honore, B., Rasmussen, H.H., Vandekerckhove, J. & Celis, J.E. (1991) Neuronal protein gene product 9.5 (IEF SSP 6104) is expressed in cultured human MRC-5 fibroblasts of normal origin and is strongly down-regulated in their SV40 transformed counterparts. *FEBS Lett.* **280**, 235–240.

Hough, R., Pratt, G. & Rechsteiner, M. (1987) Purification of two high molecular weight proteases from rabbit reticulocyte lysate. *J. Biol. Chem.* **262**, 8303–8313.

Huibregtse, J.M., Scheffner, M. & Howley, P.M. (1991) A cellular protein mediates association of p53 with the E6 onco-protein of human papillomavirus types 16 or 18. *EMBO J.* **10**, 4129–4135.

Huibregtse, J.M., Scheffner, M. & Howley, P.M. (1993). Cloning and expression of the cDNA for E6-AP, a protein that mediates the interaction of the human papillomavirus E6 oncoprotein with p53. *Mol. Cell Biol.* **13**, 775–784.

Hunt, L.T. & Dayhoff, M.O. (1977) Amino-terminal sequence identity of ubiquitin and the non-histone component of nuclear protein A24. *Biochem. Biophys. Res. Commun.* **74**, 650–655.

Ikai, A., Nishigai, M., Tanaka, K. & Ichihara, A. (1991) Electron microscopy of 26 S complex containing 20 S proteasome. *FEBS Lett.* **292**, 21–24.

Isenman, L.D. & Dice, J.F. (1989) Secretion of intact proteins and peptide fragments by lysosomal pathways of protein degradation. *J. Biol. Chem.* **264**, 21 591–21 596.

Ishida, N., Tanaka, K., Tamura, T., Nishizawa, M., Okazaki, K., Sagata, N. & Ichihara, A. (1993) Mos is degraded by the 26S proteasome in a ubiquitin-dependent fashion. *FEBS Letters* **324**, 345–348.

Ishiura, S., Nomura, Y., Tsukahara, T. & Sugita, H. (1989) Addition of ATP increases the apparent molecular mass of the multicatalytic proteinase, ingensin. *FEBS Lett.* **257**, 123–126.

Isoe, T., Naito, M., Shirai, A., Hirai, R. & Tsuruo, T. (1992) Inhibition of different steps of the ubiquitin system by cisplatin and aclarubicin. *Biochim. Biophys. Acta* **1117**, 131–135.

Ivy, G.O., Kitani, K. & Ihara. Y. (1989) Anomalous accumulation of tau and ubiquitin immunoreactivities in rat brain caused by protease inhibition and by normal aging: A clue to PHF pathogenesis. *Brain Res.* **498**, 360–365.

Jabben, M., Shanklin, J. & Vierstra, R.D. (1989) Ubiquitin-phytochrome conjugates. Pool dynamics during *in vivo* phytochrome degradation. *J. Biol. Chem.* **264**, 4998–5005.

Jentsch, S. (1992) Ubiquitin-dependent protein degradation: A cellular perspective. *Trends Cell. Biol.* **2**, 98–103.

Jentsch, S., McGrath, J.P. & Varshavsky, A. (1987) The yeast DNA repair gene RAD6 encodes a ubiquitin-conjugating enzyme. *Nature* **329**, 131–134.

Jentsch, S., Seufert, W. & Hauser, H.-P. (1991) Genetic analysis of the ubiquitin system. *Biochem. Biophys. Acta* **1089**, 127–139.

Kanayama, H., Tanaka, K., Aki, M., Kagawa, S., Miyaji, H., Satoh, M., Okada, F., Sato, S., Shimbara, N. & Ichihara, A. (1991) Changes in expressions of proteasome and ubiquitin genes in human renal cancer cells. *Cancer Res.* **51**, 6677–6685.

Kanayama, H.O., Tamura, T., Ugai, S., Kagawa, S., Tanahashi, N., Yoshimura, T., Tanaka, K. & Ichihara. A. (1992) Demonstration that a human 26S proteolytic complex consists of a proteasome and multiple associated protein components and hydrolyzes ATP and ubiquitin-ligated proteins by closely linked mechanisms. *Eur. J. Biochem.* **206**, 567–578.

Kay, G.F., Ashworth, A., Penny, G.D., Dunlop, M., Swift, S., Brockdorff, N. & Rastan, S. (1991) A candidate spermatogenesis gene on the mouse Y-chromosome is homologous to ubiquitin-activating enzyme. E1. *Nature* **354**, 486–489.

Klein, U., Gernold, M. & Kloetzel, P.M. (1990) Cell-specific accumulation of *Drosophila* proteasomes (MCP) during early development. *J. Cell Biol.* **111**, 2275–2282.

Kloetzel, P.M., Falkenburg, P. E., Hossl, P., and Glatzer, K. H. (1987) The 19S ring-type particles of *Drosophila*. Cytological and biochemical analysis of their intracellular association and distribution. *Exp. Cell Res.* **170**, 204–213.

Knight, E., Jr., Fahey, D., Cordova, B., Hillman, M., Kutney, R., Reich, N. & Blomstrom, D. (1988) A 15-kDa Interferon-induced protein is derived by COOH-terminal processing of a 17-kDa Precursor. *J. Biol. Chem.* **263**, 4520–4522.

Kolman, C.J., Toth, J. & Gonda, D.K. (1992) Identification of a portable determinant of cell cycle function within the carboxyl-terminal domain of the yeast CDC34 (UBC3) ubiquitin-conjugating (E2) enzyme. *EMBO J.* **11**, 3081–3090.

Kuehn, L., Dahlmann, B. & Kopp, F. (1990) Effect of the multicatalytic proteinase (prosome) on translational activity in rabbit reticulocyte lysates. *FEBS Lett.* **261**, 274–278.

Kulka, R.G., Raboy, B., Schuster, R., Parag, H.A., Diamond, G., Ciechanover, A. & Marcus, M. (1988) A Chinese hamster cell cycle mutant arrested at G2 phase has a temperature-sensitive ubiquitin-activating enzyme, E1. *J. Biol. Chem.* **263**, 15 726–15 731.

Kwak, S., Masaki, T., Ishiura, S. & Sugita, H. (1991) Multicatalytic proteinase is present in Lewy bodies and neurofibrillary tangles in diffuse Lewy body disease brains. *Neurosci. Lett.* **128**, 21–24.

Laszlo, L., Doherty, F.J., Osborn, N.U. & Mayer, R.J. (1990) Ubiquitinated protein conjugates are specifically enriched in the lysosomal system of fibroblasts. *FEBS*, **261**, 365–368.

Latchman, D.S., Estridge, J.K. & Kemp, L.M. (1987) Transcriptional induction of the ubiquitin gene during *Herpes simplex* virus infection is dependent upon the viral immediate-early protein ICP4. *Nucleic Acids Res.* **15**, 7283–7293.

Lenk, S.E., Dunn, W.A., Jr., Trausch, J.S., Ciechanover, A. & Schwartz, A.L. (1992) Ubiquitin-activating enzyme, E1, is associated with maturation of autophagic vacuoles. *J. Cell Biol.* **118**, 301–308.

Levinger, L. & Varshavsky, A. (1982) Selective arrangement of ubiquitinated and D1 protein-containing nucleosomes within the Drosophila genome. *Cell* **28**, 375–385.

Li, X.C., Gu, M.Z. & Etlinger, J.D. (1991) Isolation and characterization of a novel endogenous inhibitor of the proteasome. *Biochemistry* **30**, 9709–9715.

Li, X.S. & Etlinger, J.D. (1992) Ubiquitinated proteasome inhibitor is a component of the 26 S proteasome complex. *Biochemistry* **31**, 11 964–11 967.

Loeb, K.R. & Haas, A.L. (1992) The interferon-inducible 15-kDa ubiquitin homologue conjugates to intracellular proteins. *J. Biol. Chem.* **267**, 7806–7813.

Loetscher, H., Schlaeger, E.J., Lahm, H.W., Pan, Y.C., Lesslauer, W. & Brockhaus, M. (1990) Purification and partial amino acid sequence analysis of two distinct tumour necrosis factor receptors from HL60 cells. *J. Biol. Chem.* **265**, 20 131–20 138.

Lorca, T., Galas, S., Fesquet, D., Devault, A., Cavadore, J.C. & Doree, M. (1991) Degradation of the proto-oncogene product p39mos is not necessary for cyclin proteolysis and exit from meiotic metaphase: Requirement for a Ca^{2+}-calmodulin dependent event. *EMBO J.* **10**, 2087–2093.

Lowe, J., Fergusson, J., Kenward, N., Laszlo, L., Landon, M., Farquhar, C., Brown, J., Hope, J. & Mayer, R.J. (1992) Immunoreactivity to ubiquitin-protein conjugates is present early in the disease process in the brains of scrapie-infected mice. *J. Pathol.* **168**, 169–177.

Lowe, J., McDermott, H., Landon, M., Mayer, R.J. & Wilkinson, K.D. (1990) Ubiquitin carboxyl-terminal hydrolase (PGP 9.5) is selectively present in ubiquitinated inclusion bodies characteristic of human neurodegenerative diseases. *J. Pathol.* **161**, 153–160.

Matsui, S., Sandberg, A.A., Negoro, S., Seon, B.K. & Goldstein, G. (1982) Isopeptidase: A novel eukaryotic enzyme that cleaves isopeptide bonds. *Proc. Natl Acad. Sci. USA* **79**, 1535–1539.

Matsui, S.I., Seon, B.K. & Sandberg, A.A. (1979) Disappearance of a structural chromatin protein A24 in mitosis: Implications for molecular basis of chromatin condensation. *Proc. Natl Acad. Sci. USA* **76**, 6386–6390.

Matsumoto, Y., Yasuda, H., Marunouchi, T. & Yamada, M. (1983) Decrease in uH2A (protein A24) of a mouse temperature-sensitive mutant. *FEBS Lett.* **151**, 139–142.

Matthews, W., Driscoll, J., Tanaka, K., Ichihara, A. & Goldberg, A.L. (1989) Involvement of the proteasome in various degradative processes in mammalian cells. *Proc. Natl Acad. Sci. USA* **86**, 2597–2601.

Mayer, A.N. & Wilkinson, K.D. (1989) Detection, resolution, and nomenclature of multiple ubiquitin carboxyl-terminal esterases from bovine calf thymus. *Biochemistry* **28**, 166–172.

Mayer, R.J., Arnold, J., Laszlo, L., Landon, M. & Lowe, J. (1991) Ubiquitin in health and disease. *Biochim. Biophys. Acta* **1089**, 141–157.

Mayer, R.J., Laszlo, L., Landon, M., Hope, J. & Lowe, J. (1992) Ubiquitin, lysosomes, and neurodegenerative diseases. *Ann. NY Acad. Sci.* **674**, 149–160.

McGrath, J.P., Jentsch, S. & Varshavsky, A. (1991) UBA 1: An essential yeast gene encoding ubiquitin-activating enzyme. *EMBO J.* **10**, 227–236.

McGuire, M.J. & DeMartino, G.N. (1989) The latent form of macropain (high molecular weight multicatalytic protease) restores ATP-dependent proteolysis to soluble extracts of BHK fibroblasts pretreated with anti-macropain antibodies. *Biochem. Biophys. Res. Commun.* **160**, 911–916.

McGuire, M.J., Reckelhoff, J.F., Croall, D.E. & DeMartino, G.N. (1988) An enzyme related to the high molecular weight multicatalytic proteinase, macropain, participates in a ubiquitin-mediated, ATP-stimulated proteolytic pathway in soluble extracts of BHK 21/C13 fibroblasts. *Biochim. Biophys. Acta* **967**, 195–203.

Meyer, E.M., West, C.M., Stevens, B.R., Chau, V., Nguyen, M. T. & Judkins, J.H. (1987) Ubiquitin-directed antibodies inhibit neuronal transporters in rat brain synaptosomes. *J. Neurochem.* **49**, 1815–1819.

Meyers, G., Tautz, N., Dubovi, E.J. & Thiel, H.J. (1991) Viral cytopathogenicity correlated with integration of ubiquitin-coding sequences. *Virology* **180**, 602–616.

Miller, H.I., Henzel, W.J., Ridgeway, J.B., Kuang, W., Chisholm, V. & Liu, C. (1989) Cloning and expression of a yeast ubiquitin-protein cleaving activity in *Escherichia coli*. *Biotechnology* **7**, 698–704.

Mitchell, M.J., Woods, D.R., Tucker, P.K., Opp, J.S. & Bishop, C.E. (1991) Homology of a candidate spermatogenic gene from the mouse Y-chromosome to the ubiquitin-activating enzyme E1. *Nature* **354**, 483–486.

Mori, S., Heldin, C.H. & Claesson Welsh, L. (1992) Ligand-induced poly-ubiquitination of the platelet-derived growth factor beta-receptor. *J. Biol. Chem.* **267**, 6429–6434.

Mori, S., Heldin, C.H. & Claesson Welsh, L. (1993) Ligand-induced ubiquitination of the platelet-derived growth factor β-receptor plays a negative regulatory role in its mitogenic signaling. *J. Biol. Chem.* **268**, 577–583.

Nishitani, H., Goto, H., Kaneda, S., Yamao, F., Seno, T., Handley, P., Schwartz, A.L. & Nishimoto, T. (1992) tsBN75 and tsBN423, temperature-sensitive X-linked mutants of the BHK21 cell line, can be complemented by the ubiquitin-activating enzyme E1 cDNA. *Biochem. Biophys. Res. Commun.* **184**, 1015–1021.

Nishizawa, M., Okazaki, K., Furuno, N., Watanabe, N. & Sagata, N. (1992) The 'second-codon rule' and autophosphorylation govern the stability and activity of Mos during the meiotic cell cycle in *Xenopus* oocytes. *EMBO J.* **11**, 2433–2446.

Nowak, T.S., Jr., Bond, U. & Schlesinger, M.J. (1990) Heat shock RNA levels in brain and other tissues after hyperthermia and transient ischaemia. *J. Neurochem.* **54**, 451–458.

Ohtsubo, M. & Nishimoto, T. (1988) The gene coding a ubiquitin-activating enzyme may locate on X-chromosome. *Biochem. Biophys. Res. Commun.* **153**, 1173–1178.

Orino, E., Tanaka, K., Tamura, T., Sone, S., Ogura, T. & Ichihara, A. (1991) ATP-dependent reversible association of proteasomes with multiple protein components to form 26S complexes that degrade ubiquitinated proteins in human HL-60 cells. *FEBS Lett.* **284**, 206–210.

Ozkaynak, E., Finley, D., Solomon, M.J. & Varshavsky, A. (1987) The yeast ubiquitin genes: A family of natural gene fusions. *EMBO J.* **6**, 1429–1439.

Pan, Y. & Briggs, M.S. (1992) Hydrogen exchange in native and alcohol forms of ubiquitin. *Biochemistry* **31**, 11 405–11 412.

Paolini, R. & Kinet, J.P. (1993) Cell surface control of the multi-ubiquitination and deubiquitination of high-affinity immunoglobulin E receptors. *EMBO J.* **12**, 779–786.

Perry, G., Friedman, R., Shaw, G. & Chau, V. (1987) Ubiquitin is detected in neurofibrillary tangles and senile plaque neurites of Alzheimer disease brains. *Proc. Natl Acad. Sci. USA* **84**, 3033–3036.

Pickart, C.M. & Rose, I.A. (1985) Functional heterogeneity of ubiquitin carrier proteins. *J. Biol. Chem.* **260**, 1573–1581.

Rechsteiner, M. (1991) Natural substrates of the ubiquitin proteolytic pathway. *Cell* **66**, 615–618.

Richter Ruoff, B., Heinemeyer, W. & Wolf, D.H. (1992) The proteasome/multicatalytic-multi-functional proteinase. *In vivo* function in the ubiquitin-dependent N-end rule pathway of protein degradation in eukaryotes. *FEBS Lett.* **302**, 192–196.

Rivett, A.J. & Sweeney, S.T. (1991) Properties of subunits of the multicatalytic proteinase complex revealed by the use of subunit-specific antibodies. *Biochem. J.* **278**, 171–177.

Scheffner, M., Munger, K., Huibregtse, J.M. & Howley, P.M. (1992) Targeted degradation of the retinoblastoma protein by human papillomavirus E7-E6 fusion proteins. *EMBO J.* **11**, 2425–2431.

Scheffner, M., Werness, B.A., Huibregtse, J.M. & Levine, A. J. (1990) The E6 onco-protein encoded by human papillomavirus types 16 and 18 promotes the degradation of p53. *Cell* **63**, 1129–1136.

Scherrer, H., Seidah, N.G., Renjannet, S., Crine, P., Lis, M. & Chretien, M. (1978) Biosynthesis of a ubiquitin-related peptide in rat brain and in human and mouse pituitary tumours. *Biochem. Biophys. Res. Commun.* **84**, 874–885.

Schueler, P.A. & Elde, R.P. (1988) Differential expression of ubiquitin within the rat brain: Discretely localized increases following salt-loading. *J. Chem. Neuroanat.* **1**, 165–175.

Schwartz, A.L., Ciechanover, A., Brandt, R.A. & Geuze, H.J. (1988) Immunoelectron microscopic localization of ubiquitin in hepatoma cells. *EMBO J.* **7**, 2961–2966.

Schwartz, A.L., Trausch, J.S., Ciechanover, A., Slot, J.W. & Geuze, H. (1992) Immunoelectron microscopic localization of the ubiquitin-activating enzyme E1 in HepG2 cells. *Proc. Acad Sci USA* **89**, 5542–5546.

Seelig, A., Kloetzel, P.M., Kuehn, L. & Dahlmann, B. (1991) Molecular interaction of the proteasome (multicatalytic proteinase). Evidence that the proteasome is not a constituent of the '26S' multienzyme complex. *Biochem. J.* **280**, 225–232.

Serdaroglu, P., Askanas, V. & Engel, W.K. (1992) Immunocytochemical localization of ubiquitin at human neuromuscular junctions. *Neuropathol. Appl. Neurobiol.* **18**, 232–236.

Seufert, W. & Jentsch, S. (1990) Ubiquitin-conjugating enzymes UBC4 and UBC5 mediate selective degradation of short-lived and abnormal proteins. *EMBO J.* **9**, 543–550.

Seufert, W. & Jentsch, S. (1992) *In vivo* function of the proteasome in the ubiquitin pathway. *EMBO J.* **11**, 3077–3080.

Shanklin, J., Jabben, M. & Vierstra, R.D. (1987) Red light-induced formation of ubiquitin-phytochrome conjugates: Identification of possible intermediates of phytochrome degradation. *Proc. Natl Acad. Sci. USA* **84**, 359–363.

Shanklin, J., Jabben, M. & Vierstra, R.D. (1989) Partial purification and peptide mapping of ubiquitin-phytochrome conjugates from oat. *Biochemistry* **28**, 6028–6034.

Sharon, G., Raboy, B., Parag, H.A., Dimitrovsky, D. & Kulka, R.G. (1991) RAD6 gene product of *Saccharomyces cerevisiae* requires a putative ubiquitin-protein ligase (E3) for the ubiquitination of certain proteins. *J. Biol. Chem.* **266**, 15 890–15 894.

Siegelman, M., Bond, M.W., Gallatin, W.M., St John, T., Smith, H.T., Fried, V.A. & Weissman, I.L. (1986) Cell-surface molecule associated with lymphocyte homing is a ubiquitinated branched-chain glycoprotein. *Science* **231**, 823–829.

Silva-Pereira, I., Bey, F., Coux, O. & Scherrer, K. (1992) Two mRNAs exist for the Hs PROS-30 gene encoding a component of human prosomes. *Gene* **120**, 235–242.

Silver, E.T., Gwozd, T.J., Ptak, C., Goebl, M. & Ellison, M. J. (1992) A chimeric ubiquitin-conjugating enzyme that combines the cell cycle properties of CDC34 (UBC3) and the DNA repair properties of RAD6 (UBC2): Implications for the structure, function and evolution of the E2s. *EMBO J.* **11**, 3091–3098.

Skilton, H.E., Eperon, I.C. & Rivett, A.J. (1991) Co-purification of a small RNA species with multicatalytic proteinase (proteasome) from rat liver. *FEBS Lett.* **279**, 351–355.

Spencer, S.A., Hammonds, R.G., Henzel, W.J., Rodriguez, H., Waters, M.J. & Wood, W.I. (1988) Rabbit liver growth hormone receptor and serum binding protein. Purification, characterization, and sequence. *J. Biol. Chem.* **263**, 7862–7867.

Sullivan, M.L. & Vierstra, R.D. (1991) Cloning of a 16-kDa ubiquitin carrier protein from wheat and *Arabidopsis thaliana*. Identification of functional domains by *in vitro* mutagenesis. *J. Biol. Chem.* **266**, 23 878–23 885.

Swerdlow, P.S., Finley, D. & Varshavsky, A. (1986) Enchancement of immunoblot sensitivity by heating of hydrated filters. *Anal. Biochem.* **156**, 147–153.

Taccioli, G.E., Grotewold, E., Aisemberg, G.O. & Judewicz, N.D. (1991) The cDNA sequence and expression of an ubiquitin-tail gene fusion in *Neurospora crassa*. *Gene* **102**, 133–137.

Takeda, M., Tatebayashi, Y., Tanimukai, S., Nakamura, Y., Tanaka, T. & Nishimura, T. (1991) Immunohistochemical study of microtubule-associated protein 2 and ubiquitin in chronically aluminum-intoxicated rabbit brain. *Acta Neuropathol. (Berlin)* **82**, 346–352.

Tanaka, K. & Ichihara, A. (1988) Involvement of proteasomes (multicatalytic proteinase) in ATP-dependent proteolysis in rat reticulocyte extracts. *FEBS Lett.* **236**, 159–162.

Tanaka, K., Tamura, T., Yoshimura, T. & Ichihara, A. (1992) Proteasomes: Protein and gene structures. *New. Biol.* **4**, 173–187.

Terlecky, S.R., Olson, T.S. & Dice, J.F. (1991) A pathway of lysosomal proteolysis mediated by the 73-kilodalton heat shock cognate protein. *Acta Biol. Hung.* **42**, 39–47.

Thorburn, D.R. & Beutler, E. (1989) Decay of hexokinase during reticulocyte maturation: Is oxidative damage a signal for destruction. *Biochem. Biophys. Res. Commun.* **162**, 612–618.

Tierney, D.J., Haas, A.L. & Koop, D.R. (1992) Degradation of cytochrome P450 2E1: Selective loss after labilization of the enzyme. *Arch. Biochem. Biophys.* **293**, 9–16.

Tobias, J.W. & Varshavsky, A. (1991) Cloning and functional analysis of the ubiquitin-specific protease gene UBP1 of *Saccharomyces cerevisiae*. *J. Biol. Chem.* **266**, 12 021–12 028.

Toniolo, D., Persico, M. & Alcalay, M. (1988) A 'housekeeping' gene on the X-chromosome encodes a protein similar to ubiquitin. *Proc. Natl Acad. Sci. USA* **85**, 851–855.

Townsend, A., Bastin, J., Gould K., Brownlee, G., Andrew, M., Coupar, B., Boyle, D., Chan, S. & Smith, G. (1988) Defective presentation to class I-restricted cytotoxic T-lymphocytes in vaccinia-infected cells is overcome by enhanced degradation of antigen. *J. Exp. Med.* **168**, 1211–1224.

Trausch, J.S., Grenfell, S.J., Handley Gearhart, P.M., Ciechanover, A. & Schwartz, A.L. (1993) Immunofluorescent localization of the ubiquitin-activating enzyme, E1, to the nucleus and cytoskeleton. *Am. J. Physiol.* **264**, C93–102.

Ueno, T. & Kominami, E. (1991) Mechanism and regulation of lysosomal sequestration and proteolysis. *Biomed. Biochim. Acta* **50**, 365–371.

Uney, J.B., Anderton, B.H. & Thomas, S.M. (1993) Changes in heat shock protein 70 and ubiquitin mRNA levels in C1300 N2A mouse neuroblastoma cells following treatment with iron. *J. Neurochem.* **60**, 659–665.

van de Rijn, M., Weissman, I.L. & Siegelman, M. (1990) Biosynthesis pathway of gp90MEL-14, the mouse lymph node-specific homing receptor. *J. Immunol.* **145**, 1477–1482.

Van Nocker, S. & Vierstra, R.D. (1991) Cloning and characterization of a 20-kDa ubiquitin-carrier protein from wheat that catalyses multiubiquitin chain formation *in vitro*. *Proc. Natl Acad. Sci. USA* **88**, 10 297–10 301.

Varshavsky, A. (1992) The N-end rule. *Cell* **69**, 725–735.

Vierstra, R.D., Langan, S.M. & Schaller, G.E. (1986) Complete amino acid sequence of ubiquitin from the higher plant *Avena sativa*. *Biochemistry* **25**, 3105–3108.

Vijay-Kumar, S., Bugg, C.E., Wilkinson, K.D. & Cook, W.J. (1985) Three-dimensional structure of ubiquitin at 2.8 Å resolution. *Proc. Natl Acad. Sci. USA* **82**, 3582–3585.

Vijay-Kumar, S., Bugg, C.E., Wilkinson, K.D., Vierstra, R.D., Hatfield, P.M. & Cook, W.J. (1987) Comparison of the three-dimensional structures of human, yeast, and oat ubiquitin. *J. Biol. Chem.* **262**, 6396–6399.

Wang, G.P., Khatoon, S., Iqbal, K. & Grundke Iqbal, I. (1991) Brain ubiquitin is markedly elevated in Alzheimer's disease. *Brain Res.* **566**, 146–151.

Wiebel, F.F. & Kunau, W.H. (1992) The Pas2 protein essential for peroxisome biogenesis is related to ubiquitin-conjugating enzymes. *Nature* **359**, 73–76.

Wilk, S. and Orlowski, M. (1983) Evidence that pituitary cation-sensitive neutral endopeptidase is a multicatalytic protease complex. *J. Neurochem.* **40**, 842–849.

Wilkinson, K.D. (1988) Purification and Structural Properties of Ubiquitin. In 'Ubiquitin' (ed. M. Rechsteiner), pp5–38. New York, Plenum Press.

Wilkinson, K.D., Cox, M.J., Mayer, A.N. & Frey, T. (1986a) Synthesis and characterization of ubiquitin-ethyl ester, a new substrate for ubiquitin carboxyl-terminal hydrolase. *Biochemistry* **25**, 6644–6649.

Wilkinson, K.D., Cox, M.J., O'Connor, L.B. & Shapira, R. (1986b). Structure and activities of a variant ubiquitin sequence from bakers' yeast. *Biochemistry* **25**, 4999–5004.

Wilkinson, K.D., Deshpande, S. & Larsen, C.N. (1992) Comparisons of neuronal (PGP 9.5) and non-neuronal ubiquitin C-terminal hydrolases. *Biochem. Soc. Trans.* **20**, 631–637.

Wilkinson, K.D., Lee, K.M., Deshpande, S., Duerksen Hughes, P., Boss, J.M. & Pohl, J. (1989) The neuron-specific protein PGP 9.5 is a ubiquitin carboxyl-terminal hydrolase. *Science* **246**, 670–673.

Wilkinson, K.D. & Mayer, A.N. (1986) Alcohol-induced conformational changes of ubiquitin. *Arch. Biochem. Biophys.* **250**, 390–399.

Wilkinson, K.D., Smith, S.E., O'Connor, L., Sternberg, E., Taggart, J.J., Berges, D.A. & Butt, T. (1990) A specific inhibitor of the ubiquitin activating enzyme: Synthesis and characterization of adenosyl-phospho-ubiquitinol, a nonhydrolyzable ubiquitin adenylate analogue. *Biochemistry* **29**, 7373–7380.

Wilkinson, K.D., Urban, M.K. & Haas, A.L. (1980) Ubiquitin is the ATP-dependent proteolysis factor I of rabbit reticulocytes. *J. Biol. Chem.* **255**, 7529–7532.

Wilson, P.O.G., Barber, P.C., Hamid, Q.A., Power, B.F., Dhillon, A.P., Rode, J., Day, I.N.M., Thompson, R.J. & Polak, J.M. (1988) The immunolocalization of protein gene product 9.5 using rabbit polyclonal and mouse monoclonal antibodies. *Br. J. Exp. Path* **69**, 91–104.

Wostmann, C., Tannich, E. & Bakker-Grunwald, T. (1992) Ubiquitin of *Entamoeba histolytica* deviates in six amino acid residues from the consensus of all other known ubiquitins. *FEBS Lett.* **308**, 54–58.

Yarden, Y., Escobedo, J.A., Kuang, W.J., Yang-Feng, T.L., Daniel, T.O., Tremble, P.M., Chen, E.Y., Ando, M.E., Harkins, R.N., Francke, U., Fried, V.A., Ullrich, A. & Williams, L.T. (1986) Structure of the receptor for platelet-derived growth factor helps define a family of closely related growth factor receptors. *Nature* **323**, 226–232.

Zacksenhaus, E. & Sheinin, R. (1990) Molecular cloning, primary structure and expression of the human X-linked A1S9 gene cDNA which complements the ts A1S9 mouse L cell-defect in DNA replication. *EMBO J.* **9**, 2923–2929.

Zdebska, E., Antoniewicz, J., Nilsson, B., Sandhoff, K., Furst, W., Janik, P. & Koscielak, J. (1992) Ganglioside binding proteins of calf brain with ubiquitin-like N-terminals. *Eur. J. Biochem.* **210**, 483–489.

Zhan, S.S., Beyreuther, K. & Schmitt, H.P. (1992) Neuronal ubiquitin and neurofilament expression in different lysosomal storage disorders. *Clin. Neuropathol.* **11**, 251–255.

Zhang, N., Wilkinson, K. & Bownes, M. (1993) Cloning and analysis of expression of a ubiquitin carboxyl-terminal hydrolase expressed during oogenesis in *Drosophila melanogaster. Developmental Biology* **157**, 214–223.

Zhaung, Z.P. & McCauley, R. (1989) Ubiquitin is involved in the *in vitro* insertion of monoamine oxidase B into mitochondrial outer membranes. *J. Biol. Chem.* **264**, 14 594–14 596.

Zhuang, Z.P., Marks, B. & McCauley, R.B. (1992) The insertion of monoamine oxidase A into the outer membrane of rat liver mitochondria. *J. Biol. Chem.* **267**, 591–596.

CHAPTER 10

UBIQUITIN, THE CYTOSKELETON AND NEURODEGENERATIVE DISEASES

Dennis W. Dickson and Shu-Hui C. Yen

Department of Pathology, Albert Einstein College of Medicine, Bronx, New York 10461, USA

Table of Contents

Summary

Heat- or stress-related proteins are a common constituent of a number of structural alterations in neurons and glia in ageing and degenerative disorders of the central nervous system (CNS). Based upon immunocytochemical, and less often, biochemical studies, the most common stress protein associated with these cellular

Heat Shock Proteins in the Nervous System
ISBN 0-12-480960-X

alterations is ubiquitin. Routine application of ubiquitin immunocytochemistry to CNS tissue has increased our understanding of ageing and degenerative diseases. It has revealed a wealth of changes that were not visible with standard histological techniques. Among these ubiquitin-immunoreactive structures are lesions that are accompanied by cytoskeletal changes and those that are not. Structures that are characteristic of neurodegenerative diseases are often associated with cytoskeletal alterations. On the other hand, ubiquitin-immunoreactive structures that are a manifestation of ageing are more often not associated with cytoskeletal proteins. Among the former lesions are inclusions containing intermediate filaments and inclusions containing microtubule-associated protein tau. Neuronal intermediate filament (neurofilament) inclusions are Lewy bodies, Lewy body-like inclusions of amyotrophic lateral sclerosis and ballooned neurons. Glial intermediate filament (GFAP) inclusions are Rosenthal fibres (RF) and gliofibrillary inclusions of infancy. Microtubule-associated protein inclusions are neurofibrillary tangles (NFT) and neuropil threads, granulovacuolar bodies, Pick bodies, neuropil grains, glial tangles and glial cytoplasmic inclusions of multiple system atrophy. The most common age-related alterations revealed with ubiquitin immunocytochemistry are granular degeneration of myelin and dystrophic axons. The results suggest that stress proteins are involved in ageing and degenerative disease processes. Of these inclusions only NFT and RF have been subjected to biochemical analysis. Work on the pathogenesis of the other cellular changes is based upon descriptive methods. Given the small number of lesions, biochemical studies may be impractical or impossible. Animal or tissue culture experimental models will be needed to determine the mechanism of formation of many of the stress-related inclusions.

10.1 Introduction

10.1.1 The cytoskeleton

The cytoskeleton, composed of three major classes of structural proteins and their associated proteins, form microfilaments (5 nm in diameter), intermediate filaments (10 nm in diameter) and microtubules (22 nm in diameter). The major protein constituents are actin for microfilaments, the family of intermediate filament proteins for intermediate filaments and tubulin for microtubules. Microfilaments are instrumental for cell attachment and motility, intermediate filaments provide rigid structural support, and microtubules have both structural and dynamic properties. Comprehensive reviews should be consulted for additional information (Wang *et al.*, 1985; Soifer, 1986). Of the cytoskeletal components, intermediate filaments and microtubules are most important in neuropathology.

Intermediate filaments are developmentally-regulated and relatively cell-type-specific proteins. Despite intense research over the last several decades, the function

Table 1 Intermediate filament proteins

Class	Names	Distribution
I	Acidic keratins	Epithelia
II	Basic keratins	Epithelia
III	Desmin	Muscle
	Vimentin	Mesenchyme
	Glial fibrillary acidic protein (GFAP)	Glia
	Peripherin and NF66	Neurons
IV	Neurofilaments	Neurons
V	Lamins	Nuclear envelope
VI	Nestin	Neuroepithelial stem cells

of intermediate filaments remains unknown. The ability of intermediate filaments to bind to nuclear and plasma membranes suggests that intermediate filaments may integrate cytoplasmic space and possibly transmit signals from the extracellular space to the internal milieu (Goldman *et al.*, 1985). There are six classes of intermediate filaments (Table 1). All intermediate filaments share a basic structure characterized by an α-helical, coiled coil, rod domain and non-helical amino- and carboxy-terminal domains (Steiner and Roop, 1988; Stewart, 1990; Liem, 1990). An increasingly recognized subset of proteins associated with intermediate filaments is instrumental in mediating their actions. Included among these are several proteases and kinases.

10.1.1.1 Glial fibrillary acidic protein (GFAP)

GFAP is the major intermediate filament protein of astrocytes (Eng, 1985). In developing or cultured astrocytes vimentin is co-expressed with GFAP. GFAP expression is increased in response to many stimuli, including cellular stress. It is a phosphoprotein, and several different kinases have been shown to phosphorylate GFAP, most notably protein kinase C, c-AMP-dependent protein kinase and intermediate filament-associated kinase. Recent studies have also shown that proline-directed MAP kinases, including cdc2, phosphorylate GFAP during specific stages of the cell cycle (Skalli *et al.*, 1992). Phosphorylation decreases the ability of GFAP to assemble into filaments.

10.1.1.2 Neurofilament proteins (NF)

Neurofilaments are composed of at least three major proteins: NF-L, NF-M and NF-H (Liem, 1990). The proteins co-assemble to form intermediate filaments. The neurofilament proteins differ from each other in the extent of post-translational modification, molecular weight and the size of carboxy-terminal domains. In NF-M and NF-H these are large domains with multiple phosphorylation (lysine-serine-proline – KSP) repeats. In the NF-H there are up to 50 KSP repeats. The heat-stable microtubule-associated proteins also contain this structural motif and cross-reactivity between some antibodies to NF and MAP is due to antibody recognition of these epitopes (Ksiezak-Reding *et al.*, 1987). The functional properties of neurofilaments are modulated by their state of phosphorylation, and specific kinases are closely associated with neurofilaments. Neurofilaments have been demonstrated in vitro to bind both microfilament proteins and tubulin. The cross bridges between NF and microtubules that have been recognized at the electron microscopic level for decades are the likely correlates of this interaction.

10.1.1.3 Microtubules and microtubule associated proteins (MAPs)

Microtubules are dynamically instable structures composed of polymers α and β tubulin. Microtubules are directional polymers with plus and minus ends. In most dividing cells, one of the most important roles of microtubules is formation of the mitotic spindle. In neurons, which are post-mitotic, microtubules are key elements in the system of fast axoplasmic transport, where they form the backbone of vesicular transport (Howard *et al.*, 1989). In contrast, tubulin and other cytoskeletal proteins are transported by slow axoplasmic transport (Hollenbeck, 1990).

The functional properties of microtubules appear to be determined in part by microtubule associated proteins (MAPs). MAPs are a highly diverse group of proteins that co-purify with microtubules (Matus, 1988; Goedert *et al.*, 1991). Both high molecular weight and mid-molecular weight MAPs have been described. The major mid-molecular weight MAPs are heat-stable proteins called tau proteins. Tau is a protein with 5–6 isoforms due to both alternative gene splicing and post-translational modification (Himmler, 1989; Kosik, 1990). The major form of post-translational modification of tau is phosphorylation.

Like intermediate filament proteins, tau phosphorylation leads to decreased ability of tau to promote microtubule assembly and favours disassembly of microtubules. Several kinases have been shown to phosphorylate tau, including calcium-calmodulin kinase and protein kinase C. Particular interest in recent years has focused on F_A kinase and proline directed kinases, including cdc2 and MAP kinases as mediators of tau phosphorylation (Drewes *et al.*, 1992).

Recent evidence suggests that ubiquitin may be associated with the normal microtubular network (Murti *et al.*, 1988). Studies of cell division have also suggested that ubiquitin mediated protein degradation may be involved in mitosis (Glotzer *et al.*, 1991).

10.1.2 Cytoskeleton in stress

In stressed cells the intermediate filament network collapses into a peri- or juxtanuclear bundle, while microtubules are relatively preserved (Welch *et al.*, 1985). The redistribution of intermediate filaments is not merely due to generalized decrease in protein synthesis in the stressed cells. Possible explanations for these changes may be increased phosphorylation or proteolysis of intermediate filaments coincident with increased intracellular calcium in the stressed cell (Bagi *et al.*, 1990; Duncan and Hershey, 1989). Increased phospho-inositol in stressed cells may also contribute to protein kinase C activation and subsequent intermediate filament phosphorylation. Studies have also suggested that heat shock proteins may bind to and stabilize tubulin, which may partially explain the relative resistance of microtubules to acute cell stress (Lee *et al.*, 1992).

Biochemical studies of heat-stressed cells demonstrate altered phosphorylation of cytoskeletal proteins. In neuroblastoma cell lines tau protein is dephosphorylated by heat stress (Chiang *et al.*, 1993). These clues from basic cellular biology may have relevance to the pathological alterations that characterize neurodegeneration.

10.1.2.1 Protein degradation in stress

It is not the purview of this chapter to review the systems of protein degradation active in stressed cells. In broad outline they include lysosomal and non-lysosomal systems (Dice, 1987). One of the major non-lysosomal systems is mediated by ubiquitin (Parag *et al.*, 1987); Hershko and Ciechanover, 1992), which is discussed more fully in Chapter 11 of this book. Ubiquitin is one of a family of low molecular weight proteins whose expression increases in response to diverse cellular stresses. As such, ubiquitin is considered to be a heat shock protein. Since stress is often associated with protein damage, it is reasonable to expect that synthesis of a protein such as ubiquitin, which is involved in energy-dependent degradation of abnormal or denatured proteins, would increase coincidentally. The proteins to be degraded by the ubiquitin-dependent system are covalently modified by ubiquitin. Thus, ubiquitin conjugation to proteins in certain specific states represents a cellular adaptive response to stress.

Of particular interest to neuropathologists is the fact that ubiquitin is associated with a number of cytological alterations in the ageing brain and in neurodegenerative diseases (Lowe *et al.*, 1988; Mayer *et al.*, 1991). In most cases it is unknown whether or not the association is due to covalent attachment of ubiquitin to the proteins in the cytological alterations of the stressed cells. Until further biochemical analysis is performed, much of our knowledge stems from indirect approaches, including immunocytochemistry and immunochemistry. The focus of this chapter will be to describe many of these alterations.

10.2 Neurodegenerative diseases with stress protein inclusions

One of the hallmarks of cellular pathology associated with stress is the generation of inclusion bodies. This topic has been reviewed previously in detail (Yen *et al.*, 1986). The focus of this chapter will be on those aspects of cellular inclusions related to the stress response. Inclusion bodies are composed of abnormal proteins that accumulate in the cell because of perturbations in their degradation, production or other structural abnormalities. In most cases the nature of the inciting stress that leads to inclusion body formation is unknown. Immunochemical and immunocytochemical methods have identified several classes of such inclusions, many of which are associated with stress proteins, the most common of which is ubiquitin. Each of the major components of the cytoskeleton are associated with inclusions.

10.2.1 Microfilament inclusions

Hirano bodies are eosinophilic rod bodies that accumulate in dendrites and perikarya of neurons of the hippocampus. They are composed of actin and actin binding proteins, such as tropomyosin (Goldman, 1983). Unlike almost every other cytoskeletal inclusion in neurodegenerative diseases, ubiquitin and other stress proteins have not been associated with Hirano bodies. In this respect it is unclear how the pathogenesis of Hirano bodies differs from other inclusion bodies.

10.2.2 Intermediate filament inclusions

10.2.2.1 Neurofilament inclusions

Lewy Bodies (LBs): LBs are spherical, eosinophilic neuronal inclusions in monoaminergic and cholinergic neurons of the brainstem, diencephalon, basal forebrain and cerebral cortex (review: Pollanen *et al.*, 1993). Neurodegenerative diseases associated with LBs can be considered to form a disease spectrum based upon the number and distribution of LBs. At one end of the spectrum are cases with few LBs confined to the substantia nigra and at the other end are cases with many and widespread LBs, including LBs in the cortex, a disorder referred to as diffuse Lewy body disease (DLBD) (Dickson *et al.*, 1989). LBs in the brainstem usually have a dense, hyaline, eosinophilic core and a clear halo. Ultrastructurally, brainstem-type LBs have a dense, osmiophilic core composed of granular and vesicular material and a concentric rim of radiating to haphazardly arranged 8–10 nm fibrils. Cortical LBs are smaller, eosinophilic cytoplasmic inclusions with loosely arranged fibrils and granular material, but no distinct core. Cortical LBs are most often observed in lower cortical layers in small to medium-sized nonpyramidal neurons. Cortical LBs are most numerous in the cingulate gyrus, insular cortex, amygdala and frontotemporal neocortex.

Neurofilament protein epitopes are present in both brainstem and cortical LBs. Using a panel of monoclonal antibodies to NF-L, NF-M, and NF-H, it has been demonstrated that epitopes from the full-length of all three NF subunits are present in both cortical and subcortical LBs (Hill *et al.*, 1991). An antigenic determinant thought to reside near the junction of the NF-M rod and tail domains (recognized by monoclonal antibody RM032) may detect an epitope relatively specific to LBs (Schmidt *et al.*, 1991). Stress proteins that have been immunocytochemically associated with LBs include ubiquitin (Kuzuhara *et al.*, 1988), ubiquitin hydrolase

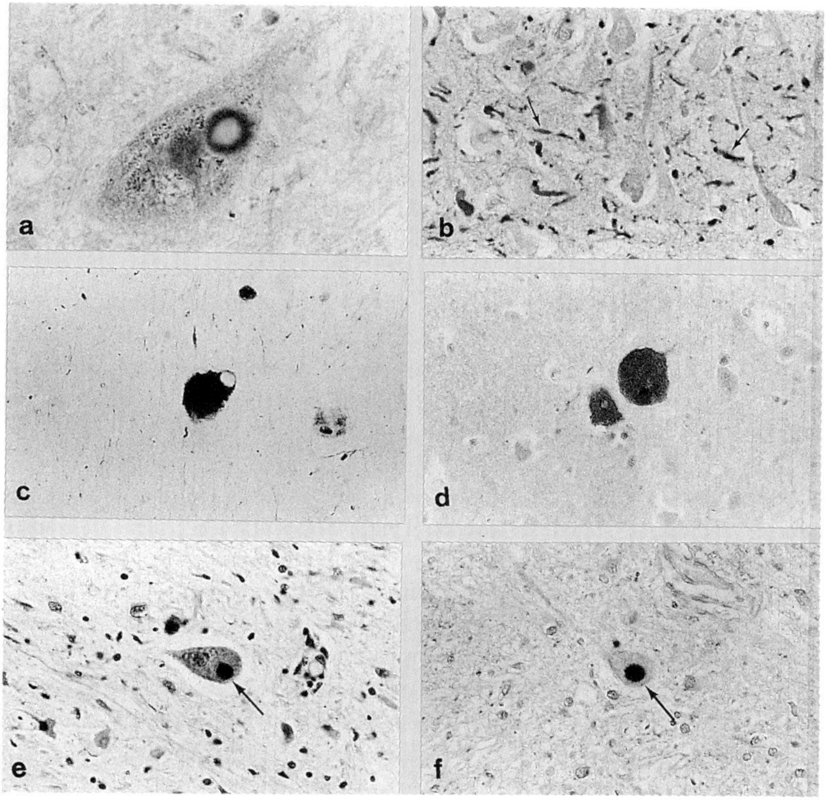

Figure 1 Neurofilament inclusions. A classic Lewy body in a neuron of the substantia nigra in Parkinson's disease immunostained for ubiquitin shows a ring-like pattern of immunoreactivity (a). Ubiquitin-staining of the CA2/3 region of the hippocampus in DLBD shows numerous neurites (arrows) (b). Ballooned neurons in corticobasal ganglionic degeneration (CBD) are immunoreactive with a phosphate-dependent anti-NF (c). Ballooned neurons from the same case are also immunoreactive for ubiquitin (d). Lewy body-like inclusions in the anterior horn cell of ALS are stained with anti-NF (e) and anti-ubiquitin (f). (e is counterstained with haematoxylin)

(Lowe *et al.*, 1992) and αB-crystallin (Lowe *et al.*, 1992), a protein that is enriched in lens tissue and that shares homology with low molecular weight heat shock protein 27 (hsp27). Several studies have now demonstrated that ubiquitin immunocytochemistry is the most sensitive method for detecting LBs (Lennox *et al.*, 1989; Figure 1a). The presence of ubiquitin in LBs has been interpreted as a reflection of either abnormalities of the ubiquitin-dependent degradation or conjugation to an abnormal substrate protein. Ubiquitinated proteins are degraded by a multimeric proteinase, 26S proteasome (Goldberg and Rock, 1992). Antibodies to multicatalytic protease have also stained LBs (Kwak *et al.*, 1991). This suggests that ubiquitination is tied to proteolysis in LBs. Ultrastructural studies have not, however, identified proteasomes associated with LBs. Ubiquitin is also removed from conjugated proteins by a C-terminal hydrolase. Antibodies to this molecule, also known as PGP9.5, also label LBs (Lowe *et al.*, 1990). These studies of fixed tissue suggest that ubiquitin immunoreactivity in LBs reflects a dynamic process of conjugation, hydrolysis and protein degradation. The role of αB-crystallin in LBs is poorly understood (Lowe *et al.*, 1992). Although αB-crystallin is a major lens protein, it has also been shown to be a small heat shock protein (Klemenz *et al.*, 1991). In normal cells it is most often detected in glia (see below) rather than neurons (Iwaki *et al.*, 1990). Given the presence of both cytoskeletal elements and stress proteins, it is tempting to speculate that LBs represent collapsed cytoskeleton in response to cellular stress. Studies of Parkinson's disease have implicated reactive oxygen species, generated in response to environmental toxins, as potential sources of cellular stress (Jenner and Marsden, 1993).

Hippocampal dystrophic neurites: The hippocampus is vulnerable to a number of stresses, including ischemia (Nowak, 1990). Several studies have demonstrated that heat shock proteins are induced in the hippocampus in response to a variety of such insults (Abe *et al.*, 1991). It is with some interest that neurons of the hippocampus are also vulnerable to degenerative diseases. In Alzheimer's disease and Pick's disease, neurons of the hippocampus are particularly vulnerable to neurofibrillary degeneration and Pick-body formation (see below). In DLBD neuronal processes rather than cell bodies appear to be most vulnerable. One particular region of the hippocampus (CA2/3 region) regularly develops neuritic degeneration in DLBD and less often in Parkinson's disease (Dickson *et al.*, 1991). The hippocampal neurites in DLBD are long, tortuous and sometimes segmented (Figure 1b). They are stained with antibodies to ubiquitin and inconsistently with neurofilament monoclonal antibodies. At the ultrastructural level they contain intermediate filaments. They have an immunoreactivity profile similar, but not identical to LBs. For example, they are inconsistently or unreactive with antibodies to ubiquitin C-terminal hydrolase (PGP9.5) or αB-crystallin (unpublished observations). Similar neuritic degeneration is also found in other regions of the brain in DLBD, including the basal forebrain, substantia nigra, pedunculopontine nuclei, raphe nuclei and dorsal motor nuclei of the vagus. The extent of ubiquitin-immunoreactive neuritic degeneration in the hippocampus correlates with the density of cortical LBs. It remains a mystery why these particular subsets of neuronal processes are vulnerable to degeneration in DLBD.

Ballooned neurons (BN): BN, also referred to as Pick cells, are one of the hallmarks of Pick's disease, but they are also detected in a number of neurodegenerative disorders, including corticobasal ganglionic degeneration (CBD) and even Alzheimer's disease. Pick's disease, like Alzheimer's disease, is a degenerative disease that produces dementia and very often severe deterioration of the personality. Pick's disease is associated with profound focal atrophy of the frontal and temporal lobes. In Pick's disease, BN are found in atrophic cortices and usually in the lower cortical layers; in other disorders BN are most often detected in the amygdala, claustrum, insular cortex and cingulate gyrus. The distribution of neurons vulnerable to BN is similar to those that are most vulnerable to LBs. CBD is a disorder that is associated with focal cortical degeneration, but also marked involvement of deep grey matter, brainstem and cerebellum. It may be associated with dementia, focal cortical deficits and Parkinsonism. BN in CBD are numerous and found in widespread areas of the neocortex, especially the parietal lobe (Gibb *et al.*, 1989).

BN are weakly argyrophilic and intensely reactive with neurofilament antibodies (Dickson *et al.*, 1986; Figure 1c). Ultrastructural studies show them to be composed of granular material and disordered intermediate filaments. Phosphorylated NF epitopes are preferentially located in BN. In normal neurons phosphorylated NF epitopes are usually distributed in the axoplasm and non-phosphorylated epitopes are present in the somatodendritic domain (Sternberger and Sternberger, 1982). The abnormal profile in BN may represent redistribution of phosphorylated NF or, alternatively, increased phosphorylation of NF in the perikarya. The reactivity of BN with NF antibodies is more consistent than reactivity of LBs with the same antibodies. In contrast, BN are less consistently reactive with antibodies to ubiquitin (Lee *et al.*, 1989; Figure 1d). Both BN and LB are reactive with antibodies to αB-crystallin and ubiquitin C-terminal hydrolase (Lowe *et al.*, 1992; Lowe *et al.*, 1990; Kato *et al.*, 1992).

Lewy body-like inclusions in amyotrophic lateral sclerosis (ALS): Lewy body-like inclusions are occasionally detected in anterior horn cells in sporadic and familial motor neuron disease (Leigh *et al.*, 1988, 1989; Kato *et al.*, 1989). Similar to LBs, they are round to irregular shaped eosinophilic cytoplasmic inclusions that are reactive with antibodies to phosphorylated NF epitopes and ubiquitin (Figure 1e,f). Ultrastructural studies have shown these inclusions to be composed of aggregates of intermediate filaments. They are distinct from other inclusions in anterior horn cells in ALS, including Bunina bodies and skein-like inclusions. The latter are also reactive with ubiquitin, but not NF, and at the ultrastructural level are paracrystalline aggregates of microfilaments.

10.2.3 Glial filament inclusions

Rosenthal fibres: RF are one of the best studied stress-protein inclusions in astrocytes. They are found in a variety of conditions associated with chronic glial scars and even in low grade, slowly growing astrocytic tumours. They are numerous in a rare leukodystrophy, Alexander's disease. RF are eosinophilic carrot-shaped inclusions in fibrous astrocytes especially in astrocyte end-feet around blood vessels and at the pial surface. At the ultrastructural level, RF are granulofilamentous inclusions. Studies with GFAP immunocytochemistry demonstrate that the filaments at the periphery of the dense osmiophilic centre are glial filaments, but the centre of the RF is negative for GFAP.

Given the abundance of RF in Alexander's disease, it has been possible to isolate and characterize the major protein constituents of RF. Work in Goldman's laboratory demonstrated that the inclusions contained GFAP, ubiquitin and αB-crystallin (Iwaki *et al.*, 1989; Goldman and Corbin, 1991). Antibodies to ubiquitin immunostain RF (Figure 2a). Biochemical studies show that αB-crystallin in RF is ubiquitinated. Although αB-crystallin was first described as a lens protein, it is now known that αB-crystallin is also a stress protein that shares sequence homology with 27-kDa HSP. Antibodies to αB-crystallin have immunodecorated not only RF, but other stress-protein inclusions in neurons, which share the common feature of being composed predominantly of intermediate filaments. Recent evidence suggests that in addition to αB-crystallin, hsp27 is also present in RF (Goldman, personal communication). Molecular analysis has failed to reveal an abnormality in any of the proteins in RF, but they may be more phosphorylated than normal cellular proteins.

Gliofibrillary inclusions: Infants who have suffered hypoxic stress, such as in metabolic disorders or hypoxic encephalopathy, sometimes develop lipidized astrocytes with eosinophilic cytoplasmic inclusions. These inclusions have some similar properties with RF, but they differ in other respects. They are reactive with GFAP, ubiquitin (Figure 2b) and αB-crystallin, but at the ultrastructural level are composed of disordered aggregates of intermediate filaments without dense osmiophilic granular material. It is tempting to speculate that astrocytes have a limited repertoire with respect to response to injury. In response to acute stress, especially in the developing brain, gliofilamentous inclusions form, whereas in response to chronic stress due to metabolic or degenerative diseases RF are formed. The results reiterate the fact that similar stress-related proteins can be incorporated into a variety of inclusions.

10.2.4 Microtubule associated protein (Tau) inclusions

Neurofibrillary tangles (NFT) and Neuropil Threads (NT): NFT are a hallmark of Alzheimer's disease (AD) and progressive supranuclear palsy (PSP or Steele Richardson Olszewski syndrome). Neuropil threads are similar structures within neuronal processes rather than in the cell bodies (Braak *et al.*, 1986). Most of the studies have shown few differences between the immunochemical characteristics of NFT and NT, and they will be discussed together. These inclusions are intensely argyrophilic. NFT are variably flame or globose in shape depending upon the cell in which they are found. They are relatively insoluble structures that can be isolated intact with relatively simple methods of tissue dissociation. At the ultrastructural level the major component of the NFT are twisted tubules, referred to as paired helical filaments (PHF). They have a diameter of 22 nm and a periodicity of 160 nm. Several recent studies have described methods of obtaining relatively pure populations of PHF. Direct biochemical analysis of these preparations has been possible. The major protein component in PHF is tau protein that has several

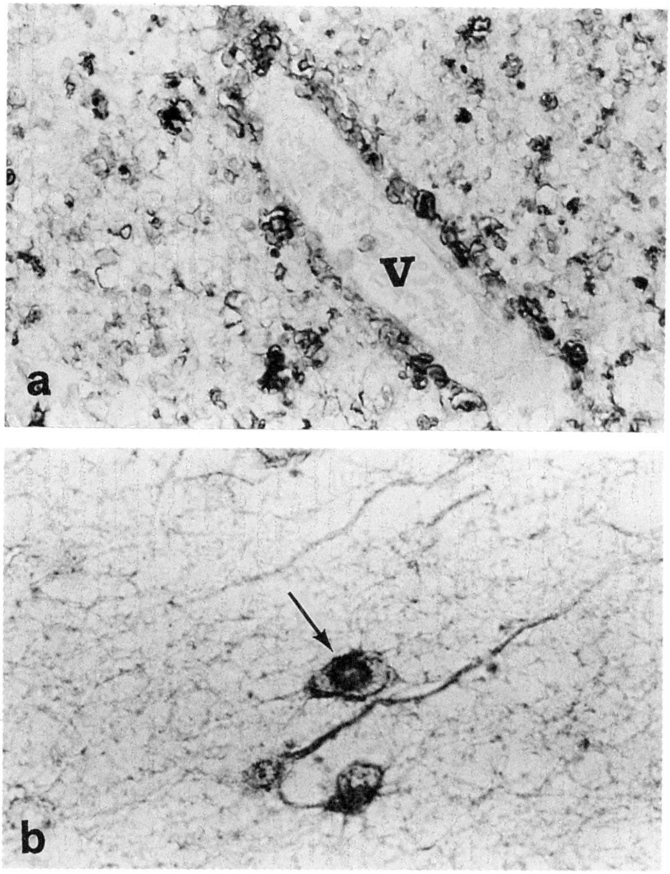

Figure 2 Glial filament inclusions. Numerous Rosenthal fibres are reactive with anti-ubiquitin in this case of Alexander's disease (a). Note in particular many inclusions around a blood vessel (v) in astrocyte end-feet. Round ubiquitin-immunoreactive inclusions are also present in these astrocytes in the white matter of a child with mitochondrial encephalopathy and anoxic encephalopathy (b).

differences from normal tau protein, also referred to as PHF-tau or A68 (Grundke-Iqbal *et al.*, 1986; Wolozin and Davies 1987; Lee *et al.*, 1991; Liu *et al.*, 1991).

NFT that occur in basal ganglia and brain stem nuclei in PSP differ from those in AD in that they are composed of straight filaments of 15–18 nm in diameter rather than PHF (Dickson *et al.*, 1985). Biochemical analysis of PSP brains shows subtle differences in the proportion of the three major isoforms of tau in PSP compared to AD (Flament *et al.*, 1991). Recent biochemical studies of a case of CBD shows similar biochemical changes in tau (Ksiezak-Reding, 1993), suggesting that there may be overlap between CBD and PSP. Whereas NFT are present in the pyramidal neurons of the cortex, hippocampus, amygdala, basal forebrain and

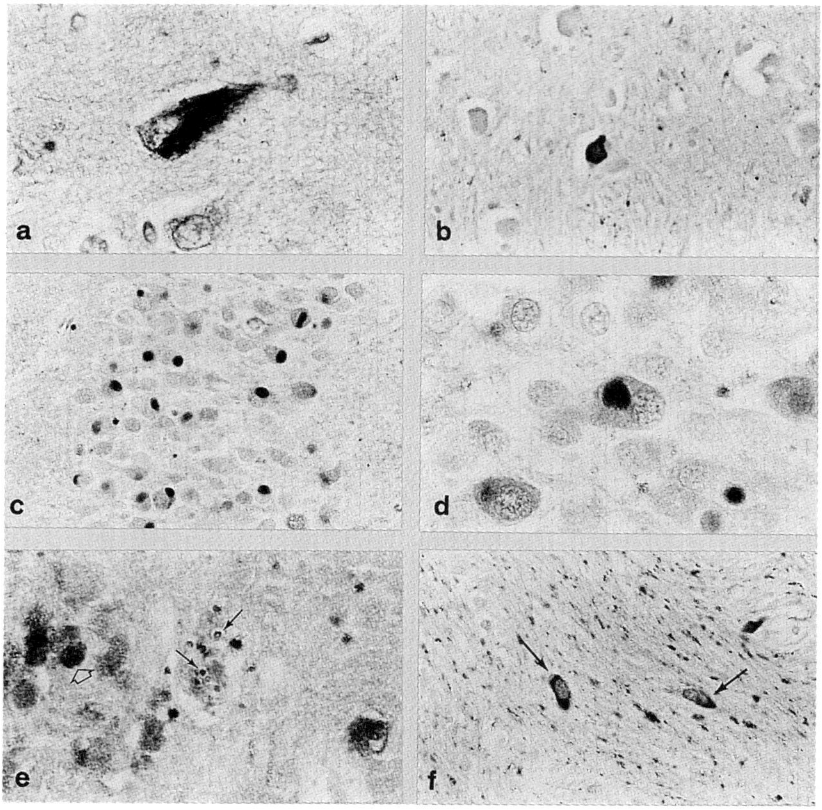

Figure 3 Microtubule-associated protein tau inclusions. Neurofibrillary tangles in Alzheimer's disease (a) and PSP (b) are immunostained with anti-ubiquitin. Flame-shaped NFT in (a) is from hippocampus and the globose NFT in (b) is from the inferior olivary nucleus. In Pick's disease numerous inclusions (Pick bodies) are present in the dentate fascia of the hippocampus (c). They are reactive with anti-ubiquitin (c,d). Granulovacuolar bodies (arrows) within the cytoplasm of a neuron in the hippocampus of Alzheimer's disease are reactive for ubiquitin. Note also the dystrophic neurites (see below) surrounding a senile plaque (arrowhead). In multisystem atrophy glial cells have ubiquitin-immunoreactive inclusions (arrows) (f). The white matter also has extensive granular degeneration of myelin (see below).

monoaminergic neurons of the brainstem in AD, in PSP they have a different distribution, affecting small neurons in the basal ganglia, thalamus, brainstem and cerebellum. NFT in both conditions are reactive with antibodies to ubiquitin (Mori *et al.*, 1987; Perry *et al.*, 1987; Manetto *et al.*, 1988; figure 3a,b). It has been stated that NFT are less reactive with ubiquitin in PSP compared to AD (Cruz-Sanchez *et al.*, 1992).

Immunocytochemical studies have also shown that NFT contain epitopes in common with normal cytoskeletal proteins, including neurofilament, MAP2 and tau proteins (Yen *et al.*, 1987). The biochemical basis for this may be the fact that of these molecules share microtubule binding domains or common phosphorylated epitopes. The evidence is most compelling for antibodies to NF. Those NF antibodies that recognize a cross-reacting epitope in tau immunostain NFT, whereas those that do not cross-react with tau generally do not react with NFT (Ksiezak-Reding *et al.*, 1987). In contrast, almost every antibody to tau protein tested recognizes NFT. These results clearly separate the filaments in LBs from those in NFT.

Phosphorylation may play a critical role in the formation of PHF. There are multiple phosphorylation sites in tau, some of which appear to be essential for PHF formation and others that are not necessary for PHF formation (Liu *et al.*, 1993). Protein kinases that phosphorylate tau to a form that resembles that seen in PHF (PHF-tau) include proline-directed kinases, such as MAP kinase and cdc2 kinase (Drewes *et al.*, 1992). Regulation of cdc2 kinase is through binding to one of a family of cyclins. It is of some interest that cell levels of cyclin are regulated by ubiquitin-dependent proteolysis (Glotzer *et al.*, 1991). Furthermore, there is evidence that ubiquitination of proteins is itself regulated by phosphorylation (Kong and Chock, 1992).

It is also of interest that the stress response is associated with stabilization of microtubules, possibly by interaction of heat shock proteins with tubulin (Gupta, 1990) and that tau and heat shock proteins co-purify in the stressed cell. The heat stress or stress response is also associated with a number of physiological changes in cellular metabolism including inhibition of protein synthesis and altered phosphorylation. It is tempting to speculate that PHF formation may be a reaction of the cytoskeleton to cellular stress. Recent biochemical studies of *post mortem* AD provide support for this general conclusion (Perez *et al.*, 1991).

Neuropil Grains and Glial Tangles: In addition to BN, CBD is characterised by several, additional cytoskeletal alterations that are best appreciated with immunocytochemical techniques with antibodies to either tau protein or ubiquitin (Dickson and Mattiace, 1992). Grain-like neurites were first described as argyrophilic structures that defined a neurodegenerative disease associated with dementia (Braak and Braak, 1987), but the most common disorder associated with neuropil grains is CBD. Grains are short stubby processes that are immunoreactive for tau protein; in CBD they are diffusely scattered in the cortical neuropil, but also characteristically clustered in plaque-like aggregates that are not associated with amyloid deposits (Mattiace *et al.*, 1991b; Figure 4b). Most of the grains are also reactive for ubiquitin (Figure 4a). Immunocytochemistry also shows that CBD brains contain abnormal tau reactivity in processes in the white matter and within glial cells in the white matter (Figure 4c,d). Biochemical studies of CBD reveal PHF-tau (A68) abnormalities in both grey and white matter (Ksiezak-Reding *et al.*, 1993). At the ultrastructural level neuropil grains are composed of straight filaments similar to those in PSP. Somewhat similar neuritic clusters have been described in PSP (Hauw *et al.*, 1990). Some PSP cases also have tau-reactive axons in the basal ganglia

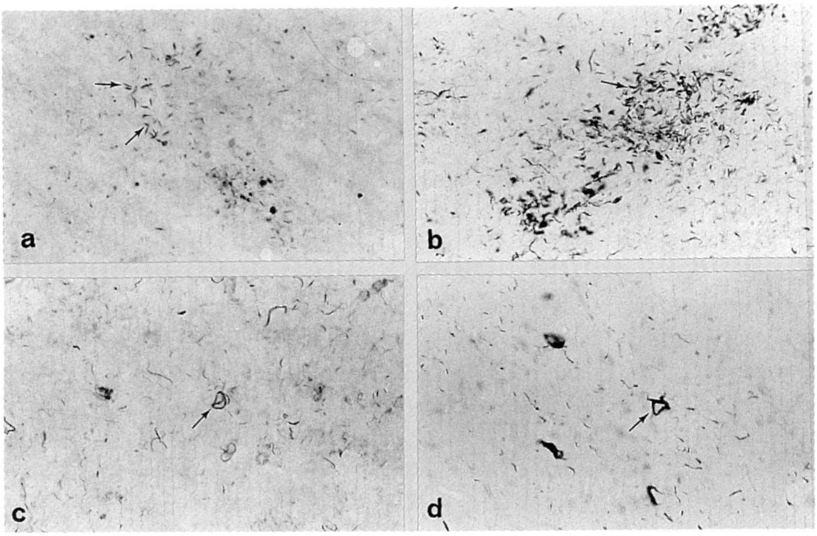

Figure 4 Neuropil grains and glial tangles in CBD. Anti-ubiquitin staining of the cortical grey and white matter of a case of CBD shows clusters of grain-like structures (arrows in a) and glial tangles (arrows in b). Similar structures are also reactive with tau antibodies in adjacent sections from the same case (a,b – grey matter; c,d – white matter).

as in CBD. The two disorders thus overlap pathologically, biochemically and sometimes clinically (Masliah *et al.*, 1991). Further studies are needed to define the pathological distinctions between PSP and CBD. For example, the BN, which are the most characteristic pathological hallmark of CBD, are rarely described in PSP.

Pick bodies (PB): In addition to BN (Pick cells), Pick's disease is also characterized by argyrophilic inclusion bodies (Pick bodies) in small neurons of the cerebral cortex and dentate fascia of the hippocampus. PB have almost identical antigenic properties to PHF (Love *et al.*, 1988). At the ultrastructural level they are composed of loosely packed and irregularly arranged, twisted tubules with a longer periodicity and wider diameter than PHF (Yen *et al.*, 1986). Few biochemical studies of the cytoskeleton and stress proteins have been performed in Pick's disease brain, but PHF-tau (A68) has been detected in much lower amounts than in AD (Wolozin and Davies, 1987). Pick bodies are consistently stained with ubiquitin antibodies (Figure 3c,d).

Granulovacuolar bodies (GVB): Neurons in the hippocampus develop dense granular cytoplasmic inclusions surrounded by clear vacuoles (GVB) as a manifestation of ageing and AD. Early histochemical studies suggested that GVB were derived from lysosomes, but little progress was made in understanding their biochemical composition until the advent of immunocytochemical techniques. With immunocytochemistry GVB have been shown to be immunoreactive with phosphorylated neurofilament epitopes that are shared with tau protein (Dickson *et al.*, 1987). Recent studies have confirmed the presence of tau epitopes in GVB (Dickson *et al.*, 1993). In addition, GVB are consistently reactive with antibodies to ubiquitin. (Figure 3e). Unlike PHF, not all tau antibodies show reactivity with GVB. The fact that only limited tau epitopes are present in GVB, as well as their frequent appearance in neurons with PHF or vulnerable to PHF formation, suggests that GVB may contain tau proteins that are partially degraded by autophagy. That ubiquitin may be associated with lysosomal degradation is of some interest and is considered in greater detail in other chapters of this book. It is worth noting that ubiquitinated proteins may be degraded in lysosomes by autophagy (review: Mayer *et al.*, 1991). Evidence for a specific pathway for ubiquitin mediated lysosomal degradation is not known.

Glial Cytoplasmic Inclusions (GCI): Multisystem atrophy is a relatively uncommon neurodegenerative disorder, with clinical features of cerebellar dysfunction, extrapyramidal movement abnormality and autonomic nervous system dysfunction. It is characterized by neurodegeneration of select populations of neurons in the putamen, substantia nigra, pons, inferior olive, cerebellar cortex and spinal cord. The neurodegeneration is not associated with any identifiable neuronal alterations; however, recently, GCI have been found to be a pathological hallmark of this disorder (Papp *et al.*, 1989; Horoupian and Dickson, 1991). The glial inclusions are immunoreactive for tau protein and are also consistently reactive for ubiquitin (Figure 3f). Double labelling studies with markers for oligodendroglia suggest that GCI in multisystem atrophy are within the cytoplasm of oligodendroglia. At the ultrastructural level GCI are tubulofilamentous structures. The explanation for the predilection for oligodendroglia in this neurodegenerative disease remains a mystery. It should also be emphasized that unlike most of the other tau-positive inclusions described above, GCI are in glia. In non-diseased states tau has infrequently been demonstrated in astrocytes, but not in oligodendroglia. The presence of tau and ubiquitin immunoreactive inclusions in oligodendroglia is truly extraordinary.

10.2.5 Non-cytoskeletal ubiquitin-immunoreactive inclusions

Neuronal inclusions in dementia lacking distinctive histology: A relatively common non-Alzheimer degenerative dementia goes by the rather unappealing designation of dementia-lacking-distinctive-histology (DLDH). This disorder is a cortical degeneration whose clinical features vary, depending upon the distribution of the brunt of the cortical pathology. Patients with frontal lobe ('frontal lobe dementia'), parietal lobe ('dysphasic dementia') or temporal lobe ('ALS dementia') degeneration have been described. Involvement of deep grey matter is inconsistent, but frequently detected. The disorder takes it name from the fact that with standard histopathological methods there is only non-specific neuronal loss, gliosis and spongiosis, especially of the upper cortical layers. Using a battery of special

Table 2 Inclusion bodies in neuropathology

Inclusion body	Characteristic disease	Location	Cell Type	EM Findings	Structural protein	Stress protein
Lewy bodies (LB)	Parkinson's disease; Diffuse Lewy body disease (DLBD)	Substantia nigra, brainstem monoaminergic, basal forebrain	Neurons	Granulofilamentous Intermediate filaments	NF	Ubiquitin; αB-crystallin
Dystrophic neurites	DLBD	CA2/3, basal forebrain, brainstem nuclei	Neuronal processes	Intermediate filaments	NF	Ubiquitin
Cortical Lewy bodies	DLBD	Limbic & paralimbic cortices, amygdala	Neurons	Intermediate filaments	NF	Ubiquitin; αB-crystallin
Ballooned neurons (BN)	Corticobasal ganglionic degeneration (CBG), Pick's disease	Limbic & paralimbic cortices, amygdala	Neurons	Filamentous; granular material	NF	Ubiquitin; αB-crystallin
Lewy body-like inclusions in ALS	ALS – sporadic and familial	Anterior horn cells	Neurons	Intermediate filaments	NF	Ubiquitin; αB-crystallin
Rosenthal fibres	Alexander's disease	Subpial and perivascular; white matter	Astrocytes	Granulofilamentous; glial filaments	GFAP	Ubiquitin; αB-crystallin hsp27
Gliofibrillary inclusion	Infantile hypoxic encephalopathies	White matter	Astrocytes	Glial filaments	GFAP	Ubiquitin; αB-crystallin
Neurofibrillary tangles (NFT) and neuropil threads	Alzheimer's disease (AD); Progressive supra-nuclear palsy (PSP)	Cortex, hippocampus, basal forebrain, brainstem nuclei	Neurons	Paired helical filaments (PHF); straight filaments (SF)	PHF-tau	Ubiquitin
Pick bodies	Pick's disease	Hippocampal dentate fascia, cerebral cortex	Neurons	PHF with wider diameter and longer periodicity	PHF-tau	Ubiquitin
Neuropil grains	Corticobasal ganglionic degeneration (CBD); Dementia with grains	Cortex, hippocampus	Neurons	SF	PHF-tau	Ubiquitin

Table 2 *continued*

Gliolfibrillary tangles	CBG; PSP	White matter	Astrocytes	SF	PHF-tau	Ubiquitin
Glial cytoplasmic inclusions (GCI)	Multisystem atrophy	White matter	Oligodendro-cytes	Tubulofilamentous	PHF-tau	Ubiquitin
Granulovacuolar bodies (GVB)	Ageing; AD	Hippocampus	Neurons	Dense granule in vacuole	Tau, NF	Ubiquitin
Ubiquitin reactive inclusions	Dementia lacking distinctive histopathology (DLDH)	Cortex (small neurons in layer 2); dentate fascia	Neurons	Unknown	Unknown	Ubiquitin
Marinesco bodies	Ageing	Substantia nigra	Neurons	Paracrystalline array of microfilaments	Unknown	Ubiquitin
Neuroaxonal dystrophy	Ageing	Dorsal column nuclei, substantia nigra, globus pallidus	Neurons distal axons	Filaments, dense bodies, vesicular material	Unknown	Ubiquitin
Granular degeneration myelin	Ageing	Myelin of brain and spinal cord	Myelin lamellae	Dense bodies in myelin lamellae	Unknown	Ubiquitin
Granular Dystrophic neurites	Ageing senile plaques; Amyloid plaques in prion disease	Associated with senile plaques	Neuronal processes	Dense bodies, vesicular material	Unknown	Ubiquitin

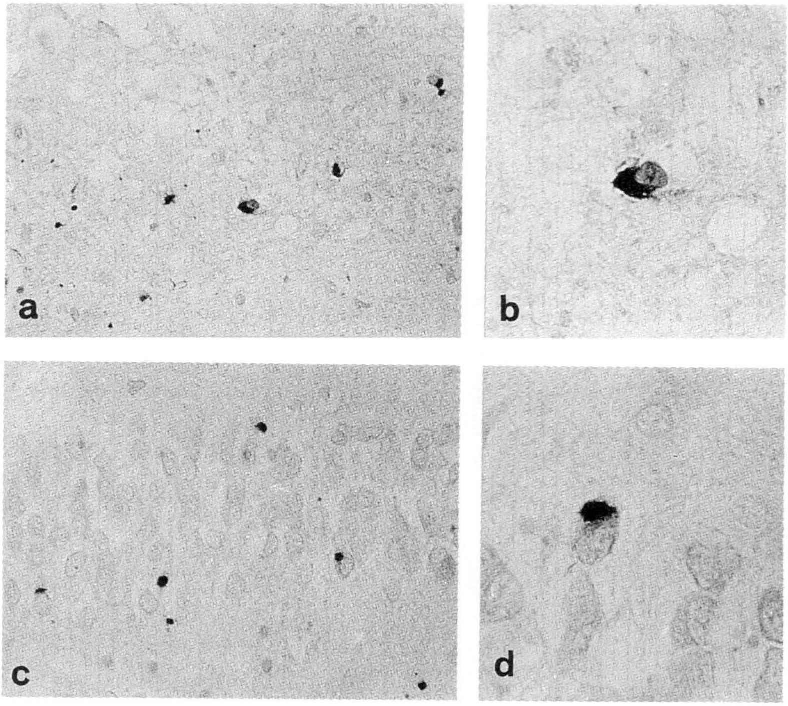

Figure 5 Ubiquitin-immunoreactive inclusions in dementia lacking distinctive histopathology (DLDH). The upper cortical layer has a fine vacuolar change and scattered neurons have ubiquitin-immunoreactive inclusions (a) that are seen in higher magnification in (b). Similar inclusions are present in the dentate fascia (c) seen at higher magnification in (d). These inclusions have a similar distribution to Pick bodies (see Figure 3c,d) but are not reactive with tau antibodies like Pick bodies.

neurohistological techniques no distinctive pathologic inclusions are detected. On the other hand, recent studies of such cases with antibodies to ubiquitin have revealed an oval to crescent shaped inclusion in small non-pyramidal neurons in the upper cortical layer and in the dentate fascia (Wightman *et al.*, 1992; Figure 5). The inclusions are not reactive with other antibodies, but extensive studies have not been performed to date. Ultrastructural studies have also not yet been reported.

Marinesco bodies: Intranuclear eosinophilic inclusions that are most often detected in aged brains in neurons of the substantia nigra are composed of intranuclear paracrystalline aggregates of microfilaments that are not reactive with antibodies to neurofilaments. Marinesco bodies are consistently stained with antibodies to ubiquitin (Figure 6a). Since some stress proteins are known to translocate from cytosol to the nucleus the presence of stress proteins in intranuclear inclusions is not entirely surprising. Ubiquitin is frequently detected in the nucleus in nuclear histones, but there is no evidence that Marinesco bodies contain histones.

Neuroaxonal dystrophy: Dystrophic axons are a common age-associated pathology. They are nearly universally present in the dorsal-column nuclei and can be detected as early as the second or third decade of life. They are increased in frequency in diabetes. Other common locations include the substantia nigra and globus pallidus. They are immunoreactive with ubiquitin (Dickson *et al.*, 1991; Figure 6b). A location where they are relatively common includes the cortex and the deep grey matter, especially the amygdala and the entorhinal cortex. This is a pathology that was not apparent until ubiquitin immunocytochemistry became widely used. The ultrastructural appearance, like the light microscopic appearance of dystrophic axons, is heterogeneous. They are eosinophilic and granular to hyaline in appearance. At the EM level they contain filaments, membranous organelles and dense bodies. They are also invariably reactive with ubiquitin. Immunolocalization studies of ubiquitin have not been performed. It is unclear to which element ubiquitin is conjugated in dystrophic axons.

Granular degeneration of myelin: Myelin invariably undergoes degeneration associated with ageing. This has been noted in electron microscopic studies. Until the advent of ubiquitin immunocytochemistry, the extent and distribution of this pathology was unsuspected (Dickson *et al.*, 1991; Mattiace *et al.*, 1991a; Figure 6c). The clinical significance of granular degeneration of myelin is also unknown, but it is tempting to speculate that myelin pathology may underlie some of the age-associated slowing that occurs in central motor and sensory information processing.

Granular dystrophic neurites in senile plaques: Senile plaques are complex structures that are associated with extracellular amyloid deposits and variable glial and neuronal alterations. They are very common in the ageing brain and are essential features of AD. They are also detected in some of the spongiform encephalopathies. The role of stress proteins in these disorders is discussed more fully in Chapter 11, but for the sake of completeness they are discussed briefly at this point. Amyloid is a class of relatively insoluble and inert material with common physical and staining properties (β-pleated sheet and birefringence to Congo red). The major protein in the amyloid deposits is often derived from cells of the immune system, most often mononuclear phagocytes.

With ubiquitin immunocytochemistry it is possible to detect dystrophic neuronal processes associated with dense amyloid deposits, regardless of the nature of the amyloid (Suenaga *et al.*, 1990a,b). For example, dystrophic neuronal processes are associated with senile plaques in the cortex and cerebellum in AD (Figure 7a,b), pathological ageing (Figure 7c) and Gerstmann-Sträussler-Scheinker syndrome (a familial spongiform encephalopathy) (Figure 7d). The dystrophic neurites are coarsely granular and not reactive with tau antibodies, except for some neurites in AD. These findings suggest that amyloid or some associated factor is toxic to

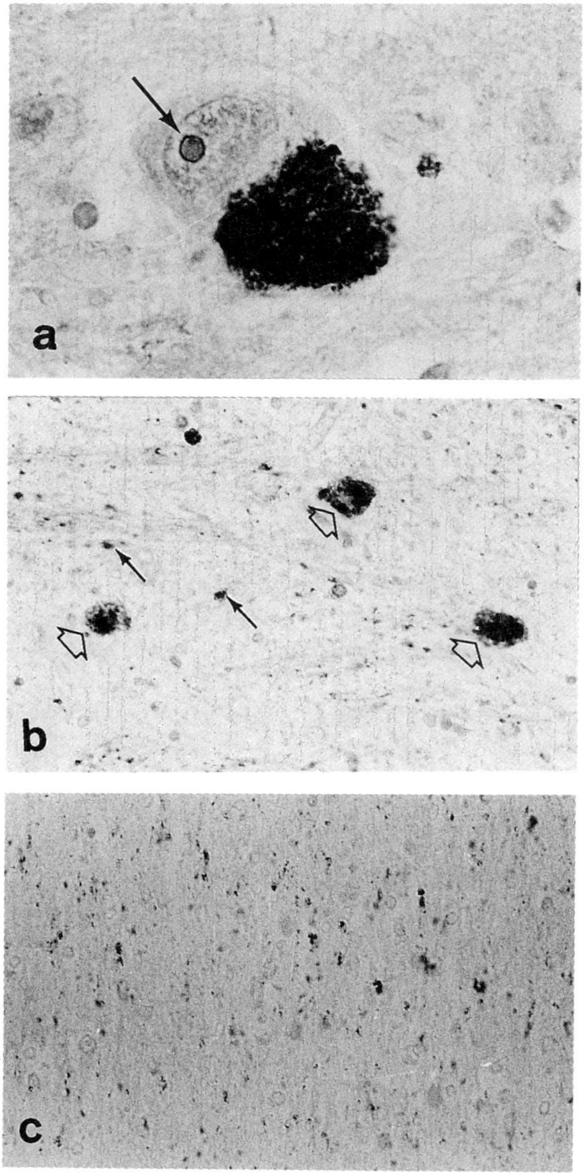

Figure 6 Ubiquitin-immunoreactive inclusions of uncertain nature. A Marinesco body in the nucleus of a pigment bearing neuron of the substantia nigra is immunoreactive for ubiquitin (a). Dystrophic axons in the globus pallidus (arrowheads) of this elderly person are also reactive with ubiquitin with coarse granular staining of the inclusion. The smaller granular staining (arrows) is granular degeneration of myelin, which is better appreciated in the white matter in (c).

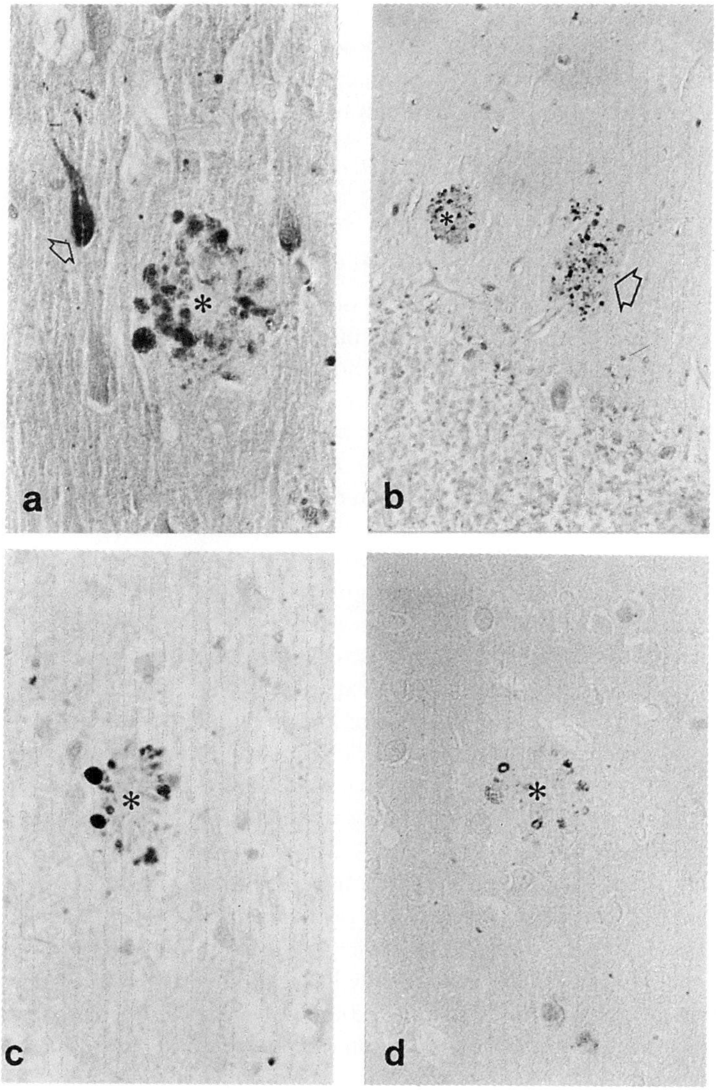

Figure 7 Ubiquitin-immunoreactive dystrophic neurites associated with amyloid deposits in senile plaques. In the hippocampus and the cerebellum of Alzheimer's disease-dense core amyloid deposits (*) are surrounded by ubiquitin-immunoreactive dystrophic processes. Note staining of NFT (arrow) with anti-ubiquitin in (a). In the cerebellum even diffuse amyloid deposits (arrow) are frequently associated with similar ubiquitin-immunoreactive dystrophic neurite (b). An elderly patient with Creutzfeldt–Jakob disease with Aβ amyloid deposits (*) (c) and a young man with Gerstmann–Sträussler–Scheinker syndrome and prion protein amyloid deposits (d) are also surrounded by ubiquitin-immunoreactive processes. These dystrophic neurites are not reactive with tau antibodies and correspond to membranous dense bodies at the ultrastructural level.

neuronal processes. At the ultrastructural level ubiquitin-immunoreactive dystrophic neurites in all these conditions are composed of swollen processes filled with membranous dense bodies possibly of lysosomal origin (Dickson *et al.*, 1990; Jicha and Dickson, 1993) that contain ubiquitin immunoreactive proteins.

10.3 Discussion

It is important to reiterate that much of our current knowledge of these cellular alterations is based upon immunocytochemistry simply because the inclusions are present in numbers too few to adequately isolate for biochemical studies. There are only a few exceptions. The PHF in AD are very abundant and AD is also a common disorder. Intensive research efforts in recent years have perfected methods to obtain relatively pure preparations of PHF that have permitted molecular analysis (Liu *et al.*, 1991). Other much less common disorders with many NFT, such as PSP, have been subjected to more limited analysis. Progress has also been made in biochemical characterization of Rosenthal fibres (RF), since they are very abundant in Alexander's disease. Although Alexander's disease is rare, there are sufficient number of lesions that RF have proven to be amenable to biochemical analysis. The evidence clearly indicates that RF are astrocytic stress-related inclusions. It is of course unknown if the observations in Alexander's disease have relevance to other disorders in which RF are detected. There are preliminary reports of limited success in partial purification of Lewy bodies. The major protein detected in these preparations cross-reacted with a polyclonal antibody to tau protein and neurofilament. It is unknown whether this actually represents tau, neurofilament or some yet to be identified related protein.

In the case of RF, biochemical studies have demonstrated that proteins in RF are conjugated to ubiquitin. This has not been convincingly demonstrated for the other inclusions. In order to understand the pathogenesis of these lesions it will be important to learn if ubiquitin is covalently linked to the major structural proteins. The possible mechanisms to be considered are that ubiquitin-mediated proteolysis is overwhelmed by the excessive production of abnormal structural proteins or that the ATP-dependent degradation of proteins is impaired due to defective elements (e.g. proteasome disorders) or this complex proteolytic pathway. A common denominator of cell stress is the production of denatured proteins that may be the target for ubiquitin-mediated degradation. The nature of the abnormalities of the cellular metabolism that generate the abnormal proteins that accumulate in these lesions remains to be determined

In a more practical sense the use of ubiquitin to routinely evaluate CNS tissue has redefined our way of looking at the nervous system. A wealth of age-associated histopathology is present in the normal aged brain, including granular degeneration of myelin, granular cytoplasmic and nuclear inclusions and dystrophic neuronal processes in both gray and white matter and occurring in association with age-associated amyloid deposition. The extensive ubiquitin-immunoreactivity in the ageing brain suggests that histopathological manifestations of CNS ageing should be thought of as a stress response. Much of this histopathology is occult with routine histopathological methods. In most cases age-associated pathology can be distinguished from disease-related processes. In ageing, ubiquitin immunoreactive structures are most often not associated with cytoskeletal elements, whereas the cytoskeleton is perturbed in some respect in lesions associated with disease.

If we accept the fact that stress reaction is involved in the formation of, or response to, neuronal and glial inclusions, much is still left with respect to understanding their pathogenesis. Why, for example, are glial inclusions characteristic of multisystem degeneration, which had been considered to be a neurodegenerative disease for many years? Why are neuronal and not glial inclusions characteristic of other neurodegenerative diseases? What determines which particular set of neurons are affected in each disorder? What is the nature of the stress that leads to production of a particular type of inclusion? The interaction of stress proteins and the cytoskeleton and further elucidation of the physiological changes associated with stress, including the effects on altered intracellular calcium metabolism and the cascade of events that follow perturbations in intracellular calcium, including phosphorylation and proteolysis, may eventually shed light on the cytoskeletal inclusions in neurological diseases. The ultimate goal is to understand the nature of the stress that sets into motion the pathological chain of events that produces the characteristic neuronal and glial alterations that we recognize as inclusion bodies.

Acknowledgments

Drs A. Hirano, K. Klein and T. Zimmerman submitted case material for study. Jorge Bermudez and Lola Leon provided technical support. Greg Jicha, Linda Mattiace and Meng-Liang Zhao performed some of the immunocytochemistry. Their contributions are gratefully acknowledged. Supported by National Institutes of Health grants AG06803, AG01136 and AG04145.

References

Abe, K., Tanzi, R.E. & Kogure, K. (1991) Induction of HSP70 mRNA after transient ischemia in gerbil brain. *Neurosci. Lett.* **125**, 166–168.

Bagi, G. & Hidvégi, E.J. (1990) Protein phosphorylation and kinase activities in tumour cells after hyperthermia. *Int. J. Radiat. Biol.* **58**, 633–650.

Braak, H., Braak, E., Gundke-Iqbal, I. & Iqbal, K. (1986) Occurrence of neuropil threads in the senile human brain and in Alzheimer's disease: A third location of paired, helical filaments outside of neurofibrillary tangles and neuritic plaques. *Neurosci. Lett.* **65**, 351–355.

Braak, H. & Braak, E. (1987) Argyrophilic grains: Characteristic pathology of cerebral cortex in cases of adult onset dementia without Alzheimer changes. *Neurosci. Lett.* **76**, 124–127.

Chiang, M.F., Liu, W.-K. & Yen, S.-H. (1993) Reversible heat stress-related loss of phosphorylated Alzheimer type epitopes in tau proteins of human neuroblastoma cells. *J. Neurochem.* **13**, 4854–4860.

Cruz-Sanchez, F.F., Rossi, M.L., Cardozo, A., Deacon, P. & Tolosa, E. (1992) Clinical and pathological study of two patients with progressive supranuclear palsy and Alzheimer's changes. Antigenic determinants that distinguish cortical and subcortical neurofibrillary tangles. *Neurosci. Lett.* **136**, 43–46.

Dice, J.F. (1987) Molecular determinants of protein half-lives in eukaryotic cells. *FASEB J.* **1**, 349–357.

Dickson, D.W., Wertkin, A., Kress, Y., Ksiezak-Reding, H. & Yen, S.-H. (1990) Ubiquitin immunoreactive structures in normal human brains: Distribution and developmental aspects. *Lab. Invest.* **63**, 87–99.

Dickson, D.W., Yen, S.-H., Suzuki, K.I., Davies, P., Garcia, J.H. & Hirano, A. (1986) Ballooned neurons in select neurodegenerative diseases contain phosphorylated neurofilament epitopes. *Acta Neuropathol. (Berl.)* **71**, 216–223.

Dickson, D.W., Ksiezak-Reding, H., Crowe, A & Yen, S.-H. (1987) A monoclonal antibody that recognizes a phosphorylated epitope in Alzheimer neurofibrillary tangles, neurofilament and tau proteins immunostains granulovacuolar degeneration. *Acta Neuropathol. (Berl.)* **73**, 254–258.

Dickson, D.W., Crystal, H., Mattiace, L., Kress, Y., Schwagerl, A., Davies, P. & Yen, S.-H. (1989) Diffuse Lewy body disease: Light and electron microscopic immunocytochemistry of senile plaques. *Acta Neuropathol. (Berl.)* **78**, 572–584.

Dickson, D.W., Werkin, A., Mattiace, L.A., Fier, E., Kress, Y., Davies, P. & Yen, S.-H. (1990) Ubiquitin immunoelectron microscopy of dystrophic neurites in cerebellar senile plaques of Alzheimer's disease. *Acta Neuropathol. (Berl.)* **79**, 486–493.

Dickson, D.W., Ruan, D., Crystal, H., Mark, M.H., Davies, P., Kress, Y. & Yen, S.-H. (1991) Hippocampal degeneration differentiates diffuse Lewy body disease (DLBD) from Alzheimer's disease: Light and electron microscopic immunocytochemistry of CA2–3 neurites specific to DLBD. *Neurology* **41**, 1402–1409.

Dickson, D.W., Kress, Y., Crowe, A. & Yen, S.-H. (1985) Monoclonal antibodies to Alzheimer neurofibrillary tangles. 2. Demonstration of a common antigenic determinant between ANT and neurofibrillary degeneration in progressive supranuclear palsy. *Am. J. Pathol.* **120**, 292–303.

Dickson, D.W., & Mattiace, L.A. (1992) Immunocytochemical studies distinguish corticobasal degeneration from progressive supranuclear palsy. *J. Neuropathol. Exp. Neurol.* **51**, 321.

Dickson, D.W., Liu, W.-K., Kress, Y., Ku, J., DeJesus, O. & Yen, S.-H. (1993) Phosphorylated tau immunoreactivity of granulovacuolar bodies (GVB) of Alzheimer's disease: Localization of two amino terminal tau epitopes in GVB. *Acta Neuropathol. (Berl.)* **85**, 463–470.

Drewes, G., Lichtenberg-Kraag, B., Doring, F., Mandelkow, E.M., Biermnat, J., Goris, J., Doree, M. & Mandelkow, E. (1992) Mitogen activated (MAP) kinase transforms tau protein into an Alzheimer-like state. *EMBO J.* **11**, 2131–2138.

Duncan, R.F. & Hershey, J.W.B. (1989) Protein synthesis and protein phosphorylation during heat stress, recovery and adaptation. *J. Cell Biol.* **109**, 1467–1481.

Eng, L.F. (1985) Glial fibrillary acidic protein (GFAP): The major protein of glial intermediate filaments in differentiated astrocytes. *J. Neuroimmunol.* **8**, 203–214.

Flament, S., Delacourte, A., Verny, M., Hauw, J.-J. & Javoy-Agid, F. (1991) Abnormal tau proteins in progressive supranuclear palsy: Similarities and differences with the neurofibrillary degeneration of the Alzheimer type. *Acta Neuropathol. (Berl.)* **81**, 591–596.

Gibb, W.R.G., Luthert, P.J. & Marsden, C.D. Corticobasal degeneration. (1989) *Brain* **112**, 1171–1192.

Glotzer, M., Murray, A.W. & Kirschner, M.W. (1991) Cyclin is degraded by the ubiquitin pathway. *Nature* **349**, 132–138.

Goedert, M., Crowther, R.A. & Garner, C.C. (1991) Molecular characterization of microtubule-associated proteins tau and MAP2. *Trends Neurosc.* **14**, 193–199.

Goldberg, A.L. & Rock, K.L. (1992) Proteolysis, proteasomes and antigen presentation. *Nature* **357**, 375–379.

Goldman, R., Goldman, A., Green, K., Jones, J., Lieska, N. & Yang, H.-Y. (1985) Intermediate filaments: Possible functions as cytoskeletal connecting links between the nucleus and the cell surface. In *Intermediate Filaments* (eds. Wang, E., Fischman, D., Liem, R.K.H. & Sun, T.-T.), pp. 1–17. *Annals of the New York Academy of Sciences*, Vol. 445, New York.

Goldman, J.E. (1983) The association of actin with Hirano bodies. *J. Neuropathol. Exp. Neurol.* **42**, 146–152.

Goldman, J.E. & Corbin, E. (1991) Rosenthal fibres contain ubiquitinated αB-crystallin. *Am. J. Pathol.* **139**, 933–938.

Grundke-Iqbal, I., Iqbal, K., Quinlan, M., Tung, Y.-C., Zaidi, M.S. & Wisniewski, H.M. (1986) Microtubule-associated protein tau. A component of Alzheimer paired helical filaments. *J. Biol. Chem.* **261**, 6084–6089.

Gupta, R.S. (1990) Microtubules, mitochondria, and molecular chaperones: A new hypothesis for *in vivo* assembly of microtubules. *Biochem. Cell. Biol.* **68**, 1352–1363.

Hauw, J.J., Verny, M., Delaere, P., Cervera, P., He, Y. & Duyckaerts, C. (1990) Constant neurofibrillary changes in the neocortex in progressive supranuclear palsy. Basic differences with Alzheimer's disease and aging. *Neurosci. Lett.* **119**, 182–186.

Hershko, A. & Ciechanover, A. (1992) The ubiquitin system for protein degradation. *Annu. Rev. Biochem.* **61**, 761–807.

Hill, W.D., Lee, V.M.-Y., Hurtig, H.I., Murray, J.M. & Trojanowski, J.Q. (1991) Epitopes located in spatially separate domains of each neurofilament subunit are present in Parkinson's disease Lewy bodies. *J. Comp. Neurol.* **309**, 150–160.

Himmler, A., Drechesel, D., Kirschner, M.W. & Martin, D.W. Jr. (1989) Tau consists of a set of proteins with repeated C-terminal microtubule-binding domains and variable N-terminal domains. *Moleo. Cell. Biol.* **9**, 1381–1388.

Hollenbeck, P.J. (1990) Cytoskeleton on the move. *Nature* **343**, 408–409.

Horoupian, D.S. & Dickson, D.W. (1991) Striatonigral degeneration, olivoponto-cerebellar atrophy and 'atypical' Pick disease. *Acta Neuropathol. (Berl.)* **81**, 287–295.

Howard, J., Hudspeth, A.J. & Vale, R.D. (1989) Movement of microtubules by single kinesin molecules. *Nature* **342**, 154–158.

Iwaki, T., Kume-Iwaki, A., Liem, R.K.H. & Goldman, J.E. (1989) αB-crystallin is expressed in non-lenticular tissues and accumulates in Alexander's disease brain. *Cell* **57**, 71–78.

Iwaki, T., Kume-Iwaki, A. & Goldman, J.E. (1990) Cellular distribution of αB-crystallin in non-lenticular tissues. *J. Histochem. Cytochem.* **38**, 31–39.

Jenner, P. & Marsden, C.D. (1993) MPTP-induced Parkinsonism: A model of Parkinson's disease and its relevance to the disease process. In *Parkinson's Disease and Movement Disorders* (eds. Jankovic, J. and Tolosa, E.) pp. 55–76. Second Edition. Baltimore, Williams & Wilkins.

Jicha, G. & Dickson, D.W. (1993) Gerstmann-Sträussler syndrome (GSS): Evidence for microglial participation in amyloid deposition. *Neurology* **43**, A252.

Kato, S., Hirano, A., Umahara, T., Kato, M., Herz, F. & Ohama, E. (1992) Comparative immunohistochemical study on the expression of αB-crystallin, ubiquitin and stress response protein 27 in ballooned neurons in various disorders. *Neuropathol. Appl. Neurobiol.* **18**, 335–340.

Kato, T., Katagiri, T., Hirano, A., Kawanami, T. & Sasaki, H. (1989) Lewy-like hyaline inclusions in sporadic motor neuron disease are ubiquitinated. *Acta Neuropathol. (Berl.)* **77**, 391–399.

Klemenz, R., Frohli, E., Steiger, R.H., Schafer, R. & Aoyama, A. (1991) αB-crystallin is a small heat shock protein. *Proc. Natl Acad. Sci USA* **88**, 3652–3656.

Kong, S.K. & Chock, P.B. (1992) Protein ubiquitination is regulated by phosphorylation. An *in vitro* study. *J. Biol. Chem.* **267**, 14 189–14 192.

Kosik, K.S. Tau protein and Alzheimer's disease. *Curr. Opin. Cell Biol.* **2**, 101–104.

Ksiezak-Reding, H., Dickson, D.W., Davies, P. & Yen, S.-H. (1987) Recognition of tau epitopes by anti-neurofilament antibodies that stain Alzheimer neurofibrillary tangles. *Proc. Natl Acad. Sci. USA* **84**, 3410–3414.

Ksiezak-Reding, H., Morgan, K., Weidenheim, K., Mattiace, L.A. & Dickson, D.W. (1993) Paired helical filaments (PHF) in corticobasal degeneration. *J. Neuropathol. Exp. Neurol.* **52**, 283.

Kuzuhara, S., Mori, H., Izumiyama, N., Yoshimura, M. & Ihara, Y. (1988) Lewy bodies are ubiquitinated: A light and electron microscopic immunocytochemical study. *Acta Neuropathol. (Berl.)* **75**, 345–353.

Kwak, S., Masaki, T., Ishiuri, S. & Sugita, H. (1991) Multicatalytic proteinase is present in Lewy bodies and neurofibrillary tangles in diffuse Lewy body-disease brains. *Neurosci. Lett.* **128**, 21–24.

Lee, V.M.-Y., Balin, B., Otvos, L. Jr. & Trojanowski, J.Q. (1991) A68: A major subunit of paired helical filaments and derivatized forms of normal tau. *Science* **151**, 678–680.

Lee, W.C., Lin, K.Y., Chen, K.D. & Lai, Y.K. (1992) Induction of HSP70 is associated with vincristine resistance in heat shocked 9L rat brain tumour cells. *Br. J. Cancer* **66**, 653–659.

Lee, S., Park, Y.D., Yen, S.-H., Ksiezak-Reding, H., Goldman, J.E. & Dickson, D.W. (1989) Infantile motor neuron disease: A case report with electron microscopic and ubiquitin and neurofilament immunohistochemical studies. *Neuropediatrics* **20**, 107–111.

Leigh, P.N., Anderton, B.H., Dodson, A., Gallo, J.-M., Swash, M. & Power, D.M. (1988) Ubiquitin deposits in anterior horn cells in motor neurone disease. *Neurosci. Lett.* **93**, 197–203.

Leigh, P.N., Dodson, A., Swash, M., Brion, J.-P. & Anderton, B. H. (1989) Cytoskeletal abnormalities in motor neuron disease: An immunocytochemical study. *Brain* **112**, 521–535.

Lennox, G., Lowe, J., Morrell, K., Landon, M. & Mayer, R.J. (1989) Anti-ubiquitin immunocytochemistry is more sensitive than conventional techniques in the detection of diffuse Lewy body disease. *J. Neurol. Neurosurg. Psychiatry* **52**, 67–71.

Liem, R.K.H. (1990) Neuronal intermediate filaments. *Curr. Opinion Cell Biol.* **2**, 86–90.

Liu, W.-K., Ksiezak-Reding, H. & Yen, S.-H. (1991) Abnormal tau proteins from Alzheimer's disease brains: purification and amino acids analysis. *J. Biol. Chem.* **266**, 21 723–21 727.

Liu, W.S., Dickson, D.W. & Yen, S.-H. (1993) Heterogeneity of tau protein in Alzheimer's disease: Evidence for increased expression of an isoform and preferential distribution of a phosphorylated isoform in neurites. *Am. J. Pathol.* **142**, 387–394.

Love, S., Saitoh, T., Quijada, S., Cole, G.M. & Terry, R.D. (1988) Alz-50, ubiquitin and tau immunoreactivity of neurofibrillary tangles, Pick bodies and Lewy bodies. *J. Neuropathol. Exp. Neurol.* **47**, 393–405.

Lowe, J., Blanchard, A., Morrell, K., Lennox, G., Reynolds, L., Billett, M., Landon, M. & Mayer, R.J. (1988) Ubiquitin is a common factor in intermediate filament inclusion bodies of diverse type in man, including those of Parkinson's disease, Pick's disease, and Alzheimer's disease, as well as Rosenthal fibres in cerebellar astrocytomas, cytoplasmic bodies in muscle, and Mallory bodies in alcoholic liver disease. *J. Pathol.* **155**, 9–15.

Lowe, J., McDermott, H., Landon, M., Mayer, R.J. & Wilkinson, K.D. (1990) Ubiquitin carboxyl-terminal hydrolase (PGP 9.5) is selectively present in ubiquitinated inclusion bodies, characteristic of human neurodegenerative diseases. *J. Pathol.* **161**, 153–160.

Lowe, J., McDermott, H., Pike, I., Spendlove, I., Landon, M. & Mayer, R.J. (1992) αB-crystallin expression in non-lenticular tissues and selective presence in ubiquitinated inclusion bodies in human disease. *J. Pathol.* **166**, 61–68.

Lowe, J., Errington, D.R., Lennox, G., Pike, I., Spendlove, I., Landon, M. & Mayer, R.J. (1992) Ballooned neurons in several neurodegenerative diseases and stroke contain αB-crystallin. *Neuropathol. Appl. Neurobiol.* **18**, 341–350.

Manetto, V., Perry, G., Tabaton, M., Mulvihill, P., Fried, V.A., Smith, H.T., Gambetti, P. & Autilio-Gambetti, L. (1988) Ubiquitin is associated with abnormal cytoplasmic filaments characteristic of neurodegenerative diseases. *Proc. Natl Acad. Sci. USA* **85**, 4501–4505.

Masliah, E., Hansen, L.A., Quijada, S., DeTeresa, R., Alford, M., Kauss, J. & Terry, R.D. (1991) Late onset dementia with argyrophilic grains and subcortical tangles or atypical progressive supranuclear palsy. *Ann. Neurol.* **29**, 389–396.

Mattiace, L.A., Kress, Y., Davies, P., Ksiezak-Reding, H., Yen, S.-H. & Dickson, D.W. (1991a) Ubiquitin-immunoreactive dystrophic neurites in Down's Syndrome brains. *J. Neuropathol. Exper. Neurol.* **50**, 547–559.

Mattiace, L.A., Wu, E., Aronson, M. & Dickson, D.W. (1991b) A new type of neuritic plaque without amyloid in corticonigral degeneration with neuronal achromasia. *J. Neuropathol. Exp. Neurol.* **50**, 312.

Matus, A. (1988) Microtubule-associated proteins: Their potential role in determining neuronal morphology. *Annu. Rev. Neurosci.* **11**, 29–44.

Mayer, R.J., Arnold, J., László, L., Landon, M. & Lowe, J. (1991) Ubiquitin in health and disease. *Biochim. Biophys. Acta* **1089**, 141–157.

Mori, H., Kondo, J. & Ihara, Y. (1987) Ubiquitin is a component of paired helical filaments in Alzheimer's disease. *Science* **235**, 1641–1644.

Murti, K.G., Smith, H.T. & Fried, V.A. (1988) Ubiquitin is a component of the microtubule network. *Proc. Natl Acad. Sci. USA* **85**, 3019–3023.

Nowak, T.S. Jr. (1990) Protein synthesis and the heat-shock/stress response after ischaemia. *Cerebrovasc. Brain Metab. Rev.* **2**, 345–366.

Papp, M.I. Kahn, J.E. & Lantos, P.L. (1989) Glial cytoplasmic inclusions in the CNS of patients with multiple system atrophy (striatonigral degeneration, olivopontocerebellar degeneration and Shy-Drager syndrome). *J. Neurol. Sci.* **94**, 79–100.

Parag, H.A., Raboy, B. & Kulka, R.G. (1987) Effect of heat shock on protein degradation in mammalian cells: Involvement of the ubiquitin system. *EMBO J.* **6**, 55–61.

Perez, N., Sugar, J., Charya, S., Johnson, G., Merril, C., Bierer, L., Perl, D., Haroutunian, V. & Wallace, W. (1991) Increased synthesis and accumulation of heat shock 70 proteins in Alzheimer's disease. *Mol. Brain Res.* **11**, 249–254.

Perry, G., Friedman, R., Shaw, G. & Chau, V. (1987) Ubiquitin is detected in neurofibrillary tangles and senile plaque neurites of Alzheimer disease brains. *Proc. Natl Acad. Sci. USA* **84**, 3033–3036.

Pollanen, M.S., Dickson, D.W. & Bergeron, C. (1993) Pathology and biology of the Lewy body. *J. Neuropathol. Exp. Neurol.* **52**, 183–191.

Rechsteiner, M. (1991) Natural substrates of the ubiquitin proteolytic pathway. *Cell* **66**, 615–618.

Schmidt, M.L., Murray, J., Lee, V.M.-Y., Hill, W.D., Wertkin, A. & Trojanowski, J.Q. (1991) Epitope map of neurofilament protein domains in cortical and peripheral nervous system Lewy bodies. *Am. J. Pathol.* **139**, 53–65.

Skalli, O., Chou, Y.H. & Goldman, R.G. (1992) Cell cycle-dependent changes in the organization of an intermediate filament-associated protein: Correlation with phosphorylation by $p34^{cdc2}$. *Proc. Natl Acad. Sci USA* **89**, 11 959–11 963.

Soifer, D. (1986) *Dynamic Aspects of Microtubule Biology. Annals of the New York Academy of Sciences*, Vol. **466**, New York.

Steiner, P.M. & Roop, D.R. (1988) Molecular and cellular biology of intermediate filaments. *Annu. Rev. Biochem.* **57**, 593–625.

Sternberger, L.A. & Sternberger, N.H. (1982) Monoclonal antibodies distinguish phosphorylated and nonphosphorylated forms of neurofilaments *in situ. Proc. Natl Acad. Sci. USA* **80**, 6126–6130.

Stewart, M. (1990) Intermediate filaments: Structure, assembly and molecular interactions. *Curr. Opinion Cell Biol.* **2**, 91–100.

Suenaga, T., Hirano, A., Llena, J.F., Ksiezak-Reding, H., Yen, S.-H. & Dickson, D.W. (1990a) Modified Bielschowsky and immunohistochemical studies on cerebellar plaques in Alzheimer's disease, *J. Neuropathol. Exper. Neurol.* **49**, 31–40.

Suenaga, T., Hirano, A., Llena, J.F., Ksiezak-Reding, H., Yen, S.-H. & Dickson, D.W. (1990b) Ubiquitin immunoreactivity in kuru plaques in Creutzfeldt-Jakob disease. *Ann. Neurol.* **28**, 174–176.

Wallace, W., Johnson, G., Merril, C., Sugar, J. & Refolo, L. (1992) Potential role of HSP 72/73 in the reversible phosphorylation of tau to a form A68 in heat shocked neuronal PC12 cells. *Soc. Neurosci. Abstr.* **18**, 199.

Wang, E., Fischman, D., Liem, R.K.H. & Sun, T.T. (1985) Intermediate Filaments, *Ann. NY Acad. Sci.* Vol. **445**, New York.

Welch, W.J., Feramisco, J.R. & Blose, S.E. (1985) The mammalian stress response and the cytoskeleton: Alterations in intermediate filaments. In *Intermediate Filaments* (eds. Wang, E., Fischman, D., Liem, R.K.H. & Sun, T.-T.), pp. 57–67. *Ann. NY. Acad. Sci.*, Vol. **445**, New York.

Wightman, G., Anderson, V.E., Martin, J., Swash, M., Anderton, B.H., Neary, D., Mann, D., Luthert, P. & Leigh, P.N. (1992) Hippocampal and neocortical ubiquitin-immunoreactive inclusions in amyotrophic lateral sclerosis with dementia. *Neurosci. Lett.* **139**, 269–274.

Wolozin, B. & Davies, P. (1987) Alzheimer-related neuronal protein A68: Specificity and distribution. *Ann. Neurol.* **22**, 521–526.

Yen, S.-H., Dickson, D.W., Crowe, A., Butler, M. & Shelanski, M.L. (1987) Alzheimer neurofibrillary tangles contain unique epitopes, and epitopes in common with heat-stable microtubule-associated proteins, MAP_2 and tau. *Am. J. Pathol.* **126**, 63–73.

Yen, S.-H., Dickson, D.W., Peterson, C. & Goldman, J.E. (1986) Cytoskeletal abnormalities in neuropathology. In *Progress in Neuropathology*, Vol. 6 (ed. Zimmerman, H.M.) pp. 63–90. New York, Raven Press.

CHAPTER 11

UBIQUITIN, ENDOSOMES-LYSOSOMES AND NEURODEGENERATIVE DISEASES

M. Landon, J. Lowe* and R. J. Mayer

*Departments of Biochemistry and *Pathology, University of Nottingham Medical School, Queen's Medical Centre, Nottingham NG7 2UH, UK*

Table of Contents

Heat Shock Proteins in the Nervous System
ISBN 0-12-480960-X

11.1 The context

The neuron is a very special non-dividing, polarized cell, having a requirement for a life-span at least equal to that of the animal and with much of its essential activities taking place at a distance from the cell body, where the bulk of the biosynthetic and degradative enzymes reside. Consequently, reliable transport along cytoskeletal elements is crucial to neuronal survival. Just as ageing in man leads to declining physical abilities, it is easy to believe that transport might become compromised in aged neurons. Most of the human neurodegenerative diseases occur in middle- to late-life and parallel other physical and cellular features of ageing. Neurons exist in environments where exposure to external stresses is limited; nevertheless, low-level chronic stress is believed to underlie many of the neurodegenerative disorders. In view of the importance of a functioning nervous system to the survival of an organism, one would expect neuronal protective responses to stress to be well developed since cell replacement is not an option.

The first of the two preceding chapters has established the position of ubiquitin as a cell stress (heat shock) protein: the second has given a comprehensive survey of ubiquitin in relation to the neuropathological features of the diseases that will also be covered here, albeit from a different viewpoint. Some of the diseases have entirely unknown causes, while others are clearly inherited, often in an autosomal dominant manner, with the affected gene identified in certain cases; some are both heritable and idiopathic; the prion diseases are transmissible and of especial interest since the causative agent is apparently a host protein, as evidenced by close linkage to mutations in heritable forms that are also transmissible. Here we wish to look at the common features of the molecular pathology of these diseases that generate the 'how' and 'why' questions for which we still have few definitive answers.

Finally, in setting the scene, the classical concept of the lysosome as a single entity needs to be displaced by a complex picture of a family of acid compartment organelles, which develop by budding and fusion events that are not fully understood. Here, the terms endosomes-lysosomes or lysosome-related will be used to embrace all members of this family.

1.2 Mechanisms of neurodegeneration

Molecular genetics is providing some unexpected developments in linking mutant genes with particular inherited disorders. Long-standing suspicions that major diseases, such as Alzheimer's disease (AD) are heterogeneous have been clearly substantiated by linkage with distinct gene products; early onset of pathology is explained by abnormal forms of Alzheimer amyloid precursor protein (APP) coded on chromosome 21 (Chartier-Harlin *et al.*, 1991; Goate *et al.*, 1991; Murrell *et al.*, 1991), by mutations in an as yet unidentified gene on chromosome 14 (Schellenberg *et al.*, 1992) or by over-expression of normal APP in Down's syndrome (trisomy of chromosome 21); late-onset inherited disease appears to be a consequence of an abnormality on chromosome 19 (Pericak-Vance *et al.*, 1991), recently identified as allelic variation at the apolipoprotein E *locus* and probably also of relevance in sporadic cases (Corder *et al.*, 1993). These alternative causations are only the current starting points and our understanding of the influences of environmental factors, and, in the longer term, possibilities of directed development of novel therapies, can only be enhanced by increasing knowledge of, for example, the precise cellular role of APP.

Other neurodegenerative diseases are also linked with proteins of unknown function that leave large gaps in our understanding of how the mutant form might lead to the pathology. Thus, just as mutant molecules of APP place amyloid formation at the centre of certain types of familial AD, the familial forms of human prion diseases, Creutzfeldt-Jakob disease and Gerstmann-Straussler syndrome, are now known to reflect a range of point mutations or inserts in the prion protein (PrP) gene (Prusiner, 1992) and post-translationally modified PrP is at the focus of biochemical attention in these disorders (Prusiner, 1991; Prusiner, 1992). Despite recognition of the nature and importance of variant proteins, there is a long way to go in completing our understanding of how the diseases progress.

In the case of familial motor neurone disease, a mutant form of superoxide dismutase makes an obvious connection with free radical-induced stress (Rosen *et al.*, 1993), long thought to be involved in neurodegeneration. Mitochondrial malfunction, through MPTP metabolites (Langston *et al.*, 1983) or possibly consequent upon changes in mitochondrial DNA (Linnane *et al.*, 1989), appears to underlie some forms of Parkinson's disease and shows that the neuron can be vulnerable to insufficiencies in oxidative metabolism.

Many of the pathological processes in the whole range of neurodegenerative diseases can clearly give rise, either directly or indirectly, to abnormal proteins; the role played by ubiquitin in the removal of abnormal proteins provides the basis of this chapter. Although we cannot claim to have a complete picture, our present understanding of the ramifications of ubiquitin in neurodegeneration can be summarised as cytoprotection; we will use this term to bring together what still appear to be disparate threads in an unfinished story.

11.3 The beginnings – ubiquitin and Alzheimer's disease

Three independent reports of the presence of ubiquitin in the neurofibrillary tangles (NFT) of AD appeared in 1987. A mixture of protein chemical (Mori *et al.*, 1987) and immunohistochemical evidence, the latter also showing that plaque neurites contain ubiquitin (Cole and Timiras, 1987; Perry *et al.*, 1987), led to further studies using anti-ubiquitin immunohistochemistry that greatly expanded on these key initial observations and indicated previously unsuspected relationships between degenerative diseases.

Before looking at the developments from this starting point, it is worth commenting on some of the immunochemical aspects of the work and it is also, perhaps, appropriate to look at what the implications were thought to be for the presence of ubiquitin in NFT. At that time NFT were still poorly defined chemically, though structurally known to consist predominantly of the paired helical filaments (PHF) first described by Kidd (1963). PHF are found only in human brain tissue and this uniqueness had a constraining effect on trying to understand their place in disease processes. A number of antisera and monoclonal antibodies had been raised against part-purified PHF and it was a search for the antigen for one such monoclonal antibody that led to the protein–chemical identification of ubiquitin (Mori *et al.*, 1987). Comparison of the monoclonal antibody with antisera known to preferentially recognise ubiquitin conjugates, rather than the free form, was indicative of conjugates in PHF; the use of antisera with the same conjugate specificity drew similar conclusions in the two independent immunohistochemical studies (Cole and Timiras, 1987; Perry *et al.*, 1987). There are now commercial antisera that preferentially recognise either free or conjugated ubiquitin. It is important to understand that essentially all of the data discussed here have been obtained by the use of antisera that recognise ubiquitin in its conjugated form (Haas and Bright, 1985).

A quite reasonable early conclusion was that the ubiquitin in PHF might be conjugated to the microtubule-associated protein tau, for whose presence there was immunochemical evidence (Brion *et al.*, 1985). A very close relationship between PHF and ubiquitin is demonstrable by immunogold electron microscopy (Figure 1A), although, interestingly, the target of ubiquitination has only recently been defined (Morishima-Kawashima *et al.*, 1993; also see below); as will become apparent, this is still not the case for most other ubiquitinated inclusions. The relationship between cell stress (heat shock) and ubiquitin was known, as was the role of ubiquitination of abnormal proteins and their degradation via non-lysosomal proteolysis. Well-characterized lysosomal connections for ubiquitin lay ahead; the appearance of ubiquitin outside the cytosolic compartment was somewhat closer in that, while extending knowledge of the cellular repertoire of ubiquitinated inclusions, it was noted that a proportion of granulovacuolar features of hippocampal cells in AD were ubiquitin positive (Lowe *et al.*, 1988). This so-called granulovacuolar degeneration was originally described by Simchowicz in 1911, but first investigated comprehensively by Woodard (1962), and consists of granular inclusions

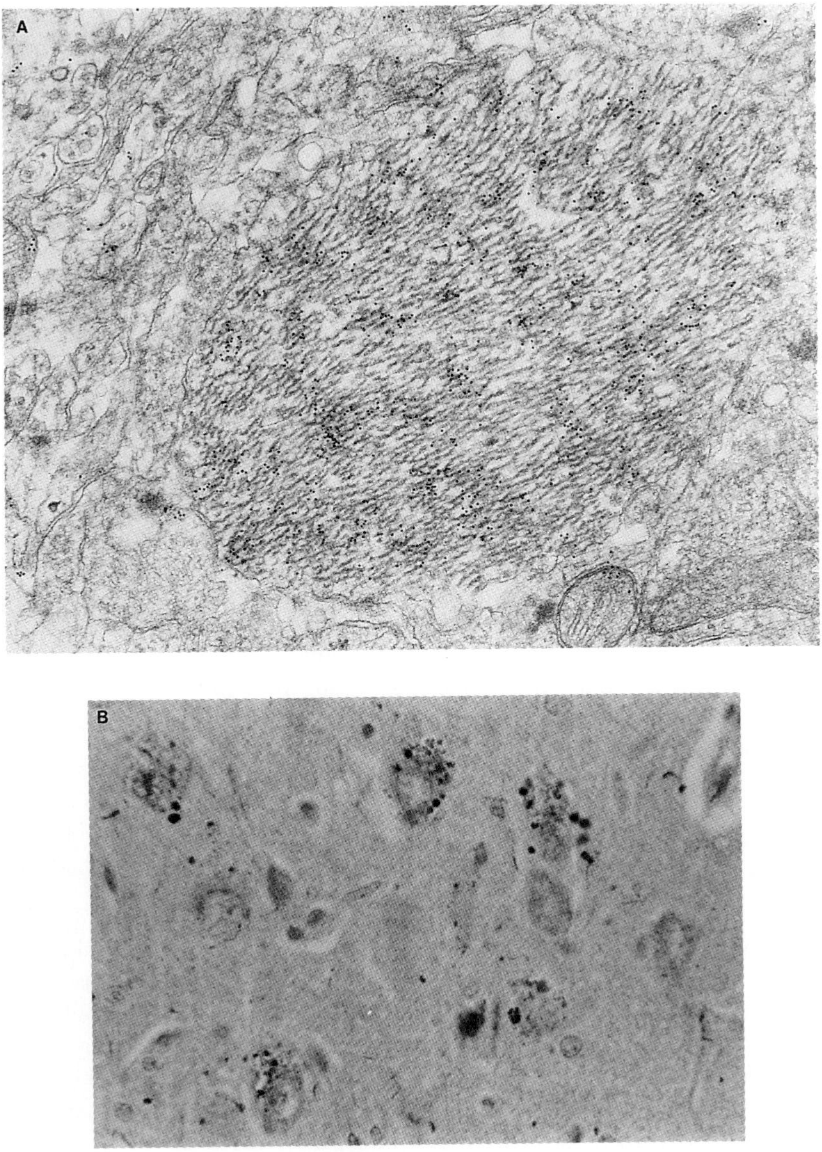

Figure 1 (A) Alzheimer's disease brain tissue biopsy sample – immunogold electron micrograph of paired helical filaments labelled for ubiquitin-protein conjugates. Mag. X47 850; (B) Granulovacuolar degeneration in Alzheimer's disease. Anti-ubiquitin-protein conjugate immunostaining.

within vacuoles (Figure 1B), thus representing an intracellular compartment. The appearance of ubiquitin-positive inclusions in a membrane-limited compartment in a cell contrasted both with the ubiquitin-positive non-membrane limited inclusions seen elsewhere and, moreover, with the perceived normal cytosolic location and roles of ubiquitin. The main emphasis of this chapter thus contrasts with the previous one where the relationship between ubiquitin and cytoskeletal elements in cellular inclusions provided the major focus.

11.4 Ubiquitin in association with neurodegenerative and other diseases – a developing story

A flurry of publications provided evidence for the presence of ubiquitin conjugates in well-characterised inclusions and allowed identification of others that had been poorly defined previously or were totally novel (Lowe *et al.*, 1988; Manetto *et al.*, 1988; Mayer *et al.*, 1991; Lowe *et al.*, 1993). The previous chapter has discussed the relationship between ubiquitin and cytoskeletal elements in inclusions in neuronal and glial cells; ubiquitin-conjugate immunoreactivity is also seen with a cytoskeletal involvement in cells outside the brain in relation to defined stresses, such as in alcoholic liver disease where Mallory bodies are found within hepatocytes (an animal model of this is produced by feeding griseofulvin to mice; see Section 11.5), and finally, and most importantly in relation to this chapter, in cells in the absence of an obvious link with the cytoskeleton.

11.5 The ageing connection

As has been noted above, ubiquitin was first observed in a pathological situation outside the cytosol in granulovacuolar degeneration. With widespread use of anti-ubiquitin immunocytochemistry it quickly became apparent that punctate reactivity was detectable in apparently normal human brain tissue and increased in an age-dependent manner (Leigh *et al.*, 1989; Pappolla *et al.*, 1989; Dickson *et al.*, 1990b; Migheli *et al.*, 1992). Dot-like immunoreactivity, as it has been named, is a consequence (and marker) of neuritic dystrophy, most visible in the white matter and, much less-commonly, in the neuropil in the deeper cortical layers of the grey matter (Figure 2A); some also reflects focal enlargement of the myelin sheath. As well as in man, it is seen in animals such as dogs (Migheli *et al.*, 1992) and mice (Lowe *et al.*, 1992; Figure 2B), establishing that it is related to the lifetime of the animal and not age *per se*. When visualised in the electron microscope, the dot-like structures can be seen to be lysosome-related dense, multilamellar or multivesicular bodies (Dickson *et al.*, 1990a,b; Migheli *et al.*, 1992). Unambiguous linkage with ageing is complicated in the same way as finding the pathological features of AD

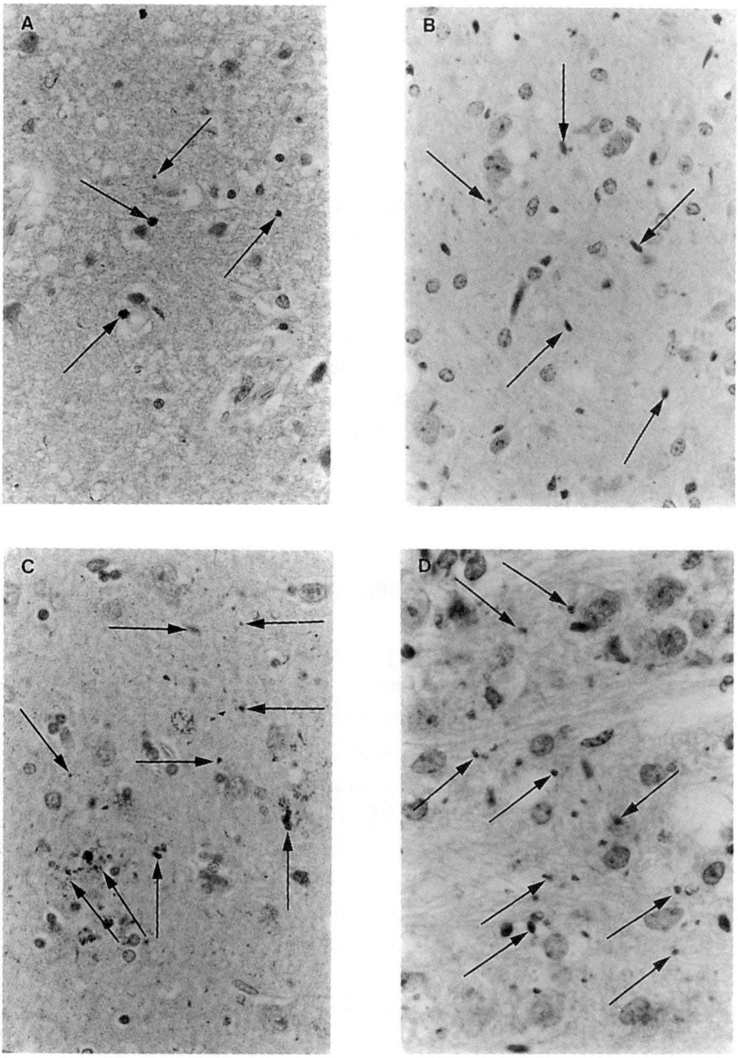

Figure 2 Anti-ubiquitin-protein conjugate dot-like immunoreactivity in: (A) Normal aged human brain; (B) Normal aged mouse brain; (C) Alzheimer's disease brain; (D) ME7-scrapie-infected mouse brain.

at a low level in most aged brains; indeed, the dot-like reactivity is increased in AD where dystrophic neurites are widespread (Figure 2C), sometimes associated with PHF and often seen peripheral to amyloid-containing plaques. Dot-like reactivity is also increased in the brains of scrapie-infected mice (Figure 2D), a disease context that is expanded upon in the next section below. With these observations in mind, one might see dot-like ubiquitin-positive dystrophic neurites as an inevitable consequence of ageing, as well as being the early pathological event in a variety of disease processes that involve difficulties in the lysosomal handling of proteins. This latter aspect derives from data that are brought together and discussed in later parts of this chapter.

11.6 Ubiquitin-positive immunoreactive structures are not simply representative of terminal events

There are great difficulties in interpreting the chronology of pathological features that precede neuronal death in chronic neurodegenerative diseases in man. The observations made on post-mortem material merely represent a snap-shot of the process, with preceding events inaccessible. For studies on AD it is possible to exploit the parallels with Down's syndrome (trisomy of chromosome 21), where all cases will develop the characteristic pathology of plaques and NFT in their fourth decade (Oliver and Holland, 1986); there are, of course, still limitations due to the paucity of staged examples and human variability. In this respect the prion diseases are unique, encompassing both human and animal direct counterparts, with the latter providing material for studies of all stages of the disease process in a reproducible host.

Ubiquitin immunoreactivity has, in particular, been associated with NFT of AD; these structures are known from studies on Down's syndrome to occur later than plaques (Mann and Esiri, 1989); a similar view prevails from the perceived central role of amyloid in AD (Hardy and Allsop, 1991) now reinforced for some cases of early-onset familial AD with the delineation of mutations in the APP (Chartier-Harlin *et al.*, 1991; Goate *et al.*, 1991; Murrell *et al.*, 1991). Other observations are not incompatible with ubiquitin-positive inclusions representing late events in a prolonged disease process. This background has generated a common view of ubiquitin being involved in reactions that are both secondary and late. Where ubiquitin immunoreactivity is seen earlier, as in relation to ageing, it has been interpreted by others as a distinct phenomenon.

There are, as noted above, substantial advantages in studying prion diseases in animals; all stages of the disease are available in properly preserved tissue, thus allowing ultrastructural investigations. Ubiquitin immunoreactivity is seen in terminal mouse scrapie at the light microscope level in three forms: in dot-like structures diffusely present throughout the neuropil (Figure 2D); in a similar form, but localised around amyloid plaques; and as granular material within or adjacent to neuronal cell bodies (Lowe *et al.*, 1990a). The dot-like lesions can be picked up very early in the disease process and their ultrastructural appearance (Lowe *et al.*, 1992) is similar to the equivalent deposits in terminal cases (Laszlo *et al.*, 1992; Figure 3A). Is scrapie typical of the other diseases where early events are inaccessible? Clearly, at present it is not possible to respond unequivocally; however, in view of the many similarities between roles for ubiquitin in scrapie and other diseases, a positive answer seems much more likely than one in the negative.

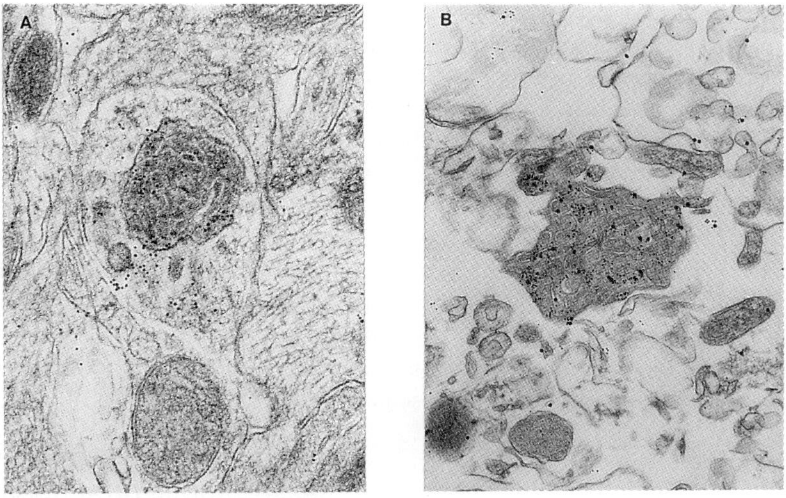

Figure 3 Immunogold-labelling electron micrographs of: (A) ME7-scrapie-infected mouse brain tissue labelled for ubiquitin-protein conjugates. Mag. X48 620; (B) Subcellular fraction isolated from ME7-scrapie-infected mouse brain and double labelled for ubiquitin-protein conjugates (10 nm gold particles) and for PrP^{SC} (20 nm gold particles). Mag X38 610.

11.7 Ubiquitin and endosomes-lysosomes

11.7.1 Studies on cells

Free ubiquitin had been identified in the lysosomes of hepatoma cells by immunogold electron microscopy (Schwartz *et al.*, 1988). However, our interest in protein turnover and the unexpected presence of ubiquitin-conjugate immunoreactivity in granulovacuolar degeneration prompted us to investigate the possibility of entry of such conjugates into the lysosomal compartment of perturbed or normal cells. In the first studies on fibroblasts treated with the thiol protease inhibitor E64, which is known to gain entry to lysosomes via cellular pinocytosis where it severely compromises proteolysis, it was apparent that ubiquitinated-proteins were detectable within those organelles by immunocytochemistry and by subcellular fractionation coupled with Western blotting (Doherty *et al.*, 1989). Normal fibroblasts, as assessed by similar methodologies, were also found to have ubiquitin-protein conjugates enriched within lysosomal compartments, a result that was subsequently confirmed by immunogold electron microscopy (Laszlo *et al.*, 1990). The high level of enrichment seen argues against non-specific entry through autophagy, a known mechanism (Finley and Chau, 1991), explaining the appearance of the conjugates within such organelles. Further studies have added to the weight of data in favour of ubiquitin being associated in some way with lysosomes; thus, lysosome-related or lysosomal links with ubiquitin have been seen in other cell types investigated, including polymorphonuclear neutrophils (Laszlo *et al.*, 1991) and Epstein-Barr virus-transformed lymphoblastoid cells (Laszlo *et al.*, 1991); the latter will be discussed further below. There has for some time been a sufficient body of evidence to adduce that ubiquitin is central to lysosomal as well as non-lysosomal degradation of proteins (Mayer *et al.*, 1991).

Recent developments reinforce the ties between ubiquitin and cellular compartments. Ubiquitin activation appears to be essential for the lysosomal degradation of proteins, since lysosomes in cells with a temperature-sensitive mutation in the ubiquitin-activating enzyme cannot undergo normal maturation events at the non-permissive temperature (Gropper *et al.*, 1991). A role in the transfer of vesicular enzymes into multivesicular endosomes has also been identified (Low *et al.*, 1993).

11.7.2 Human neurodegenerative diseases and central nervous system amyloid

Amyloid is a generic term for extracellular deposits of filamentous protein material commonly defined pathologically by specific dye staining e.g. Congo red. In the amyloidoses, a range of protein precursors, commonly normal or mutant serum proteins, give rise to amyloid in disease-related or inherited situations. Most often the amyloidoses affect either the brain or other tissues with little overlap. Cerebral amyloid is found in associaton with the walls of both meningeal and cortical vessels (amyloid angiopathy), as well as within the brain parenchyma in a variety of forms termed plaques.

11.7.3 Amyloid in prion diseases

The amyloid found in the form of plaques in prion diseases is derived from the prion protein (PrP), which is one of the family of glycosyl-phosphatidylinositol anchored (glypiated), cell-membrane molecules. Differences in the properties of the normal cellular form, designated PrP^{C}, and that associated with disease, PrP^{SC} (where SC stands for scrapie, the archetype prion disease of sheep and goats) are believed to be conformationally determined (Prusiner, 1991, 1992). In man the PrP gene is found on chromosome 20 and codes for an initial transcript containing a conventional signal sequence at the amino terminus, a second one at the other end of the molecule for the attachment of the lipid anchor and sites for glycosylation. Removal of the first signal sequence, attachment of carbohydrate and the events involved in glypiation leave a glycosylated polypeptide chain of around 200 residues in human PrP^{C}. Full-length molecules occur in disease-related scrapie-associated fibrils (Hope *et al.*, 1986), but shorter polypeptides have been isolated from plaques in one form of the inherited human disease Gerstmann–Straussler syndrome (Tagliavini *et al.*, 1991). Close association between PrP and disease is strongly supported by the recent development of mice that do not have a PrP gene (Bueler *et al.*, 1992) and which show a predicted resistance to prion infection (Bueler *et al.*, 1993).

11.7.4 Amyloid in Alzheimer's disease (and Lewy body-dementia)

Although amyloid is found in vessels in almost all cases of AD, most current work is addressed towards parenchymal deposits which take a variety of forms. All of the amyloid is derived from a much larger precursor, APP, which is synthesized by alternative splicing of exons in a single gene on chromosome 21. Three major types of APP can give rise to amyloid: APP_{695} (Kang *et al.*, 1987) is the predominant form in brain; APP_{751} and APP_{770} each contain a Kunitz-type protease inhibitor domain (Kitaguchi *et al.*, 1988; Ponte *et al.*, 1988; Tanzi *et al.*, 1988). All appear to be cell surface proteins with a putative single membrane-spanning helix towards the carboxyl terminus of the polypeptide chain, allowing most of the protein to sit on the external side of the plasma membrane (Kang *et al.*, 1987). The bulk of the amyloid material outside the vessels is a 42–43 amino acid fragment of APP terminating at its carboxyl residue within the membrane-spanning region (Kang *et al.*, 1987); that associated with vessels is 39 amino acids long (Prelli *et al.*, 1988), lacking some of the hydrophobic residues at the carboxyl terminus of the longer molecule. Both forms have been termed β-protein or β/A4; Aβ (where A stands for amyloid) is the recently recommended name within amyloid nomenclature.

The original procedures that were employed to isolate plaque cores often used as starting material, brains that were rich in plaques with few NFT, since the latter would contaminate purified plaque cores; in retrospect, such brains would most likely be from cases of Lewy body-dementia. There is good reason to believe that the same Aβ-pathological material occurs in both diseases.

11.7.5 Roles for endosomes-lysosomes in prion diseases

As has been outlined above, one of the cardinal features of prion pathology is the accumulation of PrP^{SC}, which is intimately related to amyloid deposition. PrP^{SC} also increases in amount in the absence of overt amyloid in the prion diseases where plaque formation is absent. In such cases, a second cardinal feature comes to the fore as spongiosis, the appearance of vacuoles within neurons and their expansion into larger regions where membrane-limited spaces leave the tissue looking like a sponge, from which the alternative earlier name, spongiform encephalopathy, is derived. Although there are occasions when amyloid accumulation and spongiform change appear to be mutually exclusive (a characteristic that we will return to below), on the basis of pathological parsimony, it has appeared likely that derivation of PrP^{SC} from PrP^{C} underlies both of these key features of the prion diseases; as will become apparent, ties between possible sites of conversion of PrP^{C} to PrP^{SC} and ubiquitin seem to provide the connection.

Much work has been performed on trying to identify that aspect of the PrP^{SC}

molecule that distinguishes it from its normal cellular counterpart. At present the consensus is that the absence of detectable post-translational chemical modification leaves conformation as the distinguishing feature (Prusiner, 1991, 1992). The structural requirements within PrP^{C} for conversion to PrP^{SC} have been investigated in scrapie-infected neuroblastoma cells in the absence of significant parts of the polypeptide; such studies have eliminated the need for glycosylation (Taraboulos *et al.*, 1990), lipid anchor and attachment signal as well as other parts of the amino acid chain (Rogers *et al.*, 1993). A form of human GSS is associated with PrP truncation (Kitamoto *et al.*, 1993a,b), supporting the possibility of shorter molecules leading to conformational alterations or, alternatively, pointing to proteolytic cleavage as a key post-translational change for PrP^{SC} accumulation. Studies on scrapie-infected cells have also addressed both the timing of and possible sites for PrP^{SC} formation. Following insertion of PrP^{C} into the membrane, PrP^{SC} is formed either at the cell surface or in endosomes-lysosomes (Caughey *et al.*, 1989, 1991; McKinley *et al.*, 1991; Borchelt *et al.*, 1992) after uptake, probably via caveolae, the newly-identified uptake system for glypiated molecules; the majority of the PrP^{SC} accumulates within cells, mainly within cytosolic compartments identified as endosomes-lysosomes (Taraboulos *et al.*, 1990).

How does the situation in cells compare with observations on scrapie-infected mouse brain tissue? Dot-like lesions, immunoreactive for ubiquitin, can be picked up very early in the disease process, at a time comparable to that at which PrP^{SC} is first detectable by Western blotting (Lowe *et al.*, 1992). These observations allow us to state that, within the current limits of detection, ubiquitin-related pathology parallels the appearance of PrP^{SC}: clearly there is a very close relationship between the primary disease event in scrapie- and ubiquitin-mediated processes.

Some questions then arise: how closely is ubiquitin linked to the production of PrP^{SC}? If there is close linkage, what are the processes that provide the linkage? Answers to these questions can come from ultrastructural studies and, in contrast to human tissue, well-preserved material can be viewed at all stages of mouse scrapie. Immunogold electron microscopy has provided evidence for the appearance of ubiquitin conjugates (Figure 3A), hsp70 and PrP^{SC} in a range of lysosome-related compartments, also marked by β-glucuronidase (Laszlo *et al.*, 1992).

One should note at this point that there is no evidence for the covalent attachment of ubiquitin to PrP; indeed, Western blotting of scrapie-associated fibril preparations shows no reaction with anti-ubiquitin antisera (Lowe *et al.*, 1990a). However, tracing the specific protein to which ubiquitin is attached in inclusions has proved difficult and it would be unwise to suggest that conjugation of ubiquitin to PrP does not take place.

In current studies on subcellular fractions obtained from scrapie-infected mouse brain ubiquitin-conjugates are seen to co-localise with PrP^{SC} in isolated endosome-lysosome organelles (Figure 3B) that have the same appearance as structures seen in tissue sections. These observations reinforce proposals that PrP^{SC} is formed in endosomes-lysosomes, with the organelle acting as a 'bioreactor' (Laszlo *et al.*, 1992). The largely extracellular location of PrP^{SC} could arise by ejection from lysosomes, an established process, or as a consequence of disruption of organelle structure through overloading. Re-uptake by neurons or within neurons into endosome-lysosomes would allow constant interaction between PrP^{C} and PrP^{SC} in a location that would be conducive to conformation changes. Release of hydrolytic enzymes from disrupted organelles has been pictured and used to explain spongiosis (Laszlo *et al.*, 1992).

Perhaps the balance between ejection and disruption determines differences in pathological features; ejection favouring plaque accumulation while spongiosis is accounted for by more intracellular disruption. Alternatively, such differences in pathological features might be consequent upon effects in different cell types: or within them, where reactions in neuronal processes favour amyloid deposition, and organelle disruption in the cell body leads to increased spongiosis. Finally, while speculating about explanations for disease variability, it is worth commenting on the current attention being paid to the possible amyloidogenicity of fragments of PrP. As was briefly alluded to above, proteolytic fragmentation of PrP could be the post-translational modification underlying formation of PrP^{SC}; many amyloids arise through proteolytic cleavage of normal soluble molecules and coexistence of different sized molecules is seen in some amyloid deposits (Westermark *et al.*, 1976; Cornwell *et al.*, 1977). Perhaps 'strains' of agent (also termed isolates) reflect different sizes of fragment. Whatever renders these peptides amyloidogenic, there is still every reason to see the endosome-lysosome as a 'bioreactor' site for essential truncation, if that is a necessary prelude to refolding.

11.7.6 Roles for endosomes-lysosomes in Alzheimer's disease

Following the delineation of cell-surface proteolytic events related to the release of soluble forms of APP (Esch *et al.*, 1990), a great deal of interest has centred on the formation of truncated molecules that might be involved in Aβ-production (Golde *et al.*, 1992; Haass *et al.*, 1992; Seubert *et al.*, 1993). Most attention has been directed towards the amino terminus of Aβ; although just as important, cleavage to form the carboxyl terminus of Aβ has been relatively neglected, both experimentally and in the discussion of amyloid generation. While the production of the amino terminus of Aβ is a straightforward hydrolytic cleavage within an exposed part of the polypeptide chain of APP, generation of the carboxyl terminus must occur within the putative membrane-spanning segment. It seems highly unlikely that hydrolysis occurs within that intramembranous hydrophobic environment and one should surely look at locations where the protein is disentangled from the lipid bilayer for proteases that might be involved in this part of Aβ-formation.

The compartment where all integral membrane proteins are turned over is the endosome-lysosome. Thus, while proteolysis external to the cell membrane may either preclude formation of Aβ or generate the amino terminus of that molecule, it is most probable that completion of amyloid formation, through scission at the carboxyl terminus from intact or truncated APP is an internal compartmentalised event.

A view contrary to this has arisen from detection of extracellular lysosomal proteases in association with plaques (Cataldo *et al.*, 1991). Here the possibility of APP processing by released lysosomal proteases is viewed downstream of membrane disruption (through other hydrolases released concurrently), as a means of providing access to the intra-bilayer peptide bonds. It is noteworthy, however, that in the same study elevated levels of lysosomal proteases were detected within particular neurons; since internalized APP will enter endosome–lysosome compartments, the elevated levels of enzymes and presumably lysosomes, would be available for intracellular processing. Figure 4A shows plaques immunohistochemically stained for ubiquitin and, in Figure 4B, the similar pattern of staining for cathepsin D, which again emphasises the dystrophic neurites, here showing bulbous morphology, around the periphery of the plaque.

11.7.7 Ubiquitin and endosomes-lysosomes in Alzheimer's disease

Most often ubiquitin is associated with the intracellular PHF in cell bodies or in dystrophic neurites, with those around plaques being distant from bulk Aβ. Close links between ubiquitin and APP comparable to those described above for PrP using electron microscopy would be difficult to obtain without suitable fresh tissue. Spongiosis or vacuolation are not uncommon in AD (Smith *et al.*, 1987) or in Lewy body-dementia (Hansen *et al.*, 1989) and could occur through endosome-lysosome disruption mechanisms comparable to those outlined above for scrapie, but caused by APP/Aβ effects.

11.8 Dot-like ubiquitin-positive immunoreactive structures as the harbinger of cell death

As has been outlined above, dot-like ubiquitin-positive dystrophic neurites can be seen in what is thought of as normal ageing. If we consider the evidence for protein handling problems in AD and prion diseases it is possible to suggest that both diseases represent a consequence of an increase in the difficulty of degrading intrinsically hazardous proteins. A corollary to this posit is that the normal situation cannot be of zero-risk and the pathological features of ageing may be a consequence of a low-level of what would, at a higher level, be seen as a disease-related process. The relative incidences of AD and prion diseases make this argument more acceptable for AD which is much more common in the human situation; studies on animal prion diseases can present an argument of equal validity.

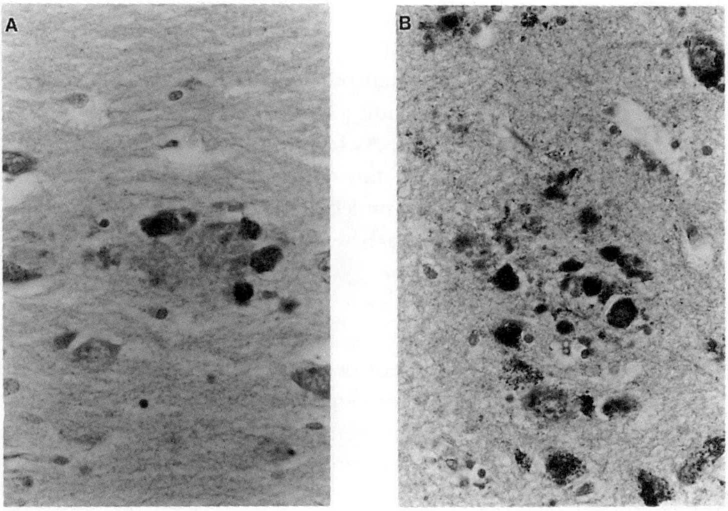

Figure 4 Plaques in Alzheimer's disease brain tissue immunostained for: (A) ubiquitin-protein conjugates and (B) cathepsin D.

11.9 Ubiquitin and inclusion body genesis

11.9.1 Ubiquitinated proteins in inclusion bodies

Although there has been a wealth of immunohistochemical observations showing conjugated ubiquitin in a wide range of inclusions, there is, in contrast, a notable paucity of chemical data identifying the target proteins to which ubiquitin is attached. The rarity of many inclusions would necessitate enormous enrichment from starting tissue in order to provide adequate quantities of material for full characterization. It is a biochemical fact that ease of purification relates almost directly to initial levels, and it is not unexpected, therefore, to find that the two instances of chemically-identified ubiquitin conjugates described below come from tissues especially rich in the pathological entities. It is, perhaps, not surprising to find that purification has been difficult when samples enriched in ubiquitin-conjugates are looked at in Western blots using anti-ubiquitin conjugate antisera and seen as smears (often of high molecular weight) rather than discrete bands. Such results are compatible with variable mono- or poly-ubiquitination, variable multiple-site attachment and partial degradation of ubiquitinated structures: such processes mitigate against the isolation of a single species for characterization.

11.9.1.1 Ubiquitinated tau

Since the ubiquitin story moved into neurodegenerative diseases through its identification in the PHF of AD, it is satisfying to have, at last, a description of the biochemical nature of the target material. As had been expected from a clarification of the large structural contribution of tau to PHF, ubiquitin is attached to that protein, mainly in the form of single molecules and at sites localized within the microtubule-binding region (Morishima-Kawashima *et al.*, 1993).

11.9.1.2 Ubiquitinated αB-crystallin

Rosenthal fibres accumulate in reactive or neoplastic astrocytes, but are more abundant still in Alexander's disease, a rare central nervous system degenerative disorder in children. Rosenthal fibres have been shown to be ubiquitin-positive as well as reactive for αB-crystallin, a major lens protein, but now also known to be more widely distributed in normal tissues; the detection of αB-crystallin in other pathological situations has been described in Chapter 10. The identification of mono- and poly-ubiquitinated forms of αB-crystallin in extracts of tissue from a case of Alexander's disease (Goldman and Corbin, 1991) represents the second situation where the target for ubiquitin within inclusions has been established.

11.9.2 Model systems

11.9.2.1 Cells

Cells stressed in a variety of ways respond by collapse of the intermediate filament network to a peri-nuclear location, while the distribution of microtubules is largely unaffected. Removal of the stress leads to recovery of the normal integrated peripheral cytoskeleton.

Epstein-Barr virus-transformed lymphoblastoid cells reflect expression of viral latent membrane-transforming protein-1 (LMP-1) by the production of intracellular aggregates that mimic some of the features of the inclusions of neurodegenerative diseases (Laszlo *et al.*, 1991). Accumulations enriched in LMP-1, vimentin, hsp70 and ubiquitin-protein conjugates can be seen by light microscopy. At the electron microscope level, the aggregates can be seen to contain ubiquitinated-proteins and hsp70 within lysosome-related structures. Once again, as with AD and prion diseases, there is a response to handling a single membrane protein, in this case viral.

11.9.2.2 Animals

Satisfactory animal models for neurodegenerative diseases are rare. A model for motor neurone disease, the motor neurone degeneration (*Mnd*) mutant mouse, has been shown to exhibit a range of ubiquitin-positive pathological entities, including abnormal structural elements; many of the features appeared throughout the life of the animal (Mazurkiewicz, 1991).

Although not directly in the remit of this chapter, the ability to induce Mallory bodies in mouse liver by feeding griseofulvin provides a useful animal system for ubiquitinated-inclusion accumulation (Ohta *et al.*, 1988). Using this model, we have been able to obtain protein chemical evidence for ubiquitin, but disappointingly could not identify the target protein to which it was attached (Bailey and Mayer, unpublished observations).

11.10 Links with other non-cytoskeletal proteins

A number of non-cytoskeletal proteins, including hsp70, αB-crystallin and PGP9.5 (one of a family of ubiquitin carboxyl-terminal hydrolases that are required in a number of hydrolytic reactions to generate monomeric ubiquitin from conjugated forms), have been associated with ubiquitinated inclusions in cells in immunohistochemical studies on diseased tissues. The selective appearances of PGP9.5 and αB-crystallin have been covered, along with cytoskeletal proteins, in Chapter 10; αB-crystallin has also been noted above as a definitive target for ubiquitination. While not usually seen as cytoskeletal, αB-crystallin is related to the hsp20-class of cell-stress genes (Ingolia and Craig, 1982), some of which have been shown to be associated with intermediate filaments in *Drosophila* (Leicht *et al.*, 1986).

Non-cytoskeletal proteins have usually been found on a selective basis, thus providing the thrust for arguments to distinguish the types of inclusion or the nature of inclusions with regard to temporal aspects of the degenerative process. In this way, the presence of PGP9.5 has been adduced to reflect an active or earlier type of aggregate, as typified by a cortical Lewy body, while its absence from most NFT gave a picture of that structure as an inactive, terminal-state inclusion (Lowe *et al.*, 1990b).

The hsp70 family includes members that are required for protein entry into lysosomes (Chiang *et al.*, 1989; Terlecky *et al.*, 1992) and it was no surprise to find the involvements in scrapie and Epstein-Barr virus-transformed lymphoblastoid cells that have been outlined above. Intriguingly, a member of the hsp70 family is coded for on chromosome 14, in the same region identified by genetic linkage studies as the probable site for most cases of early-onset familial AD (Schellenberg *et al.*, 1992).

11.11 Gene expression studies

Elevated expression of poly-ubiquitin genes has been shown in two situations. In the first, despite the complications of human post-mortem material, an approximate two-fold increase in the level of the poly-ubiquitin C message was detected in spinal cord from cases of motor neurone disease (Heggie *et al.*, 1989). In mouse scrapie, where tissue preservation is less problematic and all stages are accessible, the progressive nature of the disease is reflected by an exponential-like increase in expression of the equivalent mouse poly-ubiquitin gene, paralleled by that of hsp70, at the time when clinical symptoms become apparent (Kenward *et al.*, 1993).

11.12 Summary: ubiquitin and cytoprotection

Current widely held concepts about ubiquitinated inclusions, that they are simply terminal features of disease processes and merely reflect failure of ubiquitin-mediated protein degradation in the cytosol, are surely too narrow and need to be dispelled. As has been emphasised in this chapter, situations where stages in the pathology are accessible allow a more complex pattern to emerge. Early on in disease, a cytoprotective response for handling increased amounts of deleterious proteins is mounted; this appears to require the involvement of ubiquitin in protein degradation, either by providing the cell with more of the machinery in the form of endosomes–lysosomes, by targeting abnormal proteins into those organelles or, indeed, by a combination of these.

The response is not disease-specific, but should not be dismissed as uninformative merely on the thought that each particular neurodegenerative disease is unique and, therefore, not relatable to others. The previous chapter has emphasised that many apparently unrelated diseases are brought together by the accumulation of ubiquitinated inclusions with cytoskeletal elements. In previous proposals we have suggested that these represent purposeful cytoprotective aggregations (Mayer *et al.*, 1989; Mayer *et al.*, 1991), but links between these structures and the earlier stages of a ubiquitin-mediated cytoprotective response to stresses are currently rather tenuous. Transport requirements, inevitably involving cytoskeletal elements, may well provide a link with protein ubiquitination in the mounting of a response throughout the polarized neuron. In this respect, it is worth noting that ubiquitin has itself been associated with microtubules (Murti *et al.*, 1988; Mayer *et al.*, 1992b); one could speculate about the sites of ubiquitination of tau in PHF within microtubule-binding regions (Morishima-Kawashima *et al.*, 1993) suggesting a role for ubiquitination in cytoskeletal interactions in normal cells.

A unifying feature in AD and prion diseases is the central role of a single membrane protein in the formation of complex pathological features, including many of those covered in this and Chapter 10. Clearly the neuron and other cell types can be stressed whilst turning over membrane proteins in endosomes–lysosomes by alterations in demand brought about by apparently subtle changes in protein structure.

11.13 Future directions: the challenges ahead

11.13.1 Ubiquitin and basic cell biology

It is clear from the contents of this chapter that we are entering into areas of molecular neuropathology not easily explained by our current cell biological knowledge of ubiquitin and, in particular, its relationship with compartments within the cell. The reluctance to accept that the presence of ubiquitin within organelles might be other than fortuitous, has limited acceptance of what to us seems to be clearcut but novel; it is a novelty that others feel can be explained within older constraints. Moreover, since ubiquitin already appears to have too many roles, one instinctively favours shortening the list rather than lengthening it! Nevertheless, we feel that a broader approach to the investigation in normal cells of the potential roles of ubiquitin encompassing intracellular membrane-limited compartments will be fruitful, and retrospectively informative in relation to neurodegenerative mechanisms.

11.13.2 Ubiquitin and programmed cell death

Programmed cell death or apoptosis is a co-ordinated set of events, orchestrated from within, that lead to the complete dismantling of a cell and contrasts with necrotic cell death, where the cell in essence merely ceases to survive. Apoptosis allows for the selective removal of a cell without release of its internal contents and without affecting neighbouring cells by the initiation of an inflammatory response. Although there is no clear evidence for the morphological features that are characteristic of apoptosis in relation to the common neurodegenerative diseases (Carson and Ribiero, 1993), it is entirely possible that certain parts of the process are being expressed. It is with this in mind that attention has turned to apoptosis as an explanation of the cellular response to whatever chronic insult it is that underlies a particular neurodegenerative disease. Our own interest in apoptosis was stimulated by the involvement of ubiquitin in the developmental programmed removal of motor neurons in a moth (Schwartz *et al.*, 1990).

11.13.3 Disease heterogeneity

Ubiquitin immunocytochemistry has provided a distinguishing methodology for cortical Lewy bodies and their pathological context (Lennox *et al.*, 1989). In the preceding chapter the uses of antisera against the ubiquitin carboxyl-terminal hydrolase, PGP9.5, and αB-crystallin have been mentioned in further differentiation of ubiquitinated inclusions. It is likely that additional tools of a similar nature will be developed by, for example, investigations of other enzymes involved in ubiquitin activation or production via other ubiquitin carboxyl-terminal hydrolases. We have looked immunohistochemically at the distribution of ubiquitin-activating enzyme, E1, and find it selectively present in the same bulbous form of plaque neurites that are pictured in Figure 4 (Pirim, Layfield, Arnold, Mayer and Landon, unpublished observations).

Many of the neurodegenerative diseases affect people in the later decades of their lives; some of the inherited forms of prion diseases are characterised by ages at death that differ very significantly in relation to subtle changes in PrP (Baker *et al.*, 1991). We still have little understanding of the type of process that can produce clinical features only after 40 or 50 years and all of our experimental approaches look for results within the rather shorter period of a grant!

Huntington's disease has very recently been shown to be another degenerative disorder where ubiquitin-positive dot-like immunoreactivity is found (Cammarata *et al.*, 1993) and has now been demonstrated to involve abnormal genes containing expanded numbers of trinucleotide repeats (The Huntington's Disease Collaborative Research Group, 1993). In disorders of this nature the phenomenon of anticipation, where the illness becomes progressively more severe in succeeding generations, is accounted for by increases in the size of the expansion. The identification of spinocerebellar ataxia type 1, myotonic dystrophy, fragile X syndrome and spinobulbar muscular atrophy as triplet repeat neurological disorders (Ross *et al.*, 1993) suggests that there are probably others waiting to be linked with this mechanism. There will surely be some understanding of these timing mechanisms accruing as we acquire details of the proteins involved and their cellular functions.

11.13.4 Novel therapies

AD and prion diseases have the common involvement of abnormalities in membrane protein processing; endosome-lysosomes are likely locations for the processing and as such provide targets for drugs aimed at interfering with the battery of proteases within (Mayer *et al.*, 1992a). Most often reviewers feel uncomfortable about commenting on the possibilities of drug treatments for what are currently untreatable conditions. We hope that optimism over targeting drugs to endosome-lysosomes is not misplaced; as always, time will tell.

Acknowledgements

We wish to acknowledge financial support from the AFRC, The Wellcome Trust, The Parkinson's Disease Society, The Motor Neurone Disease Association, The Hadwen Trust and Research into Ageing. We also wish to thank L. Laszlo for the electron micrographs and C. Davies of the University of Manchester for the AD biopsy sample.

References

Baker, H.F., Poulter, M., Crow, T.J., Frith, C.D., Lofthouse, R., Ridley, R.M. & Collinge, J. (1991) Amino acid polymorphism in human prion protein and age at death in inherited prion disease. *Lancet* **337**, 1286.

Borchelt, D.R., Taraboulos, A. & Prusiner, S.B. (1992) Evidence for synthesis of scrapie prion proteins in the endocytic pathway. *J. Biol. Chem.* **267**, 16 188–16 199.

Brion, J.P., Passareiro, H., Nunez, J. & Flament-Durand, J. (1985) Mise en evidence immunologique de la proteine tau au niveau des lesions de degenerescence neurofibrillaire de la maladie d'Alzheimer. *Arch. Biol.* **95**, 229–235.

Bueler, H., Aguzzi, A., Sailer, A., Griener, R.A., Autenried, P., Aguet, M. & Weissmann, C. (1993) Mice devoid of PrP are resistant to scrapie. *Cell* **73**, 1339–1347.

Bueler, H., Fischer, M., Lang, Y., Bluethmann, H., Lipp, H.P., DeArmond, S.J., Prusiner, S.B., Aguet, M. & Weissmann, C. (1992) Normal development and behaviour of mice lacking the neuronal cell-surface PrP protein. *Nature* **356**, 577–582.

Cammarata, S., Caponnetto, C. & Tabaton, M. (1993) Ubiquitin-reactive neurites in cerebral cortex of Huntington's chorea: A pathological correlate of dementia. *Neurosci. Lett.* **156**, 96–98.

Carson, D.A. & Ribiero, J.M. (1993) Apoptosis and disease. *Lancet* **341**, 1251–1254.

Cataldo, A.M., Paskevich, P.A., Kominami, E. & Nixon, R.A. (1991) Lysosomal hydrolases of different classes are abnormally distributed in brains of patients with Alzheimer disease. *Proc. Natl Acad. Sci. USA* **88**, 10 998–11 002.

Caughey, B., Race, R.E., Ernst, D., Buchmeier, M.J. & Chesebro, B. (1989) Prion protein-biosynthesis in scrapie-infected and uninfected neuroblastoma cells. *J. Virol* **63**, 175–181.

Caughey, B., Raymond, G.J., Ernst, D. & Race, R.E. (1991) N-terminal truncation of the scrapie-associated form of PrP by lysosomal protease(s): Implications regarding the site of conversion of PrP to the protease-resistant state. *J. Virol*, **65**, 6597–6603.

Chartier-Harlin, M.-C., Crawford, F., Houlden, H., Warren, A., Hughes, D., Fidani, L., Goate, A., Rossar, M., Roques, P., Hardy, J. & Mullan, M. (1991) Early-onset Alzheimer's disease caused by mutations at codon 717 of the β-amyloid precursor protein gene. *Nature* **353**, 844–846.

Chiang, H.-l., Terlecky, S., Plant, C. & Dice, J.F. (1989) A role for a 70 kilodalton heat shock protein in lysosomal degradation of intracellular proteins. *Science* **246**, 282–284.

Cole, G.M. & Timiras, P.S. (1987) Ubiquitin-protein conjugates in Alzheimer's lesions. *Neurosci. Lett.* **79**, 207–212.

Corder, E.H., Saunders, A.M., Strittmatter, W.J., Schmechel, D.E., Gaskell, P.C., Small, G.W., Roses, A.D., Haines, J.L. & Pericak-Vance, M.A. (1993) Gene dose of apolipoprotein E type 4 allele and the risk of Alzheimer's disease in late-onset families. *Science* **261**, 921–923.

Cornwell, G.G., Husby, G., Westermark, P., Natvig, J.B., Michaelsen, T.E. & Skogen, B. (1977) Identification and characterization of different amyloid fibril proteins in tissue sections. *Scand. J. Immunol.* **6**, 1071–1080.

Dickson, D.W., Wertkin, A., Kress, Y., Ksiezak-Reding, H. & Yen, S.-H. (1990a) Ubiquitin-immunoreactive structures in normal human brains. *Lab. Invest.* **63**, 87–99.

Dickson, D.W., Wertkin, A., Mattiace, L.A., Fier, E., Kress, Y., Davies, P. & Yen, S.H. (1990b) Ubiquitin-immunoelectron microscopy of dystrophic neurites in cerebellar senile plaques of Alzheimer's disease. *Acta Neuropathol. Berl.* **79**, 486–493.

Doherty, F.J., Osborn, N.U., Wassell, J.A., Heggie, P.E., Laszlo, L. & Mayer, R.J. (1989) Ubiquitin-protein conjugates accumulate in the lysosomal system of fibroblasts treated with cysteine proteinase inhibitors. *Biochem. J.* **263**, 47–55.

Esch, F.S., Keim, P.S., Beattie, E.C., Blacher, R.W., Culwell, A.R., Oltersdorf, T., McClure, D. & Ward, P.J. (1990) Cleavage of amyloid β-peptide during constitutive processing of its precursor. *Science* **248**, 1122–1124.

Finley, D. & Chau, V. (1991) Ubiquitination. *Annu. Rev. Cell Biol.* **7**, 25–69.

Goate, A., Chartier-Harlin, M.-C., Mullan, M., Brown, J., Crawford, F., Fidani, L., Giuffra, L., Haynes, A., Irving, N., James, L., Mant, R., Newton, P., Rooke, K., Roques, P., Talbot, C., Pericak-Vance, M., Roses, A., Williamson, R., Rossor, M., Owen, M. & Hardy, J. (1991) Segregation of a missense mutation in the amyloid precursor protein gene with familial Alzheimer's disease. *Nature* **349**, 704–706.

Golde, T.E., Estus, S., Younkin, L.H., Selkoe, D.J. & Younkin, S.G. (1992) Processing of the amyloid precursor protein to potentially amyloidogenic derivatives. *Science* **255**, 728–730.

Goldman, J.E. & Corbin, E. (1991) Rosenthal fibres contain ubiquitinated αB-crystallin. *Am. J. Pathol* **139**, 933–938.

Gropper, R., Brandt, R.A., Elias, S., Bearer, C.F., Mayer, A., Schwart, A.L. & Ciechanover, A. (1991) The ubiquitin-activating enzyme, E1, is required for stress-induced lysosomal degradation of cellular proteins. *J. Biol. Chem.* **266**, 3602–3610.

Haas, A.L. & Bright, P.M. (1985) The immunochemical detection and quantitation of intracellular ubiquitin-protein conjugates. *J. Biol. Chem.* **260**, 12 464–12 473.

Haas, C., Koo, E.H., Mellon, A., Hung, A.Y. & Selkoe, D.J. (1992) Targeting of cell-surface β-amyloid precursor protein to lysosomes: Alternative processing into amyloid-bearing fragments. *Nature* **357**, 500–503.

Hansen, L.A., Masliah, E., Terry, R.D. & Mirra, S.S. (1989) A neuropathological subset of Alzheimer's disease with concomitant Lewy body disease and spongiform change. *Acta Neuropathol. Berl.* **78**, 194–201.

Hardy, J. & Allsop, D. (1991) Amyloid deposition as the central event in the aetiology of Alzheimer's disease. *Trends Pharmacol. Sci.* **12**, 383–388.

Heggie, P., Burdon, T., Lowe, J., Landon, M., Lennox, G., Jefferson, D. & Mayer, R.J. (1989) Ubiquitin gene expression in brain and spinal cord in motor neurone disease. *Neurosci. Lett.* **102**, 343–348.

Hope, J., Morton, L.J.D., Farquhar, C.F., Multhaup, G., Beyreuther, K. & Kimberlin, R. (1986) The major polypeptide of scrapie-associated fibrils (SAF) has the same size, charge-distribution and N-terminal protein-sequence as predicted for the normal brain protein (PrP). *EMBO J.* **5**, 2591–2597.

Ingolia, T.D. & Craig, E.A. (1982) Four small *Drosophila* heat shock proteins are related to each other and to mammalian α-crystallin. *Proc. Natl Acad. Sci. USA* **79**, 2360–2364.

Kang, J., Lemaire, H.J., Unterbeck, A., Salbaum, M., Masters, C.A., Grzeschik, K.H., Multhaup, G., Beyreuther, K. & Muller-Hill, B. (1987) The precursor of Alzheimer's disease amyloid A4 protein resembles a cell-surface receptor. *Nature* **325**, 733–736.

Kenward, N., Hope, J., Landon, M. & Mayer, R.J. (1994) Expression of poly-ubiquitin and heat shock protein 70 genes increases in the later stages of disease progression in scrapie-infected mouse brain. *J. Neurochem.* in press.

Kidd, M. (1963) Paired helical filaments in electron microscopy of Alzheimer's disease. *Nature* **197**, 192–193.

Kitaguchi, N., Takahashi, Y., Tokushima, Y., Shiojiri, S. & Ito, H. (1988) Novel precursor of Alzheimer's disease amyloid protein shows protease inhibitory activity. *Nature* **331**, 530–532.

Kitamoto, T., Iizuka, R. & Tateishi, J. (1993a) An amber mutation of prion protein in Gerstmann-Straussler syndrome with mutant PrP plaques. *Biochem. Biophys. Res. Commun.* **192**, 525–531.

Kitamoto, T., Ohta, M., Doh-ura, K., Hitoshi, S., Terao, Y. & Tateishi, J. (1993b) Novel missense variants of prion protein in Creutzfeldt-Jakob disease or Gerstmann-Straussler syndrome. *Biochem. Biophys. Res. Commun.* **191**, 709–714.

Langston, J. W., Ballard, P., Tetrud, J. W. & Irwin, I. (1983) Chronic parkinsonism in humans due to a product of meperidine-analogue synthesis. *Science* **219**, 979–980.

Laszlo, L., Doherty, F.J., Osborn, N.U. & Mayer, R.J. (1990) Ubiquitinated-protein conjugates are specifically enriched in the lysosomal system of fibroblasts. *FEBS Lett.* **261**, 365–368.

Laszlo, L., Doherty, F.J., Watson, A., Self, T., Landon, M., Lowe, J. & Mayer, R.J. (1991) Immunogold localization of ubiquitin-protein conjugates in primary (azurophilic) granules of polymorphonuclear neutrophils. *FEBS Lett.* **279**, 175–178.

Laszlo, L., Lowe, J., Self, T., Kenward, N., Landon, M., McBride, T., Farquhar, C., McConnell, ı., Brown, J., Hope, J. & Mayer, R.J (1992) Lysosomes as key organelles in the pathogenesis of prion encephalopathies. *J. Pathol.* **166**, 333–341.

Laszlo, L., Tuckwell, J., Self, T., Lowe, J., Landon, M., Smith, S., Hawthorne, J.N. & Mayer, R.J. (1991) The latent membrane protein-1 in Epstein-Barr virus-transformed lymphoblastoid cells is found with ubiquitin-protein conjugates and heat shock protein 70 in lysosomes oriented around the microtubule organizing centre. *J. Pathol.* **164**, 203–214.

Leicht, B.G., Beissmann, H., Palter, K.B. & Bonner, J.J. (1986) Small heat shock proteins of *Drosophila* associate with the cytoskeleton. *Proc. Natl Acad. Sci. USA* **83**, 90–94.

Leigh, P.N., Probst, A., Dale, G.E., Power, D.P., Brion, J.P. Dodson, A. & Anderton, B.H. (1989) New aspects of the pathology of neurodegenerative disorders as revealed by ubiquitin antibodies. *Acta Neuropathol. Berl.* **79**, 61–72.

Lennox, G., Lowe, J., Morrell, K., Landon, M. & Mayer, R.J. (1989) Anti-ubiquitin immunocytochemistry is more sensitive than conventional techniques in the detection of diffuse Lewy body disease. *J. Neurol. Neurosurg. Psychiat.* **52**, 67–71.

Linnane, A.W., Marzuki, S., Ozawa, T. & Tanaka, M. (1989) Mitochondrial DNA-mutations as an important contributor to aging and neurodegenerative diseases. *Lancet* **i**, 642–645.

Low, P., Doherty, F.J., Sass, M., Kovacs, J., Mayer, R.J. & Laszlo, L. (1993) Immunogold localization of ubiquitin-protein conjugates in Sf9 insect cells: Implications for the biogenesis of lysosome-related organelles. *FEBS Lett.* **316**, 152–156.

Lowe, J., Blanchard. A., Morrell, K., Lennox, G., Reynolds, L., Billett, M., Landon, M. & Mayer, R.J. (1988) Ubiquitin is a common factor in intermediate filament inclusion bodies of diverse type in man, including those of Parkinson's disease, Pick's disease, and Alzheimer's disease, as well as Rosenthal fibres in cerebellar astrocytomas, cytoplasmic bodies in muscle, and Mallory bodies in alcoholic liver disease. *J. Pathol.* **155**, 9–15.

Lowe, J., Fergusson, J., Kenward, N., Laszlo, L., Landon, M., Farquhar, C., Brown, J., Hope, J. & Mayer, R.J. (1992) Immunoreactivity to ubiquitin-protein conjugates is present early in the disease process in the brains of scrapie-infected mice. *J. Pathol.* **168**, 169–177.

Lowe, J., Mayer, R.J. & Landon, M. (1993) Ubiquitin in neurodegenerative diseases. *Brain Pathol.* **3**, 55–65.

Lowe, J., McDermott, H., Kenward, N., Landon, M., Mayer, R.J., Bruce, M., McBride, P., Somerville, R.A. & Hope, J. (1990a) Ubiquitin-conjugate immunoreactivity in the brains of scrapie-infected mice. *J. Pathol.* **162**, 61–66.

Lowe, J., McDermott, H., Landon, M., Mayer, R.J. & Wilkinson, K.D. (1990b) Ubiquitin carboxyl-terminal hydrolase (PGP9.5) is selectively present in ubiquitinated-inclusion bodies characteristic of human neurodegenerative diseases. *J. Pathol.* **161**, 153–160.

Manetto, V., Perry, G. Tabaton, M., Mulvihill, P., Fried, V.A., Smith, H.T., Gambetti, P. & Autilio-Gambetti, L. (1988) Ubiquitin is associated with abnormal cytoplasmic filaments characteristic of neurodegenerative diseases. *Proc. Natl. Acad. Sci. USA* **85**, 4501–4505.

Mann, D.M.A. & Esiri, M.M. (1989) The pattern of acquisition of plaques and tangles in the brains of patients under 50 years of age with Down's syndrome. *J. Neurol. Sci.* **89**, 169–179.

Mayer, R.J., Arnold, J., Laszlo, L., Landon, M. & Lowe, J. (1991) Ubiquitin in health and disease. *Biochim. Biophys. Acta* **1089**, 141–157.

Mayer, R.J. Landon, M. Laszlo, L., Lennox, G. & Lowe J. (1992a) Protein processing in lysosomes: The new therapeutic target in neurodegenerative disease. *Lancet* **340**, 156–159.

Mayer, R.J., Laszlo, L., Middleton, A., Landon, M., Hope, J. & Lowe, J. (1992b) Ubiquitin, lysosomes and neurodegenerative diseases. *Biochem. Soc. Trans.* **20**, 645–648.

Mayer, R.J., Lowe, J. & Landon, M. (1991) Ubiquitin and the molecular pathology of chronic degenerative diseases. *J. Pathol.* **163**, 279–281.

Mayer, R.J., Lowe, J., Lennox, G., Landon, M., MacLennan, K. & Doherty, F.J. (1989) Intermediate filament-ubiquitin diseases: Implications for cell sanitization. *Biochem. Soc. Symp.* **55**, 193–201.

Mazurkiewicz, J.E. (1991) Ubiquitin deposits are present in spinal motor neurons in all stages of the disease in the motor neuron degeneration (*Mnd*) mutant of the mouse. *Neurosci. Lett.* **128**, 182–186.

McKinley, M.P., Taraboulos, A., Kenaga, L., Serban, D., Stieber, A., DeArmond, S.J., Prusiner, S.B. & Gonatas, N. (1991) Ultrastructural localisation of scrapie prions in cytoplasmic vesicles of infected cultured cells. *Lab. Invest.* **65**, 622–630.

Migheli, A., Attanasio, A., Pezzulo, T., Gullotta, F., Giordana, M.T. & Schiffer, D. (1992) Age-related ubiquitin deposits in dystrophic neurites: An immunoelectron microscopic study. *Neuropathol. Appl. Neurobiol.* **18**, 3–11.

Mori, H., Kondo, J. & Ihara, Y. (1987) Ubiquitin is a component of paried helical filaments in Alzheimer's disease. *Science* **235**, 1641–1644.

Morishima-Kawashima, M., Hasegawa, M., Takio, K., Suzuki, M., Titani, K. & Ihara. Y. (1993) Ubiquitin is conjugated with amino-terminally processed tau in paired helical filaments. *Neuron* **10**, 1151–1160.

Murrell, J., Farlow, M., Ghetti, B. & Benson, M.D. (1991) A mutation in the amyloid precursor protein associated with hereditary Alzheimer's disease. *Science* **254**, 97–99.

Murti, K.G., Smith, H.T. & Fried, V.A. (1988) Ubiquitin is a component of the microtubule network. *Proc. Natl Acad. Sci. USA* **85**, 3019–3023.

Ohta, M., Marceau, N., Perry, G., Manetto, V., Gambetti, P., Autilio, G.L., Metuzals, J., Kawahara, H., Cadrin, M. & French, S.W. (1988) Ubiquitin is present on the cytokeratin intermediate filaments and Mallory bodies of hepatocytes. *Lab. Invest.* **59**, 848–856.

Oliver, C. & Holland, A.J. (1986) Down's syndrome and Alzheimer's disease: A review. *Psychol. Med.* **16**, 307–322.

Pappolla, M.A., Omar, R. & Saran, B. (1989) The 'normal' brain. 'Abnormal' ubiquitinilated deposits highlight an age-related protein change. *Am. J. Pathol.* **135**, 585–591.

Pericak-Vance, M.A., Bebout, J.L., Gaskell, P.C., Yamaoka, L.H., Hung, W.Y., Alberts, M.J., Walker, A.P., Bartlett, R.J., Haynes, C.A., Welsh, K.A., Earl, N.L., Heyman, A., Clark, C.M. & Roses, A.D. (1991) Linkage studies in familial Alzheimer disease: Evidence for chromosome 19 linkage. *Am. J. Hum. Genet.* **48**, 1034–1050.

Perry, G., Friedman, R., Shaw, G. & Chau, V. (1987) Ubiquitin is detected in neurofibrillary tangles and senile plaque neurites of Alzheimer disease brains. *Proc. Natl Acad. Sci. USA* **84**, 3033–3036.

Ponte, P, Gonzalez-de Whitt, P., Schilling, J., Miller, J., Hsu, D., Greenberg, B., Davis, K., Wallace, W., Lieberburg, I., Fuller, F. & Cardell, B. (1988) A new A4 amyloid mRNA contains a domain homologous to serine protease inhibitors. *Nature* **331**, 525–527.

Prelli, F., Castano, E.M., van Duinen, S.G., Bots, G.T.A.M., Luyendijk, W. & Frangione, B. (1988) Different processing of Alzheimer's disease β-protein precursor in the vessel wall of patients with hereditary cerebral haemorrhage with amyloidosis – Dutch type. *Biochem. Biophys. Res. Commun.* **151**, 1150–1155.

Prusiner, S.B. (1991) Molecular biology of prion diseases. *Science* **252**, 1515–1522.

Prusiner, S.B. (1992) Chemistry and biology of prions. *Biochemistry* **31**, 12 277–12 288.

Rogers, M., Yehiely, F., Scott, M. & Prusiner, S.B. (1993) Conversion of truncated and elongated prion proteins into scrapie isoforms in cultured cells. *Proc. Natl Acad. Sci. USA* **90**, 3182–3186.

Rosen, D.R., Siddique, T., Patterson, D., Figlewicz, D.A., Sapp, P., Hentati, A., Donaldson, D., Goto, J., O'Regan, J.P., Deng, H.-X., Rahmani, Z., Krizus, A., McKenna-Yasek, D., Cayabyab, A., Gaston, S.M., Berger, R., Tanzi, R.E., Haperin, J.J., Herzfeldt, B., Van den Bergh, R., Hung, W.-Y., Bird, T., Deng, G., Mulder, D.W., Smyth, C., Laing, N.G.,

Soriano, E., Periak-Vance, M.A., Haines, J., Rouleau, G.A., Gusella, J.S., Horwitz, H.R. & Brown, R.H. (1993) Mutations in Cu/Zn superoxide dismutase gene are associated with familial amyotrophic lateral sclerosis. *Nature* **362**, 59–62.

Ross, C.A., McInnis, M.G., Margolis, R.L. & Li, S.H. (1993) Genes with triplet repeats: Candidate mediators of neuropsychiatric disorders. *Trends Neurosci.* **16**, 254–260.

Schellenberg, G.D., Bird, T.D., Wijsman, E.M., Orr, H.T., Anderson, L., Nemens, E., White, J.A., Bonnycastle, L., Weber, J.L., Alonso, E., Potter, H., Heston, L.L. & Martin, G.M. (1992) Genetic linkage evidence for a familial Alzheimer's disease *locus* on chromosome 14. *Science* **258**, 668–671.

Schwartz, A.L., Ciechanover, A., Brandt, R.A. & Geuze, H.J. (1988) Immunoelectron microscopic localization of ubiquitin in hepatoma cells. *EMBO J.* **7**, 2961–2966.

Schwartz, L.M., Myer, A., Kosz, L., Engelstein, M. & Maier, C. (1990) Activation of polyubiquitin gene expression during developmentally programmed cell death. *Neuron* **5**, 411–419.

Seubert, P., Oltersdorf, T., Lee, M.G., Barbour, R., Blomquist, C., Davis, D., Bryant, K., Fritz, L.C., Galasko, D. Thal, L.J., Lieberburg, I. & Schenk, D.B. (1993) Secretion of β-amyloid precursor protein cleaved at the amino terminus of the β-amyloid peptide. *Nature* **361**, 260–263.

Smith, T.N., Anwer, U., DeGirolami, U. & Drachman, D.A. (1987) Vacuolar change in Alzehimer's disease. *Arch. Neurol.* **44**, 1225–1228.

Tagliavini, F., Prelli, F., Ghiso, J., Bugiani, O., Serban. D., Prusiner, S.B., Ghetti, B. & Frangione, B. (1991) Amyloid protein of Gerstmann-Straussler-Scheinker disease (Indiana kindred) is an 11-kDa fragment of prion protein with an N-terminal glycine at codon 58. *EMBO J.* **10**, 513–519.

Tanzi, R.E., McClatchey, A.I., Lamperti, E.D., Villa-Komaroff, L., Gusella, J.F. & Neve, R.L. (1988) Protease inhibitor domain encoded by an amyloid protein precursor in RNA associated with Alzheimer's disease. *Nature* **331**, 528–530.

Taraboulos, A., Rogers, M., Borchelt, D.R., McKinley, M.P., Scott, M., Serban, D. & Prusiner, S. (1990) Acquisition of protease resistance by prion proteins in scrapie-infected cells does not require asparagine-linked glycosylation. *Proc. Natl Acad. Sci. USA* **87**, 8262–8266.

Taraboulos, A., Serban, D. & Prusiner, S.B. (1990) Scrapie prion proteins accumulate in the cytoplasm of persistently infected cultured cells. *J. Cell Biol.* **110**, 2117–2132.

Terlecky, S.R., Chiang, H.L., Olson, T.S. & Dice, J.F. (1992) Protein and peptide binding and stimulation of *in vitro* lysosomal proteolysis by the 73-kDa heat shock cognate protein. *J. Biol. Chem.* **267**, 9202–9209.

The Huntington's Disease Collaborative Research Group. (1993) A novel gene containing a trinucleotide repeat that is expanded and unstable on Huntington's disease chromosomes. *Cell* **72**, 971–983.

Westermark, P., Natvig, J.B., Anders, R.R., Sletten, K. & Husby, G. (1976) Coexistence of protein AA and immunoglobulin light-chain fragments in amyloid fibrils. *Scand. J. Immunol.* **5**, 31–36.

Woodard, J.S. (1962) Clinicopathological significance of granulovacuolar degeneration in Alzheimer's disease. *J. Neuropathol. Exp. Neurol.* **21**, 85–91.

INDEX